Design Economics for the Built Environment

Design Economics for the Built Environment

Impact of Sustainability on Project Evaluation

Edited by

Herbert Robinson

United Nations African Institute for Economic Development and Planning (Senegal), a subsidiary of United Nations Economic Commission for Africa

Barry Symonds

Symonds Konsult International Ltd, UK and Rapid5D Ltd, UK

Barry Gilbertson

Barry Gilbertson Associates, UK

Benedict Ilozor

School of Engineering Technology, Eastern Michigan University, USA

WILEY Blackwell

This edition first published 2015
© 2015 John Wiley & Sons, Ltd.

Registered Office
John Wiley & Sons, Ltd, The Atrium, Southern Gate, Chichester, West Sussex, PO19 8SQ, United Kingdom

For details of our global editorial offices, for customer services and for information about how to apply for permission to reuse the copyright material in this book please see our website at www.wiley.com.

Library of Congress Cataloging-in-Publication Data applied for.

ISBN: 9780470659090

A catalogue record for this book is available from the British Library.

Set in 10/12pt Sabon by SPi Publisher Services, Pondicherry, India
Printed and bound in Malaysia by Vivar Printing Sdn Bhd

1   2015

Contents

Editors and Contributors

Editors

Prof. Herbert Robinson

Herbert Robinson, the lead editor, has over 20 years experience in research, consultancy, training and capacity building. He is currently a Regional Adviser and Head of Training Division at the United Nations African Institute for Economic Development and Planning (IDEP), which is part of United Nations Regional Economic Commission for Africa. Prior to joining the UN, he was Professor of Construction Economics and Project Management, and the Director of Research at the Department of the Built Environment at London South Bank University. He was also the Course Director for over 7 years of the highly reputable postgraduate Quantity Surveying programme at the University. He has held previous positions as a Senior Research Associate at Loughborough University, Research Scholar and as a UN National Expert. After graduating from the University of Reading (UK), he worked with leading international consultants Arup (UK) and in the UNDP Institutional Strengthening Project/World Bank funded project in The Gambia. He has a significant publication record including the co-authorship of several books such as *Infrastructure for the Built Environment: Global Procurement Strategies* and *Governance and Knowledge Management in Public Private Partnerships*. He is a member of the Editorial Boards of the *Journal of Financial Management for Property and Construction* and *Journal of Industrial Engineering*.

Barry Symonds

Barry Symonds, Senator h.c. (Biberach an der Riss), graduated from University College London with an MSc in Building Economics. He holds several professional qualifications (FRICS, FCIOB, FBEng, ACIAT), and is currently Managing Director of Symonds Konsult International Ltd (Education Consultants) and Associate Director for Rapid5D (BIM Solutions UK). He was previously the Head of Property, Surveying and Construction (2004–2010) at London South Bank University, UK. He is a chartered quantity surveyor and construction economist, and has worked as a partner and consultant for practices in the UK and New Zealand. He has held academic advisory roles and has been visiting professor, lecturer, and external examiner in the UK, peninsular Malaysia, Sabah, Sarawak, Brunei, New Zealand, China, Singapore, Hong Kong, Jamaica and Germany. He is founding member of the successful MBA in International Real Estate (established 2001) at the University of Science, Biberach an der Riss. Barry has written and presented conference papers in the EU and SE Asia. He has served on many RICS committees, and held the positions of Chair of the RICS Essex Branch and QS Division and member of RICS Divisional Council. Barry is a Liveryman of the Worshipful Company of Constructors (City of London), where he is a member of the Scholarship & Awards Committee.

Prof. Barry Gilbertson

Barry Gilbertson now serves as a non-executive director for three listed companies and one private company after 15 years as a partner at PricewaterhouseCoopers (working on cases such as Canary Wharf, Wembley Stadium, Enron and Lehman Bros). An acknowledged expert in interpreting real estate markets, Barry specialises in strategic solutions for real estate within a business context, and focuses on corporate strategy and risk. Barry has a successful track record in 'borrower advocacy' cases, restoring trust between lender and borrower. Barry was the first chartered surveyor to become a partner in a firm of chartered accountants, anywhere in the world, and the 123rd RICS President in 2004/5 (inaugurating The President's Commission on Sustainability). Barry was a member of the United Nations Real Estate Advisory Group for 5 years, a member of the Bank of England's Property Forum for 10 years and a Trustee at the College of Estate Management for 8 years. He has been Visiting Professor at the University of Northumbria in Newcastle for 11 years and Visiting Professor at the Royal Agricultural University for 1 year. Barry has also been Visiting Lecturer at 20 universities around the world a member of the Council (and Court) of the University of Bath, and lead/worked on projects, or spoken at conferences, in 31 countries.

Prof. Benedict Ilozor

Benedict Ilozor teaches construction management, architecture, facilities planning, and design & management at the School of Engineering Technology, Eastern Michigan University, where he is Research & Graduate Assistants Coordinator. His teaching and research are cutting edge, and he currently works on Bendors Air Power System for electricity generation. He previously taught in Australian universities, and was Management Discipline Coordinator for Architecture and Construction Management, and head of Facilities Management for the Built Environment Research Group at the School of Architecture & Building, Deakin University. He was also Australian Coordinator for the Master of Real Estate distance education collaboration between Deakin University and University of Greenwich. He has over 100 publications (books, refereed papers and articles) on design and construction, energy, facilities space planning & management, and organisational performance. He is Regional Editor for the *Journal of Management Development*, and he is on the editorial board of several international journals.

Contributors

John Adriaanse

John Adriaanse lectures and researches in Construction Law at London South Bank University. Before qualifying as a Barrister he had 25 years of practical experience in the construction industry in South Africa, Namibia and in the UK. His last post before joining academia was as a legal adviser for a company active in many jurisdictions which gave him exposure to different legal systems. His research interests focuses on the relationship between the Common Law and the Standard Forms of Contract, the interaction between the law of tort, the law of contract and that of restitution based on the reversal of unjust enrichment. He is the author of *Construction Contract Law*, now in its third edition. He is also a Fellow of the Chartered Institute of Arbitrators.

Aviad Almagor

Aviad Almagor is the Director of Product Design at Vico Software. For the past decade, he has been deeply involved in developing cutting edge software solutions for the building industry. He played a vital role in the development of Vico Office – the first integrated 5D environment for the construction industry. Aviad is enthusiastic about process integration and multidisciplinary collaboration as tools to improve the efficiency and quality of the construction industry. In addition to his work on software development, Aviad has over 10 years experience in architectural design using Building Information Modelling technologies. Aviad received his Bachelor of Architecture degree from the Israel Institute of Technology and his Master of Business Administration degree from Edinburgh Business School, Heriot-Watt University.

Arthlene Amos

Arthlene Amos is a Quantity Surveyor working in the oil and gas industry in Trinidad. Her present assignment is through a secondment with bpTT where she is involved in one of the major turnaround projects for bpTT and Trinidad. She was previously a lecturer in quantity surveying and construction economics at Kingston University, London. Arthlene has experience with major UK contractors working on diverse projects including rail, building and tunnelling. She was involved in a number of major joint ventures between Costain, Veolia Water and MWH in Brighton, the most significant of these were the construction of an 11 km new sewer tunnel in Brighton, UK. She was part of the team in Costain, Laing O'Rourke, Bachy and Emcor Rail Joint Venture to upgrade St Pancras Station better known as the Channel Tunnel Rail Link project, where 13 new platforms were constructed and the existing deck extended to facilitate the Eurostar trains.

Peter Barnes

Peter Barnes is a Director of Blue Sky ADR Ltd, and has been actively involved in the construction industry for almost 40 years. In the early part of his career, Peter was a chief quantity surveyor for a building contractor, and then became head of a building contractor's construction division. After that he moved into consultancy work, specialising in contract and commercial advice, and dispute avoidance and resolution in relation to the construction industry. Peter has an MSc in Construction Law and Arbitration and is a Chartered Conservationist. Peter sits on the JCT Council, and partly as a result of that, but also because of his involvement with various other related committees, has developed a particular interest in the application of Building Information Modelling (BIM) and how the use of BIM will have an ever increasing influence on building practices over the coming years. He holds the following professional qualifications: FCIOB, MICE, MRICS, FCIArb, and MCInstCES.

Dr. Ina Colombo

Ina Colombo is the Deputy Director of the International Institute of Refrigeration based in Paris, France. She was previously a senior researcher and lecturer in the field of sustainable refrigeration and the reduction of the greenhouse gases at London South Bank University. She is a Domestic Energy Assessor Consultant and has developed expertise in reducing the environmental and economic impact of using carbon dioxide refrigeration in retail applications. She specialised in the field of electromechanical engineering including building services, renewable energy and refrigeration to address climate change and global warming issues. Her project management experience includes the coordination of a new laboratory and an environmental chamber to support research projects on sustainability. Prior to joining academia, Ina was a Sustainability Consultant with Building Design Partnership (BDP- UK) and a Building Services Project Manager with AMEC (UK). She won the Ted Perry Award from the Institute of Refrigeration designed to encourage interest amongst bright and promising students for research of a practical nature related to the field of refrigeration.

Dr. Peter de Jong

Peter de Jong has been a lecturer in Building Economics since 2000 at the Department of Real Estate and Housing of the Faculty of Architecture and the Built Environment, Delft University of Technology in the Netherlands. His teaching speciality is on cost awareness in the design process as part of the BSc in Architecture and feasibility studies. His research focuses on life cycle costs, including supporting data classification and collection. Peter has taken part in classification of cost standards and the Dutch elemental method. Over the years, he has developed a strong interest in building informatics, building information modelling which has been a connecting thread in his career. His academic goal is to (re)define sustainable building economics with a strong focus on building quality.

Damilola Ekundayo

Damilola Ekundayo is a lecturer in Construction Economics/Quantity Surveying at Northumbria University, UK. He is currently undertaking a PhD in Sustainable Development. He graduated with an honours degree in Quantity Surveying from the Federal University of Technology Akure, Nigeria, and later undertook Masters studies in Construction Management at the University of Reading, UK where he obtained a Distinction and graduated as the best student. Before joining academia, he worked in the construction industry on multi-million pound projects in the UK and abroad. Damilola has worked on industry-funded projects and has co-authored several research publications and technical reports in the areas of construction economics, project management, environmental sustainability and built environment education. He is a board member of several reputable, peer-reviewed journals. Damilola is a recipient of scholarly and industry awards including the CIOB Certificate of Excellence Award. He is a Member of the CIOB, an Associate of the RICS and a Fellow of the Higher Education Academy.

Paul Farey

Paul Farey is a Director and Head of AECOM's Fiscal Incentives team in Europe (formerly Davis Langdon/NBW Crosher & James). A Quantity Surveyor by training, he worked for contracting organisations and in private practice before joining the team in 1997 because 'Quantity Surveying was a bit too exciting'.

Dual-qualified as a Chartered Surveyor and Taxation Technician he leads the firm's client tax service offer of Capital Allowances, Land Remediation Relief, VAT, R&D Credits and international tax depreciation advice. Paul has worked on numerous projects, across all sectors in the UK, Singapore and Malaysia. He is a Fellow of the Royal Institution of Chartered Surveyors (FRICS) and a regular speaker on property-related tax matters and has completed the College of Estate Management's *Sustainability for Real Estate Investment* course.

Victoria Hardy
Victoria Hardy is the CEO of the Star Island Corporation, a New Hampshire not-for-profit organisation that manages a 35 building conference complex on an island 8 miles off the coast. Prior to her appointment as CEO, Victoria served as the Academic Department Head of Design and Facilities at Wentworth Institute of Technology. Before joining Wentworth, she was the primary tenured Facility Management faculty at Ferris State University for almost 10 years. She also spent 20 years managing arts programmes and facilities at Stanford University, in the Meadowlands in New Jersey, and in Detroit, in addition to consulting in the arts and entertainment industry. Victoria Hardy was named the International Facility Management Association (IFMA) Distinguished Member of the Year for 2005 and was selected in 2001 as the IFMA Distinguished Educator. She holds a BSc, a Master's Degree in Management, and is a graduate of the Stanford Management Development Programme.

James Hayhoe
James Hayhoe studied at Southampton Business School and graduated with a first-class honours degree in Business with Entrepreneurship in 2007. It was during this time that he developed an interest in how terrorist threats were managed by businesses and in the built environment. James worked in a number of security sensitive environments which developed his understanding of risk management through passive and overt threat mitigation. As his career moved towards the construction sector, James developed his academic credentials by achieving a Master's Degree in Quantity Surveying from London South Bank University. This culminated in the submission of his MSc thesis entitled 'Cost and design considerations for tall buildings: managing the evolving threat of terrorism'. James maintains a keen interest in the evolution of high rise developments and their management of threats. He is employed as a Quantity Surveyor in London working on a wide variety of high profile commercial, residential and government funded projects across the UK.

Dr. Ann Heywood
Ann Heywood graduated from Bedford College, University of London with a first-class honours degree. She holds a PhD from the University of Salford. Ann was formerly Principal of the College of Estate Management (CEM), the leading not-for-profit supported e-learning provider for the property and construction sectors, with 3500 students in 100 countries. She was previously in private practice and was elected Green Surveyor of the Year by RICS in 1998 for her work in balancing the competing environmental, land use and financial needs of land portfolios. Before joining CEM, she worked in consultancy, specialising in sustainability and corporate social responsibility. She advised clients on the sustainability credentials of development projects, corporate responsibilities and staff training. Ann was previously managing partner of CPM, a company specialising in environmental consultancy. She successfully chaired the RICS Presidential

Commission on Sustainability (2004–2006). She was a Special Adviser, House of Commons Select Committee (1986–1989) and an Executive Board Member, Construction Industry Council (CIC) from 2009 to 2012.

Malgorzata Jacewicz

Malgorzata Jacewicz has a Master of Arts degree in Industrial Design and prototyping. She is also qualified in architecture. Since 2010, she has been CEO of Hold Foundation C.I.C. During her postgraduate research project at Technical University Delft, Hyperbody Research Group in 2008 she developed in collaboration with a community group activity based design principles of a socio-economic framework to generate positive collaborative changes and therefore reduction in the communities' impact on the environment. In 2009, *Architects' Journal* published her graduation project in architecture 'Paris Galaxy', Boidus Architectural Blogg publication 'How big can you think?'. In 2011, she expanded the socio-economic model of loop based systems as an alternative to the current one, proposing a new ownership model based on the relationship between leasing and recycling. In 2012, the Technology Strategy Board recognised her for innovation in a building design proposal 'Kit of Parts'" further investigating loop system solutions and economic viability of modular systems.

Rotimi Joseph

Rotimi Joseph is a Chartered Quantity Surveyor and Chartered Builder with over 10 year experience as a cost consultant and expert on flood damage properties reinstatement works. He is currently working for Cunningham Lindsey UK and completing his doctorate degree in the Department of Architecture and the Built Environment at the University of the West of England, Bristol. His research interest is in cost–benefit analysis of property level flood risk adaptation measures.

Dr. Jessica Lamond

Jessica Lamond is a Senior Research Fellow at the University of the West of England and member of the Centre for Floods Communities and Resilience. Jessica's research interests are in adaptation of the built environment to reduce the risk of flooding in an era of climate change including property level adaptation, sustainable urban drainage and insurance and she is co-editor/co-author of three books in the field.

Sean Lockie

Sean Lockie is the Director and Head of Sustainability and Carbon Management at Faithful+Gould. He holds a Bachelor of Arts, Bachelor of Planning and Resource Management, an MSc in Sustainability, and professional membership qualifications (MBIFM, MRICS). Sean has nearly 20 years experience in the sustainability area providing policy and project advice to a wide range of clients. He has written over 50 publications in the carbon and sustainability area. Sean is chair of the Environmental Industries Commission's Sustainable Buildings Group, an environmental columnist in *Construction News* and an advisor on the BRE Global standing panel.

Prof. David Lorenz

David Lorenz is the Director of Lorenz Property Advisors, a property valuation and strategic sustainability consulting firm located in South Germany. David has over 10 years of practical experience in valuation, consulting, estates management and property development. Since April 2012, David has also been Professor for Property Valuation and Sustainability at the Faculty of Economics and Business Engineering at the Karlsruhe Institute ofTechnology. David is a Fellow of the Royal Institution of Chartered Surveyors (RICS) and is a member of various international sustainability related research projects and working groups within the RICS; including the RICS Europe, Sustainability Task Force and the RICS Valuation Sustainability Group.

Prof. Thomas Lützkendorf

Thomas Lützkendorf is Chair of Sustainable Management of Housing and Real Estate at the Karlsruhe Institute of Technology (KIT). He is interested in the integration of sustainability issues into decision making processes for the life cycle of buildings, the relationships between buildings' environmental quality and economic advantages. Through long standing co-operation with architects and designers, he is also familiar with the topics and problems concerning the integration of sustainability issues into the design and planning process. He is a member and scientific consultant of the 'roundtable on sustainable building' at the German Federal Ministry of Transport, Building and Urban Affairs. In addition, he is a founding member of the International Initiative for a Sustainable Built Environment (iiSBE) and he is actively involved in various standardisation activities at the national, European and international level.

Dr. Shamil Naoum

Shamil Naoum is a Reader at London South Bank University. He received a BSc in Building and Construction Engineering from the University of Technology in Baghdad, an MSc in Construction Management and Economics from the University of Aston in Birmingham, and a PhD in Construction Management from Brunel University in Uxbridge. Before beginning his academic career, he worked in the construction industry as a site engineer and project manager. He is a member of the Chartered Institute of Building and the American Society of Civil Engineers. He has considerable research experience in construction management areas such as procurement methods, site productivity, human resources management and management science. He supervises PhD and postgraduate students researching construction management problems and has published papers in many international conferences and scholarly journals including: *American Society of Civil Engineers* (USA); *International Journal of Project Management* (UK); *Journal of Engineering Construction and Architectural Management* (UK); and *Chartered Institute of Building* (UK).

John Pearson

John Pearson is a Chartered Quantity Surveyor (FRICS) and has spent 36 years working and teaching within the construction industry. During the 1970s and 1980s he worked for Private Practice and for Consultant Civil Engineers, both in the UK and Finland. Since 1987 he has been a principal lecturer at Northumbria University and has held a number of responsibilities including managing Quantity Surveying research. As an active member of the UK Green Party, John is a keen public speaker on the importance of sustainable construction and is very conscious of the need to instil awareness of this in future construction professionals. At Northumbria University, he teaches in a range of subjects and makes every effort to identify the relevance of sustainability. In addition he has supervised both undergraduate and Master's Dissertations in this area. John also has a degree in Law (LLB) and a master's degree in Education (MEd).

Prof. Andrea Pelzeter

Andrea Pelzeter studied architecture at the University of Stuttgart in Germany. She worked as an architect in the field of construction and revitalisation. In 2002, she began her postgraduate studies in the field of business administration and real estate at the International Real Estate Business School (IREBS). She started as a research assistant at IREBS and in 2006 pursued her doctoral studies at the European Business School (EBS), International University Schloss Reichartshausen. Her research topic was 'Life-cycle costs of real estate: the influence of location, design and environment'. She founded her consulting agency Pelzeter Lebenszyklus-Management (Lifecycle-Management) in 2006. Since 2007 she has held a Professorship for General Business Administration, particularly Facility Management, at the Department of Cooperative Studies at the Berlin School of Economics and Law (HWR Berlin). She is the author of numerous publications on sustainable development in facilities management and building optimisation with life cycle costing.

Prof. Srinath Perera

Srinath Perera is the Chair and Professor of Construction Economics at Northumbria University, Newcastle upon Tyne. He has over 25 years' experience working as a consultant Quantity Surveyor, Project Manager and lecturer. He is a chartered surveyor and a member of both the Royal Institution of Chartered Surveyors and the Australian Institute of Quantity Surveyors. He presently leads the Construction Economics and Management Research group at the Faculty of Engineering & Environment of Northumbria University. His main research interests are in the broad field of Construction Economics covering, risk and value management, cost planning and management, innovation management; sustainability: whole life costing, cost–benefit analysis, carbon estimating; e-business: ICT in construction, e-procurement, decision support and knowledge based systems; professional education. He is currently a coordinator of the e-Business in Construction, task group TG83 of the CIB.

David Picken

David Picken is a Fellow of the Royal Institution of Chartered Surveyors and the Australian Institute of Quantity Surveyors. After university David worked as a volunteer on an aid project in Papua New Guinea. He joined a firm of consultant quantity surveyors in Adelaide in 1973, and held similar positions in the UK and Saudi Arabia. His academic career began at The Hong Kong Polytechnic University in 1979. He completed a research Masters at the University of Salford in life cycle costing in 1989, and studied value engineering practices during a placement with the Hanscomb Group in the USA in 1994. From 1995 to 2009, he taught at the School of Architecture and Building at Deakin University (Australia). His publications include textbooks for measurement practice and papers on design and construction economics in international refereed journals. His teaching was recognised by awards for excellence and outstanding achievement. He is now an adjunct teaching fellow at Bond University in Queensland. David's research interests focus on design economics, procurement and risk management.

Richard Powell

Richard Powell is a Senior Cost Manager with Turner & Townsend Cost Management, a leading international construction consultancy company. He chose quantity surveying as a career as he was attracted to having the on-site experience.

Richard successfully achieved a first-class honours degree in the Quantity Surveying Consultancy course at Kingston University. For the early part of his career Richard was mainly involved within the public sector supporting projects from school extensions to health centres. He then rose to the challenge on the prestigious Heathrow Terminal 5 project for 2 years prior to playing a key role within the cost management of a food retail account. As a chartered surveyor, he now leads the commercial management for a major retail banking client on the refurbishment of their branch network. Richard is passionate about first-class service delivery and passing his knowledge onto those undertaking their RICS APC.

Prof. David Proverbs

David Proverbs is Professor of Construction Management and presently Head of the Department of Architecture and the Built Environment and co-Director of the Centre for Floods, Communities and Resilience at the University of the West of England, Bristol. He is Chair of the Council of Heads of the Built Environment (CHOBE) in the UK, a member of the CIOB Educational Committee and a member of the RICS UK Education Standards Board. He has undertaken numerous research projects, both for industry and the government. Research funding has been secured from the research councils, and various public and private sponsors. Areas of research specialism within flood risk management issues include adaptation to flood risk, damage assessment, flood repair and flood resilience. He is Co-Editor of the *Structural Survey: Journal of Building Pathology and Refurbishment*; and the *International Journal of Sustainable Development and Planning*.

Jon Scott

Jon Scott is a Senior Cost Manager at Bruce Shaw, a multidisciplinary consultancy with both UK and international offices. Jon is a chartered surveyor; originally an Economics graduate with an MSc in Quantity Surveying. He has over 10 years' experience across a number of private firms including Cyril Sweett – a leading international construction consultancy. His experience includes a variety of sectors including residential, commercial, retail and PPP sectors, both in the UK and France. This experience includes the responsibility for cost planning from the inception of many different projects. He is currently working on a number of high specification residential projects in both London and Paris with Bruce Shaw. Jon has previously undertaken published research on Operational Private Finance Initiative projects and the payment mechanism.

Dr. Ian Selby

Ian Selby graduated from the University of Wales Aberystwyth in 1990 with a BSc (Econ.) Hons in International Politics. He then read an MPhil (1992) and PhD at University of Cambridge (1998). During the 1990s he worked for various public and private sector organisations developing research and public affairs activities, and led a research and marketing department for a major UK media organisation between 1998 and 2000. In 2000, he took up his first post in the built environment sector at the British Council for Offices, where he was responsible for establishing the research and public affairs department, which he subsequently led between 2002 and 2008 as Director of Research & Public Policy. He is currently the Research Director at The College of Estate Management. He has managed major research projects on flooding, and grey water usage in the UK housing sector. He has been a member of HMG committees, including the ODPM's Working Party on Decontamination of

Buildings, the DCLG's Working Party on Building Regulations and Energy Performance Certificates Advisory Implementation Committee. Ian is also currently an adviser to the *CRS* in Wales, and to the *Ústí Nad Labem-Libouchec Green Community Investment Project* in the Czech Republic.

John Symes-Thompson

John Symes-Thompson has built up over 30 years of experience in the commercial property investment markets, including 11 years at ING Real Estate in a fund management role, and 3 years at CBRE in investment agency. He joined CBRE in October 2005 as a Senior Director in the Capital Markets Division, but moved over to the Investment Valuation team in 2008 where he is able to bring his market experience and knowledge to the table for key institutional clients. He is currently the lead valuer for Standard Life Assurance, UBS Global Asset Management, BAE Systems Pension Trustees, Royal Bank of Scotland Pension Fund, Lothian Pension Fund, Mountgrange and Santander Pension Trustees in the UK. On the corporate side his clients include Sports Direct, BHS plc. and Arcadia Group. He has a specialist knowledge and interest in sustainability issues and is a member of the IPD ECOPAS Steering Group and the RICS Valuation Working Group on Sustainability.

Dr. Chika Udeaja

Chika Udeaja graduated as a Civil Engineer and worked briefly as a site engineer and design engineer before undertaking postgraduate studies in Concrete Structures at Imperial College London. This was followed by a brief assignment as a bridge engineer in Malaysia before he returned to the UK, to undertake a PhD in Construction Management at the London South Bank University. On completion of his PhD in 2003, he joined the University of Newcastle as a researcher, and was involved in developing CAPRIKON and other research projects. He is currently a senior lecturer in the Faculty of Engineering and Environment at Northumbria University. He teaches procurement, technology, and sustainability to future generations. His main research interests are in construction management and information technology. More recently, he has become increasingly involved in innovative product and process management looking at how modern construction management techniques and sustainable technologies can be used to deliver government and industry targets on improving efficiency and reducing carbon emissions.

Paul Ullmer

Paul Ullmer is a Quantity Surveyor with EC Harris, a leading international built asset consultancy company which is part of ARCADIS, a leading global engineering and consultancy firm, providing consultancy, design, engineering and management services.

Prof. J.W.F. Hans Wamelink

J.W.F. Hans Wamelink has been the Professor of Design and Construction Management in the Faculty of Architecture and the Built Environment, Department of Real Estate and Housing, Delft University of Technology since April 2006. The educational and research activities of his Chair intend to empower professionals and organisations in the AEC industry with new processes and business models which integrate knowledge, organisations and procurement to deliver innovative building projects, and the sustainable renewal of the built environment. The Chair takes care of the education in the bachelor degree as well as in the master's degree programmes. Apart from his role as a professor, he was owner–director of Infocus and a consultant at DHV, both companies specialised in consultancy and building management. After finishing his PhD at the Delft University of Technology he worked for 10 years as an Assistant Professor at the Faculty of Technology Management of the Eindhoven University of Technology in the Netherlands.

Dr. Lei Zhou
Lei Zhou is a lecturer in Construction Economics at the Faculty of Engineering and Environment, Northumbria University, UK. He is a columnist for *International Journal of Project Contracting & Labour Service*. He graduated from Heriot-Watt University with an honours degree in Building Economics and Quantity Surveying. He obtained an MPhil degree in sustainable construction from the University of Manchester. He further gained a PhD degree from Oxford Brookes University in the UK in 2009. He has expertise in Project Finance and Investment, Low Carbon City and Sustainable Construction, Quantity Surveying, Construction E-business and Public Project Management and Auditing.

Foreword

With continuing pressure and innovation in the built environment of today, and with more people now living in cities than in the history of mankind, getting that environment to be an exciting, vibrant, sustainable and cost effective place for communities, occupiers, as well as clients, has never been more important. Understanding design economics is critical to help deliver this vision around the globe, and to enable qualified professionals to provide effective and well considered advice in land, property and construction.

In *Design Economics for the Built Environment,* the Editors, Professor Herbert Robinson, Barry Symonds, Professor Barry Gilbertson and Professor Benedict Ilozor, are unquestionably well recognised professionals in providing such advice around the world. They, together with an expert team of academics and practitioners, bring the theory and practice alive for the reader. Collectively, they are to be congratulated on what has been a challenging task to pull together the latest thinking in such a well informed and coherent way. This is a hugely credible book, providing evidence of the importance of striking the right balance between theories and practices leading to a relevant and robust built environment of the future. It is incredibly well structured and thought through, as you would expect with such a prestigious roll call of academics and practitioners on the contributor list:

- In Part I, all the key elements necessary for effective design economics from an up-to-date view on the theories, principles, concepts and approaches to design economics. Important developments such as the new rules of measurement, new processes for productivity and efficiency, innovation and technologies including BIM, whole life sustainable costing, fiscal policies and incentives for achieving sustainability in design, effective procurement and sustainability tools (including BREEAM and LEED), sustainable communities, flooding risks and cost of mitigation all feature with many of the world's academic experts sharing their words of wisdom.
- Part II makes the theory come alive through practitioners sharing their experience through industry perspectives, practical examples and case studies. Key elements from Part I are unpacked to reinforce the theories and principles

learned and the implications of delivering value for money alongside the need to balance environmental, economic and social pressures of today's construction industry.

So wherever you are in the world, this is a lively and refreshing up-to-date view of design economics in terms of acting as a core enabler for delivering sustainable buildings and infrastructure projects. Whether you are currently studying for a related degree, are a practitioner or influencer in the field, this book will have something for you. I only wish something so well considered that conjoins the latest academic thinking with the practicalities of the built environment were available when I was studying.

It is a huge privilege to be asked to write the foreword for this book, with so many of those I have known in the industry involved in editing or contributing. All I can say is that it has to be worth a read. Design economics is the only thing that can influence the future of the built environment and with countries like India set to build the equivalent of a Chicago every year for the next 26 years, getting this right now is paramount for creating the best environment possible for the generations to come. If you are set to read this book you do so I am sure as a potential key contributor to the built environment now or for the future – there is no question this book will set you up with the latest thinking you require.

Amanda Clack
MSc BSc FRICS FAPM FIMC CMC Affiliate ICAEW
Vice President of the Royal Institution of Chartered Surveyors and
Partner at PricewaterhouseCoopers

Preface

The drive towards low carbon economy, zero carbon buildings and environmentally friendly infrastructure means that there is a growing interest to design in a way that reflects sustainability principles of balancing economic, social and environmental factors. Design economists are increasingly called upon to respond to new and complex challenges by providing solutions that deliver value for money for clients within the constraints of balancing the environmental, economic and social factors in the development process. The unifying theme throughout the book is therefore how to respond to the increasing social, economic and environmental pressure as a result of changes in regulations and clients' priorities to address emerging challenges in the built environment. Previous books on design economics are either too out-dated, or narrowly focused, on exploring the relationship between fundamental design variables relating to geometry in terms of size, shape, arrangement, height and their effects on capital costs. There have been a number of significant books written on the subject over the past decades. The *Economics of Building* by Herbert W. Robinson published in 1939 was the first book to be devoted entirely to the economic aspects of building. This book was a condensation of the author's PhD thesis written at the London School of Economics and was followed by other significant publications including Ivor Seeley's popular book on *Building Economics* in 1972.

However, it is now recognised that design economics should focus on a wide range of issues affecting construction costs and value from a user, buyer or tenant perspective within a changing policy environment and regulatory framework. This book presents new directions and perspectives reflecting the need to recognise the importance of climate change and sustainability in project design. Considerable attention is therefore given to design factors influencing sustainability and environmental externalities, life cycles of buildings with carbon emission treated as external costs, productivity and efficiency, taxation, monetary and fiscal policies, and other fiscal incentives (e.g. levies, reliefs and capital allowances), affect design and construction costs. Attention is also devoted to emerging issues such as the development of assessment frameworks to reduce the environmental costs of design, flooding risks and mitigation, cost implications of terrorism and similar explosive threats, new processes, innovation and technologies such as

Building Information Modelling, knowledge management systems, role of education and their impact on in improving productivity and efficiency of the design process to reduce both project duration and costs.

This book explores the theories, principles of design economics and how it is applied in the construction industry. It is carefully structured into two parts. Part I provides the context and discusses key theories and principles of design economics. Part II focuses on the application of the theories, principles and approaches in Part I by presenting practical examples, case studies as well as tools and frameworks used to achieve creativity resulting in sustainable design outcomes. This approach of integrating theory and principles with practice, tools and case studies provides a better understanding of the linkages between theories and principles of design economics to industry practices leading to a greater appreciation of the discipline of design economics and its increasingly important role in addressing critical economic, social and environmental challenges faced by clients of the construction industry today.

As editors, it has been a long and challenging process but a rewarding journey to put a book of this nature and complexity together. We want to take this opportunity to register our deepest appreciation to all the contributors from academia and industry. We also recognise that the book we are producing is at a time when there are unprecedented changes in the construction industry. The blend of invaluable contribution from academia and industry has made this book unique in many ways. The principles, industry case studies and practical tools incorporated are useful for final year and postgraduate students in design and architecture, construction management, facilities management, quantity surveying, engineering and project management, as well as government policy makers, consultants, contractors and advisers to client organisations. The book will enable both students and practitioners to explore and understand the multiplicity of factors that contribute to efficient design which can reduce both the capital and operating costs of buildings and infrastructure projects and minimise the environmental and social costs to society. Finally, we want to thank Madeleine Metcalfe and her team of editorial assistants and publishers at Wiley-Blackwell for their encouragement and patience in putting this book together.

Editors' and Publishers' Acknowledgement

We are grateful to Rapid 5D/Trimble for their generous support towards the colour illustrations in this book.

Part I
Theories, Principles and Approaches

Chapter 1
Economic Context, Policy Environment and the Changing Role of Design Economists

Herbert Robinson and Barry Symonds

1.1 Introduction

Construction projects are essential for industrial development, business growth and economic transformation. However, major industry reviews and current challenges in the UK construction industry have identified a number of problems forcing construction firms and their clients to reconsider the way they procure and manage projects. A number of criticisms have also been directed specifically at quantity surveying firms assuming the role of economists focusing on cost management and a range of services to provide value for money for clients of the construction industry.

This chapter provides the economic context for construction projects, examines the policy environment and underlying implications for design economists. First, the role of construction in economic development is explored followed by a discussion of the global construction market, the drive towards international trade and the internationalisation of construction services. Secondly, the policy environment is examined and policy measures to stimulate construction investment are discussed with examples of how fiscal, monetary and industry specific policies affect construction processes, resource markets which in turn affect construction costs. Thirdly, the role of the design economists in responding to the policy issues resulting in increasingly complex challenges to reduce social, environmental and economic costs through efficiency, and productivity are discussed. Throughout the chapter, reference is made to how various sections or subsequent chapters relate to the current, emerging and the future role of the design economists. There is a gradual development of the sections from applying theories and principles, and approaches to developing practical tools and techniques to respond to new challenges and policy initiatives influenced by the global and national agenda such as climate change, sustainability, resilience to withstand flooding/storm and building Information Modelling (BIM).

Design Economics for the Built Environment: Impact of Sustainability on Project Evaluation, First Edition.
Edited by Herbert Robinson, Barry Symonds, Barry Gilbertson and Benedict Ilozor.
© 2015 John Wiley & Sons, Ltd. Published 2015 by John Wiley & Sons, Ltd.

1.2   The economic context

Construction projects are capital intensive requiring resources for various activities from concept design to construction and use of the end product, whether it is for self-use as with owner occupiers, for sale or investment as with some developers, for human development, or as capital goods for use in the economic production process.

Understanding the dynamics of the construction industry is critical in tackling central economic and policy issues such as role of infrastructure (including buildings) in economic growth, employment, investment, inflation, exports and imports. There are several reasons why construction is often seen as an 'engine' for economic growth. First, *economic and trade* infrastructure, a key product of the construction industry, such as roads, ports, power supply, airports, telecommunication systems, factories, warehouses, business parks and offices are required for industrial production, manufacturing, retail and the services sector. Social infrastructure such as educational, health, sports and recreational facilities are also directly related to improving living standards, quality of life and to facilitate human capital development and productivity. For example, labour productivity will increase through provision of good education, healthcare and recreational infrastructure. Figure 1.1 illustrates the need to understand the role of construction actors (design and construction firms as producers of capital goods) and other actors (such as planning agencies and clients as facilitators and owners) in the development process.

Why is design economics important? The cost of infrastructure (or a building) matters. Understanding the cost drivers of infrastructure or building projects is key to the design of construction projects and recognising this factor is central to the discipline of design economics. First, infrastructure facilities (or buildings) are a critical input in production processes and contributes to a nation's productivity as it directly affects the cost of goods and services and their competitiveness in the global market place. Industries, manufacturing and the service sectors are therefore primarily concerned with minimising both initial capital (construction costs) and operational costs to maximise the returns on investment and increase profitability. Construction investment varies directly with business profits (or expectation of profits). As industries and businesses continue to grow through profitability and further investment, an expansion in buildings and infrastructure facilities will be needed.

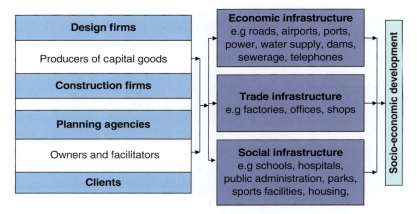

Figure 1.1   Actors in the development process.

From a business perspective, the drive to minimise capital and operational costs is important. Poorly designed and maintained infrastructure whether they are factories, warehouses and offices, power supply, roads, telecommunication, ICT systems and water supply networks increases the whole life costs of infrastructure (or buildings) which can render production processes too expensive. There is therefore a need for well-designed and properly maintained energy efficient buildings with reliable provision of infrastructure services (e.g. power, roads, water, communications) to raise both the productivity of inputs (e.g. machinery and equipment), and to lower cost of industrial production and business services. For example, to facilitate exports of goods and services, it is essential to reduce production, distribution and transaction costs by improving energy efficiency of buildings and expanding road, rail and other communication infrastructure. Construction therefore plays a crucial role in production processes and to accelerate socio-economic transformation. However, it is the nature of the design of projects, and the efficiency of construction processes that determines the unit (capital and operational) costs and performance of the end product.

Secondly, the demand for construction (whether new projects or maintenance) is derived from other goods and services such as agriculture, industry, manufacturing, services and the retail sectors which in turn depend on the state of the national economies and global markets. As a result, there is significant volatility or fluctuation in national and global construction output due to inter-linkages between various economies in the world. A mismatch between the supply and demand for construction output and services affect project costs in various ways. For example, when demand for construction exceeds supply in a particular country (i.e. the capacity of the construction supply chain), there are problems with implementing projects due to over commitment of firms or constraints in local capacity resulting in excessive delays, poor delivery (with higher future costs), and increased costs triggered by shortages of resources in the short term. As a consequence inflation rises as resource markets are overheated. The cost of construction inputs, that is labour, materials, plant and equipment will therefore rise as a result of inflationary pressure leading to increases in the overall unit cost of construction projects. Where supply exceeds demand, there will be significant underutilisation of resources and skills leading to intense competition and reduction in unit costs of construction. The globalisation of construction through the international trading system where a company from one country competes for design and/or construction services in another country helps to tackle the problems of mismatch between national and global demand and supply of construction services.

Thirdly, the construction industry generates significant employment depending on the design chosen, type of projects and method of construction determining the resource mix and utilisation. Employment is created as a result of the inputs required such as materials, equipment and plant, craft skills (e.g. carpenters, bricklayers, plumbers, steelworkers) and professional services needed (e.g. building, engineering design, architectural design, quantity surveying and real estate). Short-term jobs from concept design to construction stages and long-term employment opportunities are created due to the 'multiplier' effect in terms of new demands in other industries, manufacturing, businesses, retail and service sectors. This is as a result of the forward and backward linkages of the supply chain including maintenance services required for the life cycle of construction assets.

Fourthly, the variation in the size and nature of construction markets, composition and type of projects over time depends on the specific needs of countries, whether it is the least developing countries (LDCs), newly industrialised countries (NICs) or advanced industrial countries (AICs). The construction market is made up of construction services relating to the activities of design and construction firms and construction products and supplies such as materials and equipment used for production. The output consists of different types of buildings, infrastructure or market segments – trade infrastructure (e.g. domestic, commercial, industrial and manufacturing such as factories and warehouses), social (e.g. education, health) and economic infrastructure (e.g. roads, railways, etc.). The markets in LDCs are characterised by low levels of demand (although the need is great), weak supply chain due to constraints in design and construction capacity, skills shortages, poor quality and underdeveloped resource markets resulting in shortages of materials, equipment and plant. There is also the problem of significant volatility and unpredictability in input prices sometimes resulting in costly projects. On the other hand, the markets in AICs are characterised by high levels of demand (as effective demand is a function of population and income) and a well-developed supply chain with many national and international consulting firms, contractors, supported by international materials suppliers, equipment and plant manufacturers creating very competitive conditions.

1.3   Globalisation of construction market

The global construction market is vast accounting for approximately 13–15% of global gross domestic product (GDP). In 2012, the total output of the global construction sector was estimated at US$7.2 trillion. Currently the western economies of Europe and North America account for over 60% of global output but due to growth in the emerging economies of Brazil, Russia, India and China (the BRICS) it is predicted that output will grow more than 50% estimated to about US$12 trillion by 2020. However, the share of construction spending by world region is likely to shift significantly over time from the western economies to Asia. Figure 1.2 shows the change in Gross Value Added (GVA) in the developed and emerging nations. The indications are that the output in the western economies will fall or stagnate whilst output in emerging counties will grow enormously.

The shift from the European market to Asia is further highlighted by Euroconstruct, an organisation representing 19 of the European countries that provides statistical information and forecast trends (Euroconstruct, 2012). As can be seen from Figure 1.3, the European construction volumes are largely shrinking due to problems with sovereign debt and austerity measures to bring down the level of indebtedness. However, some growth is predicted by 2015 from Germany, the UK and Ireland, although this is unlikely to change the aggregate picture of continued stagnation throughout the EU19, due to levels of GDP growth.

Furthermore, there are also major differences between future prospects for Eastern Europe, particularly in countries such as Poland, where significant growth in construction volumes is likely due to demand and relatively small debt–GDP ratios. Indeed, it is predicted that the construction output of Russia and Poland together is set to double by 2020 (Schilling, 2013). This view is supported by Godden (2009) as illustrated in Figure 1.4.

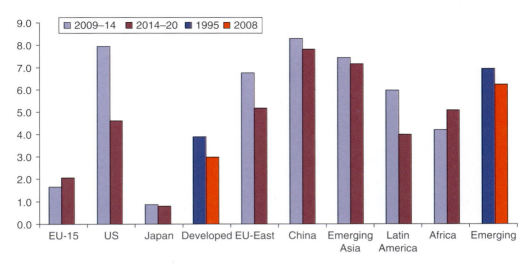

Figure 1.2   Growth in construction GVA: developed and emerging world. Source: Oxford Economics.

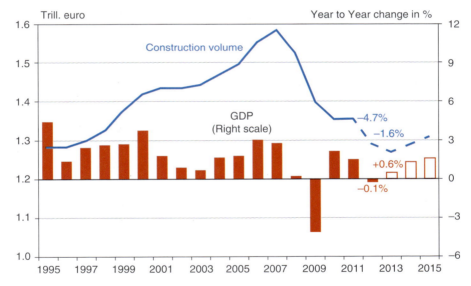

Figure 1.3   Construction volumes and GDP in Euroconstruct countries in 2012. Source: Euroconstruct (2012).

According to Davis Langdon (2012), an AECOM Group company, the general outlook based on Hazelton's (2009) research is that the stimulus for future world growth in construction output will come almost entirely from the emerging economies in Asia, in particular the economies of China and India. Citing IHS Global Insight (2011), it is predicted that whilst Western Europe and North America currently account for more than two-thirds (70%) of world output, by 2020 this will have reduced to 41% with shares from North America dropping from 25 to 17% and Western Europe from 35 to 24% (Figure 1.5).

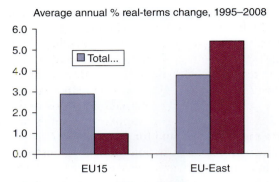

Figure 1.4   Changes in GVA between the EU15 (Western Europe) and EU-East (Eastern Europe) sectors. Source: Godden (2009).

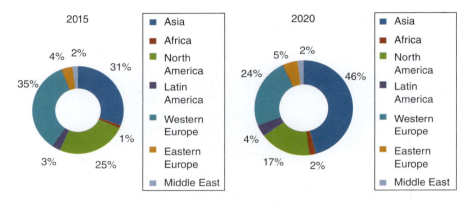

Figure 1.5   Predicted shares of global construction output in 2015 and 2020. Source: Davis Langdon (2012).

This is in sharp contrast to the share of Asia which will see a significant increase in construction output from 31 to 46% as the industrial base is blossoming and will continue to do so, leading to even stronger infrastructure demands, industrial, commercial and residential buildings to support economic growth. There is also the delocalisation of the industrial activities to Asia as manufacturers move their capital investments where labour is cheapest. Africa is expected to benefit from increased manufacturing (and associated construction) activities if governance issues and problems relating to the marginalisation of the continent in the international trading system, infrastructure deficit, stability and conflicts as well as skills shortages are addressed. The prediction for changes in the share of construction output in the other developing regions such as Africa from 1 to 2%, Latin America 3 to 4%, and Eastern Europe 4 to 5% are relatively small.

Major construction companies in Western European and North American markets are diversifying and moving to Asia as demand for design and construction services increase. International trade theories argued that countries should specialise and trade in goods and services in which they have a comparative advantage as specialisation can still result in welfare gains. For example, leading UK, Italian, USA and French design consulting firms are excellent in developing innovative design solutions at the concept stage but the detailed design development is often executed by firms from other developing and emerging countries in Asia such as India, Korea or Singapore due to the availability of the technical skills required at lower cost. For countries with competitive advantage, there is the potential for profit and to contribute to national wealth creation through increases in the GDP or gross national product (GNP) which is why some countries have specific policies and measures to support the internationalisation of design and construction firms, building materials and plant suppliers. Benefits of international trade include balancing the global demand for design and construction services, avoiding astronomical increases in construction costs in regions where there is excess demand or increase in construction activities with limited capacity. Utilising surplus capacity of design and construction firms that would otherwise be wasted in markets that are saturated are obvious advantages and benefits in international trading, as well as the diversification of project portfolio to act as a cushion in periods of economic decline. For international firms, there is also the prestige factor of gaining an international reputation which can attract iconic and landmark projects.

1.4   The policy environment and the construction industry

The global trading system and different development approaches influence the policy agenda. In classical liberal thinking, development is understood as economic growth and capital formation is seen as the key to economic development. Hence, the emphasis in the classical liberal development theory is investment on major construction projects to develop infrastructure capital. Social theories of development on the other hand focus on the importance of 'human capital', and associated investment in education, health, recreational infrastructure to improve wellbeing and welfare as the key to economic growth. This approach requires a shift from the overall rate of economic growth to considerations of social disparities, poverty reduction and inclusive growth. In the human capital approach, heavy investment in social infrastructure (and its construction) is seen as an effective means for improving living standards, quality of life and tackling social disparities and poverty. Neoclassical theory focuses on free markets as the key to economic growth, reducing the role of government to allow private investment (including infrastructure investment, design, construction, and maintenance through public private partnerships) to achieve market efficiency but the role of government is still crucial.

Formulating effective policies for construction investment requires an understanding of development theories, the role of infrastructure and construction in socio-economic development and the relationship between investment in infrastructure, construction and economic growth. According to the Bon's model, the relationship between 'share of construction' and 'GDP/capita' is an inverted U shape. The relationship between 'increases in the share of construction' and 'GDP/capita' normally follows an S shape according to the Turin/Strassman paradigm. It is therefore a challenge to develop policies to stimulate investment in construction activities and to facilitate economic growth.

Fiscal and monetary policies are essential to tackle investment in the construction sector. Fiscal policies dealing with changes in the level and composition of government expenditure in terms of sectors (e.g. education, health, infrastructure, welfare, agriculture) subsidies and taxation affect both the level of construction activity and cost of construction projects. For example, the UK approached the recent credit crunch problems by adopting fiscal policies leading to an ambitious deficit reduction plan, cutting spending on key areas relating to government expenditure affecting public sector investment and its construction activities such as Building Schools for Future (BSF) programme and the Private Finance Initiative (PFI). Monetary policies dealing with broad aggregates of money supply, interest rates and liquidity in terms of debt management also affect the level of investment. Construction investment varies inversely with interest rates. For example, monetary policies adopted in the UK during the credit crunch include lowering interest rates to an all time low and quantitative easing (QE) to enable businesses to expand by creating the conditions for affordable loan and to stimulate capital investment in manufacturing and businesses necessary to boost export performance. Such measures were also designed to facilitate the activities of small–medium enterprises (SMEs) to accelerate economic growth and employment creation. Problems in the commercial and industrial construction sectors can also be addressed through policies to enhance growth in export-led manufacturing. Other policies could include easing the availability of mortgage facilities to increase the supply and construction of housing in particular areas to respond to chronic shortages and changes in demographic patterns.

Other policies may be required to address industry specific problems affecting the demand and the supply side. Policies to stimulate demand for construction could include changes in planning regulations and land use to facilitate the expansion of housing, commercial and industrial activities. On the private sector side, this could include stimulating investment through private sector involvement (such as public private partnerships), manufacturing, industrialisation, and real estate development through tax incentives, grants and subsidies. For example, toll roads are currently being explored by the UK government to stimulate private sector investment in road construction and maintenance in an attempt to ease congestion in the national road network.

Policies on the supply side include investment in research and development (R&D), training and human capital development, critical in improving design outcomes, productivity and quality, and reducing cost of construction projects. Construction is often seen as a knowledge-based industry with significant level of tacit knowledge used in design and construction processes. Measures could therefore be directed at encouraging design and construction firms to become innovative by providing incentives to invest in R&D through the tax systems. Measures to increase efficiency and productivity of the construction process could focus on increasing the quality of labour supply or human effort required in the design and production process such as investment in skills, education, training and capacity development through the availability of education grants and sponsorships, student education loans with low interest rates and flexible payment regimes. Measures to stimulate supply and its quality include on-the-job schemes, compulsory continuous professional development (CPD), other training and certification programmes not only to increase the size of the workforce but the productivity of the entire supply chain. Nurturing of entrepreneurial talent (such as investors, developers, manufacturers) to identify development and business opportunities, take risks, generate profit and create wealth is also crucial to facilitate investment in construction.

The international agenda also influences national policy. A key international agenda relevant to the built environment is sustainability and climate change through the UN framework convention on climate change. The Kyoto Protocol adopted in 1997 setting targets to reduce greenhouse gas emissions allow developed countries to 'buy in' carbon credits from developing nations to meet their emission reduction targets through the EU Emission Trading Scheme (ETS). Under the 'cap and trade' system businesses are obliged to match their greenhouse gas emissions with equal volumes of emission allowances. According to standard economic theory 'the demand for capital is driven by the level of expected future profits and demand'. However, increased uncertainty about the future due to climate change issues can affect both the level of demand and desire for future investment in construction activities. Collective action is therefore necessary with the design economists playing an instrumental role in the built environment to tackle issues of global warming and climate change.

1.5   Current and emerging role of design economists

Understanding the economic context is a prerequisite to design appropriate policies and measures to stimulate investment in construction. Design economists respond to the economic situation and policy prescriptions by providing advice to clients on costs and value of an investment including and a range of other services to ensure value for money. Early accounts by James Nesbit (1989), a quantity surveyor and his contemporaries who were effectively the first 'design economists' focused on maximising utility (value) from a project by minimising capital cost. However, construction clients are interested in reducing both the initial capital costs and whole-life (operating) costs of their built assets to maximise the return on investment. The challenges for design economists range from the application of theories and principles to achieve greater economy, minimising whole life costs (capital cost and operational cost), to developing new areas of expertise such as carbon management and enhanced capital allowances to reduce environmental costs. Increasingly, design economists are called upon to provide advice on how to enhance the value of design, carbon management, taxation, productivity and efficiency of construction processes and built assets. There are implications in terms of the demand for certain skills which will create opportunities for the development and application of new tools to address sustainability and climate change.

Applying theories and principles for economy and value enhancement

Design economists apply theories and principles such as capital cost and whole life theories, value management, value of design and resource based theories (discussed in Chapter 2). Understanding these theories help to minimise whole life costs of buildings, and achieve optimum design outcomes through value enhancement, economies of scale, and greater planning and resource efficiency. For example, resource based theory is crucial for understanding resource markets, and the role of resource supply and demand in predicting construction costs. The cost of buildings (or infrastructure) is determined based on the cost of various resource inputs associated with design and construction processes such as materials, plant and labour including land (or space) required which can be expensive (if land resources are limited). First, to address the problem of land which adds significantly to the cost of projects, design economists would come up with a range of solutions and appraisal techniques to optimise the value of land such as residual valuation (see Chapter 5). Other approaches such as activity based design principles (discussed in Chapter 14) rather than a traditional functional design approach can be adopted to evaluate overall land requirements and to minimise land usage based on energy demand consumption analysis of various activities. Secondly, the cost of key resources such as materials, whether natural (e.g. sand, stone) or manufactured (e.g. cement, glass, doors) required for the construction process and their associated cost depends on industrial production processes influenced partly by economic policies, monetary or fiscal policies. Other resources required for construction projects affect the workforce and productivity depending on government policies, regulations and initiatives such as apprenticeship schemes, availability of education grants, subsidies for education and regulation on competition in professional services. The need for resource-based cost planning is therefore crucial for understanding resource markets and its impact on construction costs.

In making design decisions there are other factors to consider such as the optimum height to minimise cost. Understanding the relationship between construction capital cost and building height is discussed extensively in Chapter 4. It is also critical to understand the wide range of factors affecting super-tall buildings including recent design responses following the 9/11 attack to increase resilience and to cope with explosive threats (discussed in Chapter 20). Applying whole-life cost theory to minimise total construction cost and carbon emissions over the lifespan of buildings and to maximise the function of built assets is discussed in Chapters 8 and 19. Interest rates and inflation (influenced by government policies) are factored in whole life costs modelling carried out by design economists to ensure that future prediction on construction costs (both capital and operational costs such as energy, replacement, cleaning) associated with major capital investment decisions and the long-term use of built assets are appropriately determined. The application of value management theory (discussed in Chapters 22 and 23) focuses on identifying the key cost drivers and value opportunity points to facilitate value enhancement and the elimination of unnecessary costs in the design process.

However, the role of design economists in modelling, predicting and monitoring resources, costs and value management rely on accurate information and established practices. Measures to improve reliability of information and its consistency include new working methods such as the New Rules of Measurement (NRM) (discussed in Chapter 3), to ensure that cost data are accurate and to facilitate integration with BIM discussed in Chapter 21. For example, the RICS has recently published NRM 1 (new rules for standardised cost planning), NRM 2 (new measurement rules) and NRM 3 (rules for life cycle costing). The NRM is a standardised approach to enable the design economists to easily identify different components of the cost estimate such as facilitating works, building works estimate, preliminaries, overheads and profit, project/design team fees and risk allowances in order to improve predictability and reliability of construction costs. In the NRM, the cost limit may be expressed either with or without construction inflation and there are separate calculations for the provision of construction inflation. Changes in inflation and interest rates, influenced by monetary policies, affect both the cost and predictability of construction projects.

Design solutions and construction costs are influenced by fiscal and industry specific policies. An increasingly challenging role for design economists is to provide advice on taxes and fiscal incentives such as VAT, duties, taxes on income, capital investment affecting construction costs so that they are taken into account at the design stage. A comprehensive system of capital allowances and fiscal incentives developed to promote green design to reduce the carbon footprint of construction projects are discussed in Chapters 11 and 26.

Improving productivity, efficiency and leveraging new technology

Productivity and efficiency are key economic and environmental drivers central to the role of design economists in reducing costs and waste in construction projects. The determinants of productivity and efficiency are addressed using a holistic socio-techno-managerial approach (discussed in Chapter 7). Such an approach recognises that productivity can be influenced *technically* by an efficient planning and scheduling of resources, *socially*, by creating a work environment that motivates and leads people effectively, and *managerially*, by an efficient management system to communicate, co-ordinate and control design and construction activities. Design economists provide clients with advice on alternative procurement routes (discussed in Chapter 9) based on varying relationships between professional firms to improve risk management, efficiency and the predictability of building costs. The need for optimum space planning to improve organisational performance and productivity of built assets is also an important aspect of improving design outcomes (Chapter 13).

Technology is a key driver in construction processes to improve interdisciplinary collaborative practices, and the design economists have a key role in increasing the level of efficiency. Modern technology including BIM will challenge the traditional role of construction teams, and change the relationship between main parties – architects, clients, contractors, engineers, and quantity surveyors operating as design economists. The use of knowledge management tools including BIM (discussed in Chapters 16 and 21) will result in more integrated design, reduced cost, time, and result in greater efficiency and productivity of construction projects. BIM's impetus is in part enhanced by the UK Government Cabinet Office Efficiency Reform Group with various professional bodies, contracting organisations and other stakeholders to improve construction efficiency. The role of innovation in responding to sustainability challenges to increase competitiveness and profitability through greater productivity and efficiency is discussed in Chapter 24.

Responding to the international agenda

Design economists have a critical role to play in responding to international agenda such as sustainability and climate change. For example, the objectives of the UK Climate Change Act is leading to new ways of thinking about design, development and application of new tools and approaches to reduce the cost of energy supply and carbon management (see Chapter 12). Climate change has resulted in increased incidents of natural disasters and the need for better design to increase resilience and to cope with environmental emergencies such as flooding (discussed in Chapter 15). It is increasingly recognised by design economists that carbon will have to be managed in a proactive way using the cost plan and other tools as powerful communication systems. This approach will ensure that the environmental cost due to the embodied energy of building materials, construction processes, and recycling practices are adequately captured so that alternative design and materials are explored with full knowledge of their environmental cost. New materials such as low carbon concrete have been introduced to reduce environmental impact and advances in nanotechnology provide opportunities for developing other types of building materials with totally new properties. Examples of nano-sized particles that have been applied in the construction industry include titanium dioxide (TiO_2) used to break down dirt and pollution and carbon nanotubes (CNTs) used to strengthen concrete. There are several implications for the design economists. First, they need to have knowledge of specific materials with low and high embodied carbon and their cost implications. Secondly, the traditional cost plan would have to be modified to manage carbon explicitly and to inform clients about the embodied carbon of materials, and the carbon footprint of alternative designs to achieve optimum solutions. Thirdly, there will be a need to adopt a consistent approach in determining what factors to take into account in measuring and calculating embodied energy to be included in the capital cost plan. Design economists should have a better understanding of the trade-off between whole-life costs (capital costs and operating costs) and the carbon performance of different materials and alternative design (Chapter 19). New approaches include dealing with carbon emissions in whole life analysis by treating carbon emission as external costs in terms of the costs of emission certificates (Chapter 8) and the eco-cost/value ratio (EVR) tool based on life cycle analysis (LCA) to assess the ecological impact of materials and alternative design solutions (Chapter 6). Reducing fossil fuel utilisation and carbon emission by moving to alternative energy sources is also an effective solution as demonstrated in Chapter 25.

Increasingly, there are environmental costs due to legislation, planning and building regulations such as the taxation and capital allowances system. It is therefore important for design economists to take account of fiscal incentives and carbon implications in project investment decisions to mitigate the impact of climate change and sustainability. Responses to the challenges include a range of sustainable and renewable alternatives particularly in monitoring and evaluation and fiscal incentives with actual examples discussed in Chapters 11 and 26.

Changes in the priorities of clients such as promoting sustainability to achieve a higher environmental rating affect construction costs. Given the intense debate on the trade-off between carbon emissions and cost, it is increasingly relevant to reduce carbon emissions in building design, construction and operation. The application of BREEAM and LEED (similar to the Energy Star used in Australia) discussed in Chapter 12 will help to achieve low carbon design solutions. For the design economists, such tools provide a coherent and consistent framework not only to evaluate sustainability of design, but to assess the carbon and cost implications (in terms of both capital and whole life costs) of different design scenarios and to allow clients to make trade-off decisions in an informed way. For example, the additional capital costs for achieving a higher BREEAM rating could result in a reduction of whole life costs through enhanced capital allowances, lower energy costs, reduced costs associated with extended void periods, and other costs/tax advantages. There are opportunities for significantly higher rental returns as sustainable property investment is likely to influence the value of buildings (Chapters 10 and 17). Empirical evidence suggests that the economic benefits of buildings or impact on selling price/property value due to sustainable credentials range from 0.2 to 35% based on experiences cited from Germany, Switzerland, Netherlands, and USA (discussed in Chapter 10). The practical challenges in exploring the relationship between sustainable design and key factors that are likely to affect property values are also discussed from a UK practitioner's perspective (Chapter 17).

Strengthening the curriculum and training in design economics

Finally, it is expected that there will be profound changes in the future as a result of the drive for carbon-friendly design solutions, lower whole life costs, value enhancement, and the leveraging of technology for greater efficiency and productivity. Quantity surveying firms providing advice on design economics face intense competition both within and outside the construction industry. As a consequence, quantity surveying firms now provide a range of specialist technical, management and consultancy services compared with the traditional services (Chapter 16). The education and professional training of design economists will therefore be crucial in future to ensure that they continue to play a leading role in the construction process working alongside other members of the design team to achieve efficient design and cost solutions. There will be a need to review both the curriculum to capture new knowledge, and the matrix of competencies and skills discussed in Chapter 27. This is urgently required to respond to new challenges and complexity as a result of the emerging national and international agenda including the sustainability and climate change debate.

The implications are that quantity surveying firms as design economists will have to adopt new strategies to deal with the changes in the business environment. Quantity surveying firms also need to explore further options which may include repositioning from the historically narrow focus on cost management, rebranding and diversification strategies such as mergers and acquisitions, alliancing, and joint ventures that will facilitate the acquisition and leveraging of new knowledge to cope with the uncertainties arising from new challenges and technologies including BIM. The past two decades have witnessed a significant transformation of many well-known leading quantity surveying firms through mergers and acquisitions to achieve intensive growth and to enhance their competitive edge. Mergers and acquisitions provide opportunities for expanding geographical presence to attract large corporate clients, reduce costs as a result of economies of scale (and scope) which is vital to increase revenue and profit. There is the added opportunity for greater investment in knowledge tools and technology such as BIM and human interactive systems to create networks of experts and a wider pool of high quality specialist skills necessary to respond to new challenges faster.

References

Davis Langdon (2012) World Construction 2012. Report.

Euroconstruct (2012) *Construction Activity and Economic Growth in Europe.* Proceedings of the 74th Construction Market Forecast Conference, 12 December 2012, Munich, Germany.

Godden, D. (2009) *Construction-Emerging Markets to Lead the Way.* Oxford Economics Presentation, 26 November 2009.

Hazelton, S. (2009) *Global Construction Outlook.* IHS Global Insight, 23 June 2009.

Nesbit, J. (1989) *Called to Account: Quantity Surveying, 1936–1986,* Stoke Publications, London.

Schilling, D.R. (2013) Global Construction Expected to Increase by $4.8 Trillion by 2020. Industry Tap into News, 8 March 2013.

Chapter 2
Theories and Principles of Design Economics

Herbert Robinson and Barry Symonds

2.1   Introduction

A central problem in economics is scarcity of resources as human needs are unlimited whilst the means for fulfilling them are limited. Design decisions affect the built environment as there are economic, social and environmental consequences associated with construction projects, their use and performance. There is therefore a need to evaluate design not only from an economic perspective (incorporating capital and operating costs) but also in terms of the environmental and social costs to clients and other stakeholders (e.g. local people, businesses, communities and special interest groups). Factories, offices, housing, hospitals, schools, roads and airports are all essential for human development, industrial production and productivity to enhance socio-economic development. In the context of the built environment, appropriate design choices have to be made based on the resources available to meet the needs of owners and users which have to be balanced against the needs of society.

Understanding the theories and principles of design economics is fundamental in addressing this balance and achieving value for money and a cost-effective building requiring an optimum trade-off between economic, environmental and social costs. Environmental and social costs are important in design due to externalities associated with the construction process, use of buildings, growing resource problems, and carbon emissions causing global warming and climatic disruptions.

This chapter explores the theories and principles of design economics. It starts with an overview of the factors affecting design costs and benefits including externalities, followed by a discussion of capital cost theory and the whole life cost (WLC) theory to ensure that the effects of design decisions are fully considered. Value management theory focusing on maximising the function (or quality) of a design solution whether it is design space, component or materials in relation to its cost is also examined.

Design Economics for the Built Environment: Impact of Sustainability on Project Evaluation, First Edition.
Edited by Herbert Robinson, Barry Symonds, Barry Gilbertson and Benedict Ilozor.
© 2015 John Wiley & Sons, Ltd. Published 2015 by John Wiley & Sons, Ltd.

This is followed by a discussion of the 'value of design' theory to understand the relationship between economic cost (wider economic impact of attractive buildings and settings, social cost (enhanced individual, and social well-being or quality of life) and environmental cost (greater adaptability, energy efficiency, environmental sustainability, noise and pollution). The chapter concludes with the resource-based theory exploring the relationship between construction cost and resource inputs. The sum of resources and the type of resources required are a function of resource production coefficients, unit resource costs determined by types of buildings or infrastructure and the forces of demand and supply in the resource markets.

2.2   Factors affecting design costs and benefits

A facility or building's function strongly influences its design which in turn affects the construction *cost*. The *function* is important as it expresses the intended use or benefits of a project and determines the design parameters. As a result, construction costs are often expressed in *functional* units (e.g. cost per bed/seat/place/space, etc.) for offices, houses, schools, hospitals, and so on. However, construction cost is also affected by other factors such as geometry and spatial arrangements (e.g. height, *layout/groupings,* inter-linkages between buildings, common services, shared elements) and the characteristics of the site in terms of access for delivery, available services, and proximity to other buildings.

Traditional theories of design economics focused on a number of key variables and their implications in terms of capital costs. For example, the geometry of a building in terms of size, shape, arrangement and height, affects capital costs. Complex design projects characterised by difficult geometry are more expensive than simple (often repetitive) projects which benefit greatly from a reduction in unit costs as a result of the learning effects or experience curve. Complexity affects costs as projects with unusual, untried and untested design features are extremely difficult to plan, construct and manage. Uncertainty and risks are also greater in complex projects with significant cost consequences. Capital cost is also influenced by other factors such as planning requirements, building regulations and taxation and capital allowances system. However, there is increasing evidence that other design variables such as colour, lighting, sound, aroma, and landscape are also important. Some of these design factors can have a positive influence on the outcome of a project such as patients' recovery rate in hospitals, office productivity, and absenteeism. For example, in a recent study, it was reported that the provision of outdoor view reduces patients' stay time by, on average, 13.5 h and stay time by 4 h per 100 lux increase of daylight. Other examples of the economic benefits of good design include greater efficiency and productivity, savings in tax and capital allowances, reduction in staff costs, insurance costs, accident, pollution, landfill charges and energy costs resulting in a better use of the asset and return on investment.

Determining all the potential design factors influencing construction cost is therefore important in developing an effective solution but the challenge in design economics is often how to put a 'value' on certain design outcomes such as patients recovering earlier in a hospital as a result of better landscape view, sound and lighting performance. It is also problematic to put a value on savings as a result of productivity, reduction in absenteeism, reduction in pollution, carbon emissions and scenic values. The impact of design decisions can be summarised in terms of costs and benefits as shown in Table 2.1.

Table 2.1   Design costs and benefits matrix.

Costs	Benefits/value
Easy-to-price costs Economic – land, planning, design cost, construction cost	**Easy-to-price benefits** Economic – asset value, rental or sale income, normal and enhanced capital allowances
Not-so-easy to price costs Environment – pollution (emission cost), carbon cost, scenic values lost, etc. Economic – operation cost, insurance cost, loss of property value, etc.	**Not-so-easy-to-price benefits** Social – staff morale, comfort, etc. Economic – productivity, hospital recovery rates, savings in staff costs, etc. Environmental – savings in energy, emissions, reduced flooding and damage etc

In theory, the total cost associated with any project is the sum of the project's economic, social and environmental costs. Traditionally, the client is normally concerned with direct easy-to-price economic costs and benefits that are visible and associated with land purchase, planning, design and construction costs, rental, sales income and tax savings. However, other indirect not-so-easy-to-price benefits (or savings as result of reduced carbon emissions, flooding damage, productivity), social and environment costs (e.g. noise and air pollution, traffic congestion, and increased flooding risks) imposed on society, governments and other stakeholders are increasingly important. For example, energy efficiency certificates are now available to buyers/tenants which affect property value/rent.

$$\text{Total project cost} = \text{Economic cost} + \text{Social cost} + \text{Environmental cost}$$

To mitigate the effects of market failure, some social and environmental costs are passed on to clients or project owners through regulations such as charges, taxes, planning requirements or building regulations. For example, to comply with the need to reduce carbon (or environmental cost), a local authority or planning agency could request for a higher BREEAM/LEED rating for a particular development, or an increased level of flood protection or safety margin to reduce the negative consequences of a project. In the UK social costs are incorporated by statute within the Department for Communities and Local Government (2013). To this end, in England and Wales, the planning obligations (s106 T&CPA 1990 *as amended*) and payments arising from Community Infrastructure Levy (s206 Planning Act 2008) are also used to ensure that project owners contribute to the additional social and environmental costs arising as a result of a new development. This could for example mean the provision of schools, health, community and recreational facilities, bicycle lanes, and widening of some roads in a development project.

2.3   Capital cost theory

The capital cost theory was developed after the Second World War, largely in the UK. Most construction work, unlike today, was instigated by the Government sector. Post war budgets were meagre and politicians were eager to produce more for less, or to maximise the benefits given the limited resources available to achieve economic efficiency. Clients wanted to maximise utility from a project by minimising capital cost subject to certain restrictions such as building and planning regulations. This resulted in some remarkable innovative thinking at the time. There can be few better accounts of this period, than that related by James Nesbit, a quantity surveyor of that era, who has provided a history of the period 1936–1986, in a publication titled *Called to Account*. Nesbit (1989) and his contemporaries were effectively the first 'design or construction economists' and were in the forefront of the development of cost yard-sticks (a measure of acceptable value denoted in many forms, i.e. cost per square metre, cost per bed space, cost per person according to building type), elemental cost planning and building cost modelling. For example, Cartlidge (1976) noted that in 1967 the then Department of Environment in the UK issued a circular to local authorities titled *Housing Standards, Costs and Subsidies*, which together with sub-sequent revisions formed the basis of the housing cost yardstick. Similar yardsticks were developed for hospital projects by the Department of Health and Social Security and for school projects by the Department of Education and Science. These docu-ments used the elemental cost analysis of previously constructed buildings not only to measure the quantum and cost of a new building, but to derive relatively simple design related formulae (such as wall to floor ratios) to enable designers to evaluate the economic complexity (and subsequent viability) of their buildings. However, this theory is sometimes criticised due to the heavy reliance on cost per square metre capital cost guidelines for different types of buildings derived from historic data. As Sorrel (2003) noted 'the risk here is that the use of these rigid guidelines can bias clients against energy efficient buildings' required in new types of design to cope with the requirements of today's society and environmental pressure.

Since the late 1950s much has been written about design and cost planning as well as cost modelling of buildings (e.g. Seeley, 1972; Bathurst and Butler, 1973; Cartlidge, 1976; Ashworth, 1999; Ferry *et al.*, 1999; Ashworth and Hogg, 2000). These books on traditional design economics focused on factors affecting capital cost, primarily building geometry and materials. For example, there are a number of principles associated with minimising the capital costs of buildings such as external wall-to-floor ratio (known as a quantity ratio) – the lower the ratio, the more economical the design, the perimeter over plan (POP) ratio, plan shape, building size (economies of scale), planning efficiency, density, building layout, the effect of height, quality factors and site characteristics. The POP ratio is used as a measure of compactness of the design, the higher the percentage, the more effi-cient the design. This is generally true except for the circle.

$$\%\text{Compactness} = \frac{2\sqrt{\pi A}}{P} \times 100 \qquad (2.1)$$

where A is the covered area of a typical floor area and P is the perimeter enclosing that area.

The capital cost of a construction project (C) is a function of a number of design variables such as quantity ratio (Qr), size (Si), shape (Sh), height (H), materials specification (M), density (D) and planning efficiency (P).

$$C = f(Qr, Si, Sh, H, M, D, P, \ldots). \qquad (2.2)$$

For tall buildings, for example, Lee *et al.* (2011) developed the high-rise premium ratio as part of the schematic cost estimating model (SCEM) 'to identify the produc-tivity ratios of super tall buildings and to simulate construction cost as building design changes. They found that construction cost increases as unit cost rate rises due to the lower productivity ratio in projects with higher number of storeys. Table 2.2 provide examples of design variables and parameters affecting construction costs.

Table 2.2 Examples of design variables and key considerations.

Examples of design variable	Key considerations	Objective
Plan shape Some shapes are more economical than others Spatial arrangements influence building cost Complex shapes are more expensive	External wall-to-floor ratio (or quantity ratio) – the lower the ratio the less expensive the building Unusual features Roof complexities Natural lighting	Finding the plan shape which is the most economical
Building size Small buildings generally cost less but are not economical	Wall-to-floor ratio Discount on bulk purchase (economies of scale) Co-ordination and project management requirements	Finding the optimum size of a project that the team can cope with to benefit from economies of scale. Beyond this point there is diseconomies of scale
Planning efficiency Usable area varies (Net Floor Area)	Circulation space/ corridor areas/ service areas/ toilets/ lifts	To minimise non-usable space (or maximise rental income) subject to planning and building regulations
Building layout/ groupings Nature of inter-linkages between buildings reduces costs	Common services Shared elements/ external walls (e.g. terraced/ semi-detached housing)	To maximise common services and shared elements which will minimise construction cost
Height Tall structures are generally associated with higher construction costs due to vertical transportation logistics, and engineering problems	Foundation costs per square metre of GFA decreases Roof costs per square metre of GFA decreases More space for recreation/car parking Site density can be increased **However cost may start to rise due to:** Foundation loads (piling may be required) Use of plant/ equipment (hoists, tower cranes) lifts, safety considerations	Finding the optimum height of a project to reduce cost (cost per square metre) associated with tall buildings (e.g. piling, wind loading) and need for services (lifts, fire escape, etc.)
Quality factors Level of specification (materials) affects cost	Floor/wall finishes Services Fittings/technologies Environmental rating (BREEAM or LEED)	To find appropriate specification to maximise client's utility and minimise cost
Site characteristics affect building cost	Access/roads/parking Slope/ground conditions Services Location/adjacent structures	To minimise cost relating to site characteristics and surroundings

GFA, Gross Floor Area.

Cost planning techniques have now been in use for almost half a century by designers and architects, although there is limited research to establish whether cost modelling has significantly reduced building costs. However, anecdotal evidence would suggest that the building team (at least in terms of the key sectors of housing, education, health, factories and warehouses, commercial and retail) do at least consider these factors.

The UK was considered to be leading Europe (and arguably the world) in the latter half of the last century in the field of construction economics but would appear to remain one of the most expensive places to build in the EU. According to Cartlidge (2006), the cost of building hospitals per square metre is undoubtedly cheaper in France than the UK and a study produced by BWA Associates (2006) for the European Commission, cites the UK as the least efficient in Europe. Government led reports such as those by Latham (1994), and Egan (1998) in the UK have therefore focused on procurement methods in an attempt to improve efficiency. Given the current lack of funding, albeit for entirely different reasons, the UK Government is perhaps not surprisingly re-examining the issue of efficiency. The UK Government (Cabinet Office, 2011) is promoting Building Information Modelling (BIM) as an integrated management tool, simultaneously establishing 'cost targets', and developing benchmarking and performance targets together with the RICS (Martin, 2012). This will effectively establish average costs for similar buildings and examine methods of reducing those costs by 10–20 % to enable Government to achieve value for money. The cost plan (which summarises the capital cost on an elemental basis), has become arguably, the client's most useful document in terms of cost control, which even in the new age of BIM is unlikely to be replaced. However, as Crotty (2012) has muted currently cost planning is somewhat of a "black art" but BIM is likely to provide greater transparency and accuracy of information and knowledge of risks that will enable costs not only to be speedily established but interrogated with relative ease. BIM does nevertheless provide a challenge to traditional cost modelling and estimating. It is capable of providing for the first time detailed knowledge of resource costs, planning and scheduling of resources, which whilst not changing the fundamental economics of efficient building shapes, may make traditional methods of cost modelling redundant.

2.4   Whole life cost theory

WLC theory, often referred to as 'cost-in-use' theory, is an extension of the capital cost theory by including the long-term (operating) costs associated with the use of a building. It focuses on establishing a trade-off between the initial short term (capital) cost and long-term (operating) cost of alternative design solutions. To avoid inefficient use of resources, Bathurst and Butler (1973) argued that the 'full economic effect of the various design decisions taken by the architect can only be examined if capital and long-term costs can be represented together'. The design option with the lowest WLC (i.e. capital and long-term costs combined) is selected as the most efficient economically subject to certain restrictions relating to the minimum performance criteria to ensure that all options comply with minimum specification. Sometimes the term 'whole life appraisal' is used where cost and performance as well as benefits are considered.

Whole life theory attempts to establish the total cost of a facility measured over the period of interest of the owner and the objective is minimise the total cost of the design over the building's life span and to maximise the client's utility (benefits) from the facility. It is the sum of all funds expended for a facility from its conception to the end of its useful life and includes the initial capital expenditure (CapEx) for planning, design, construction and the operating expenditure (OpEx) for maintenance, energy, cleaning costs, taxation, and so on. There are economic and environmental incentives provided by governments to influence design choices or preferences in favour of energy efficient or carbon friendly design solutions through the capital and enhanced capital allowances/taxation system to reduce a project's capital and operating expenditure. WLC theory is based on quantifying all significant costs during the life of a facility using present value/ discounting technique as the costs are incurred at different time periods.

It recognises that all costs (and benefits) arising from a project are relevant for investment decisions and can be used for realistic estimating, budgeting and cash flow analysis. It is used for making choices between design alternatives to address design questions such as: Should uPVC, wooden or aluminium windows be selected? Should a particular type of roof or heating system be chosen? For example, a client might want to make a decision between carpet and wooden floor finish. The initial cost of a carpet might be lower than wooden flooring, but the running costs will be higher due to the number of replacement and higher cleaning costs associated with a carpet. The carpet may have to be replaced a few times more and cleaned more frequently compared with the wooden floor finish which may last longer (requiring less replacement) and less cleaning due to its surface.

There have been a number of studies carried out in recent years (e.g. Construction Excellence, 2004) to test the WLC theory. Its use is increasing for a number of reasons. First, both public and private clients are changing. Public sector clients are being encouraged to take a whole life approach and to discontinue the practice of separating capital and recurrent budgets. Secondly, there is an increased awareness from private sector clients in considering whole life performance in making long-term investment decisions. Other reasons for the widespread application of whole life theory includes the growing use of alternative integrated procurement systems combining design, construction and operation (e.g. DBFO, BOT, PPP/PFI), the environmental debate on energy use and long-term effects on global warming as well as the growth of the facilities management industry. WLC provides the basis for creating a sinking fund to finance the operation and planned maintenance programme for the effective management of a facility.

Key factors to consider apart from capital and operating costs are the *minimum performance specification* required to compare design alternatives. For example, for a floor finish this could be thermal properties, slip resistance, life expectancy, and appearance/aesthetics. The *period of analysis* to be used for the evaluation could be the building life, functional life, economic life, or legal life. WLC is a useful theory but there are a number of problems associated with its application as it involves *long-term forecasting* which can be difficult due to policy, economic, environmental and technological changes. There are also difficulties in obtaining reliable and consistent data due to variation in practices relating to *data collection and analysis*. Maintenance costs are very difficult to predict and even where they exist, historical data tend to be variable and problematic due to the age of buildings, changes in design and construction methods, changes in performance specification, different level of use and maintenance policies. There are also problems associated with selecting a reasonable *period of analysis* (which depends on the type of building and client), and *life cycle*, that is whether this should be based on physical, functional or economic life and how quickly is the building likely to be obsolete. A major challenge is therefore to overcome the difficulties in collecting data and predicting the life cycle or lifespan of buildings, components, systems and materials due to the techno-logical revolution, evolving practices and changes in procurement policies.

Economic factors such as discount rate, inflation, interest rates, taxation should also be determined to calculate the WLC of design alternatives. However, there are problems associated with selecting an appropriate *discount rate*. A high dis-count rate means future costs are heavily discounted which can encourage short-termism, whereas a low discount rate means future costs are highly valued. The discount rate reflects the client's long-term cost of borrowing money or the oppor-tunity cost of capital and depends on *interest rates* and *inflation* which are both difficult to predict. There are also problems associated with differential inflation as some costs such as energy tend to rise faster than others. Predicting the impact of *taxation and tax relief* can be problematic as there are two types of expenditure associated with WLC – capital expenditure and revenue expenditure. Revenue expenditure (operating costs) is tax deductible, whereas capital expenditure gen-erally is not. However, some capital expenditure qualifies for tax relief but only for some types of building and parts of buildings – for example 'machinery and plant component' but this can be very complicated. Tax relief must be included in WLC calculations for clients who pay tax.

2.5   Value management theory

Value management evolved from the work of Lawrence Miles during the Second World War. Although the terms 'value analysis' (VA) and 'value engineering' (VE) are sometimes used interchangeably, value management is increasingly used to capture both VA and VE. Design value generally means worth, significance, importance, use/usefulness or esteem associated with a particular design solution. Value therefore depends on the level of function (or quality) of a design, design space, design component or materials in relation to its cost.

$$\text{Value} = \frac{\text{Function}}{\text{Cost}} \qquad (2.3)$$

According to the function–cost ratio, value is increased either by reducing the cost of a design through identifying unnecessary costs or maximising the function (or quality) of a design for a given cost or project budget to achieve economic efficiency. Shen and Liu (2004) argued that its 'underlying hypothesis is that the cost of an element/component should match the importance of its realised function(s)'. The basic philosophy of VE is therefore to remove unnecessary cost with no loss of function and hence to increase value.

Unnecessary cost is defined as 'cost which provides neither use, nor life, quality, appearance or customer features' (Kelly and Male, 2002). Unnecessary cost can occur due to 'unnecessary' design components, materials, lifecycle, or poor build-ability. In economic terms, this reflects an inefficient use of resources which requires intervention in the form of value management. Sorrel (2003) noted that value management solutions are designed to 'optimise the level of expenditure, whilst meeting all the client's building requirements' (i.e. minimise cost of building and maximise client's utility).

The value management theory recognised that it is useful to have a 'second look' at key design decisions to explore value opportunity interventions to reduce the cost of a design solution without sacrificing the function (or quality) of the space, facility or building. Design efficiency is achieved when the benefits (from additional function) is greater than the additional costs involved. Evaluating a design solution through a value management or VE process provide benefits as different teams/stakeholders (e.g. clients, users, architects, engineers, quantity surveyors and other specialists) can examine the same design to identify waste or inefficient use of resources in terms of unnecessary functions (and cost) which is crucial in large, complex and innovative projects, particularly at the early stages. Value management can result in a reallocation of resources to improve design in other areas of the building to produce greater benefits or utility to the client.

There are several definitions of the value management or VE approach. Dell'Isola (1982) in his seminal work defined it as a 'creative organised approach whose objective is to optimise cost and/or performance of a facility or system'. Kelly *et al.* (2004) argued that it is proactive, creative, problem-solving service that involves the use of a structured, facilitated, multidisciplinary team approach to make explicit the client's value system using a variety of strategy, tools and techniques such as Pareto's law (Shen and Liu, 2004), function analysis and issue analysis. For example, Pareto's law helps to identify significant elements in a building that comprise 80% of the project cost as the focus for value management. Function analysis is a powerful tool that can be applied to design spaces (e.g. board room, bathroom, classroom, and store) or component and elements (e.g. windows, cladding, roof, floors, heating system, etc.). Money can therefore be saved by eliminating unnecessary costs associated with unused spaces in the board room, bathroom, classroom, and store and/or by selecting elements and components that are fit-for-purpose. To apply this tool, a series of questions central to the design proposed are asked such as: What does it do? What alternative will perform the same function? It is important to identify primary functions core or essential to the design and secondary functions not essential and possibly avoidable. Examples of primary functions of a window are to control ventilation, exclude moisture, transmit light, and improve security and its secondary functions are to enhance appearance, reduce sound, and assist cleaning. However, what is a secondary or primary function depends on the context of the project and the client's brief. What is a primary function in a given situation could be a secondary function in another context and vice versa. The use of function analysis should be complemented by other tools such as issue analysis to resolve high level problems in clarifying, defining, and developing a client's brief and design specification.

There are different methods employed including the 40-h workshop, VM audit, and contractors change proposal depending on the stage of design and the objectives of the value management exercise (Perera *et al.*, 2011). Examples of VE savings are shown in Table 2.3.

2.6   Value of design theory

Worpole (2000) in his book titled *The Value of Architecture: Design, Economy and the Architectural Imagination* argued that good design can contribute in terms of the 'wider economic impact of attractive buildings and settings (economic cost), ... enhanced individual, and social well-being or quality of life (social cost) and greater adaptability, energy efficiency and environmental sustainability (environmental cost)'.

The economic dimension (or project profitability or loss) of a design depends mainly on two factors – the development costs and value. The two quantifiable

Table 2.3   Examples of VE savings from selected studies.

Project	Cost	Savings (%)	Participants	Source
New university building	¥36 million	8.4	Design team and client representatives	Shen and Liu (2004)
14-mile underground railway	¥3,200 million	3.5	Contractors and value management expert	Shen and Liu (2004)
Three railway stations	¥48 million	36	Contractors and internal value management manager	Shen and Liu (2004)
New chemical factory	¥37 million	15	Original design team and client representatives	Shen and Liu (2004)
Case study A	£835, 000	6	Entire design team, contractor and client	Perera et al. (2011)
Case study B	£1.5 million	5	Entire design team and client	Perera et al. (2011)
Case study C	£2.5 million	21	Entire design team and client	Perera et al. (2011)

aspects of design are, first, the direct effect of design on costs and secondly, the impact of design on value (e.g. market rents). There are economic and financial implications of design decisions. Good design is usually sold at a premium due to benefits to occupants such as reduced energy bills, reduced disruption and associated costs of unexpected maintenance and poor services. The development value relates to sales, rental income, reduction in occupancy costs, greater productivity, better interaction and communication through flexible layouts which can be achieved through good design. The economic perspective reflects the view that development is usually undertaken to ensure that the cost of development is reasonable and there is a satisfactory return on investment or benefits to the developer or owner. The economic value of design therefore establishes the benefit to the developer in financial terms and in relations to all expenditures incurred by the developer including financing costs and interest charges.

$$Economic\ value\ of\ design(residual\ profit\ or\ loss)$$
$$=Development\ value-Development\ cost$$

In terms of the social value, Worpole (2000) noted that 'good architecture and design, can have benefits and impacts beyond aesthetics – in greater feelings of safety and security, greater legibility and assurance, and in a greater sense of locality, identity, civic pride and belonging'. He further argued that, achieving this is a 'vital part of a wider notion of quality of life...which is increasingly how towns and cities compete for inward investment and population growth'. Social dimension of design can be assessed using different methods including utility values or society's degree of satisfaction using multicriteria evaluation or panels of judges in a design competition or during post-occupancy. Slaughter (2004) in commenting on the development of Design Quality Indicators recognised that the high rise social housing in Chicago was a major source of social problems for the occupants 'creating dehumanizing and grim environments'. Gilchrist and Allouche

(2005) developed social cost indicators (in a broader sense) capturing a range of factors affecting society as a result of construction projects such as pollution, traffic, ecological and health and economic related indicators with various valuation methods to assess their social impact.

A major factor in design is to incorporate environmental considerations through better space planning, use of materials, and utilisation of buildings to reduce the embodied energy and transport related energy associated with different design solutions. The success of environmental tools such as BREEAM has also been acknowledged but it is increasingly recognised that the environmental aspect of design is intricately linked with the socio-economic dimensions (OECD, 2000; Kaatz et al., 2005; Atwood, 2008). Cooper (1999) and many other researchers recognised the tension that exists in design decisions between protecting the environment, and balancing social and economic development needs. However, there are challenges in operationalising the sustainability (or value of design) theory (Kaatz et al., 2005). An argument sometimes put forward is that the 'least sustainable [design] is the more profitable' as it avoids the environmental cost. Sir Jonathon Porritt, Chair of the Sustainable Development Commission was quoted as saying:

> You have occupiers saying we want to live in green buildings, but there aren't any. So the contractors say we can build them but developers don't want them. Developers say we want them but investors won't pay for them. Then the investors say we would pay for them but there is no consumer demand (Pickard, 2007).

The difficulties relating to uncertainty and investment risk (International Energy Agency, 2007), economic returns, environmental benefits, social preference are at the heart of the value of design debate. The trade-off between cost and value (economic), environmental and social dimension is therefore crucial in decision making. The growth of carbon financing, with the price of carbon established in market, or carbon trading reflects the increasing need to establish trade-offs in design between carbon emission (reducing environmental costs) and social and economic costs.

2.7   Carter's model

Carter (2006) argued that an integrated approach to sustainability in design has the potential to save money and increase profit margins. Carter (2006) noted that case studies have demonstrated that a growing number of developers are making a commercial strategic decision to improve the environmental and social performance of their design schemes.

$$\text{Sales income} - \text{Land value} - \text{Design and construction costs} =$$
$$\text{Profit margin} + \text{Brand value}$$

Carter (2006) proposed a relationship between sales income, land value, design/construction costs and profit margin and brand value. He noted that 'living in a zero-carbon home, and the main attraction of having cheaper utility bills, and ultimately better living conditions and standards, should enhance the "Brand Value" of a housing developer'. He further argued that, profit margin is directly linked with the brand value. Profit margin does not just enhance value of the brand, but is dependent on it. There is a direct correlation between the two and one cannot be achieved without the other. Profit margin will enhance and sustain the brand value and brand value can improve profit margins.

2.8   Resource-based theory

The total construction cost is the sum of resources required which is a function of resource production coefficients, unit cost of resources and size of the project. Production coefficients determine the resource consumption rate and unit resource costs are determined by the supply and demand in the resource markets (Robinson, 2000). Construction projects require resources such as professional input for design and management process such as design labour, construction labour, plant and materials directly used in the production process (Figure 2.1).

For example, construction labour includes steelworkers, carpenters, electricians, painters, bricklayers, masons, plumbers as well as construction managers, and materials include aggregates, glass, cement, pipes, steel reinforcement, timber and other products. The construction labour production coefficient (L^c * I) is the labour requirement for a production of a unit of construction project. A 'unit of construction' is a conceptual term expressed in various physical quantities (e.g. area, number of users, or some other measures of output). Building-type infrastructure such as schools, hospitals, police stations and houses are usually measured using the superficial area method (gross floor area). Non-building-type infrastructure, mainly civil engineering structures (e.g. roads, railways, sewerage and ports) are measured in a variety of ways – superficial area, cubic volume, linear (length or width of facility), number of users or other unit of output measures. The quantity of resources required or demand (D) is a function of the number (quantity) of infrastructure or construction projects (IQ_z) and their production coefficients (C), that is the resource requirements for a unit of infrastructure or construction project.

$$D = f(IQ_z C). \qquad (2.4)$$

The production coefficient (C) is specific to each type of resource and varies according to the type of project, development type, design and construction methods (i.e. technology adopted). Similarly, the level of resources available or supply (S) is a function of existing level of resources (E), resource growth rate (G) and

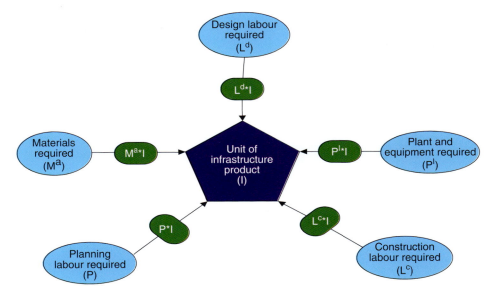

Figure 2.1   Resources required for production.

productivity rate (δ) influenced by training policy (e.g. costs and tax associated with training, availability of relevant educational courses and apprenticeship schemes) as well as improvement in procurement, technology, adoption of innovation in design and construction, and standardisation of design.

$$S = f(\delta, E, G). \tag{2.5}$$

Infrastructure or construction projects require resources and the rate of resource consumption (e.g. materials, design labour, construction labour and plant) during production depends on the type of infrastructure (or construction project), and the production coefficients. For example, the quantity of planning (professional planning input) required for a particular construction project is illustrated in 2.6.

$$RQ_j = \sum_z C_R * IQ_z. \tag{2.6}$$

In 2.6, there are j types of planning resources (R). The types of planning resources could be, for example, town planners, building control officers/planners, building inspectors, health and safety inspectors, environmental inspectors, enforcement officers and regulators. Similarly [in 2.7], there are n types of design labour (L^d) such as architects, surveyors, and various types of engineers (civil, aerodrome, transport engineers, water and sanitation, building services, electrical and power, etc.).

$$L^d Q_n = \sum_z C_{LD} * IQ_z. \tag{2.7}$$

Equation 2.8 also shows p types of construction labour (L^c) whilst 2.9 shows m types of components or materials (M^a). There are many different types of construction materials and components in the UK market. Sir John Egan (Egan, 1998) in his review of the UK construction industry titled *Rethinking Construction* noted that a house has about 40 000 different components compared with 3000 for an average car. He also cited the example of about 150 different types of toilet pans in the UK compared with six in the USA and argued for clients and designers in the UK to make much greater use of standardisation to improve efficiency and productivity.

$$L^c Q_p = \sum_z C_{LC} * IQ_z. \tag{2.8}$$

$$M^a Q_m = \sum_z C_{Ma} * IQ_z. \tag{2.9}$$

Equation 2.10 shows o types of equipment and plant resources (P). Whilst developed countries have a vast range of equipment and plant resources, the types of equipment and plant resources are often limited in many developing countries.

$$PQ_o = \sum_z C_P * IQ_z. \tag{2.10}$$

The availability of the different types of resources outlined above depends on the existing level of resources (E), resource growth rates (G) and the productivity rates (δ). The quantity of planning, design labour, construction labour resources, materials and plant resources available are illustrated in [2.11], [2.12], [2.13], [2.14] and [2.15].

$$RS_j = \delta\{E_j + (E_j * G_j)\} \tag{2.11}$$

$$L^d S_n = \delta\{E_n + (E_n * G_n)\} \tag{2.12}$$

$$L^c S_p = \delta \left\{ E_p + \left(E_p {}^* G_p \right) \right\} \tag{2.13}$$

$$M^a S_m = \delta \left\{ E_m + \left(E_m {}^* G_m \right) \right\} \tag{2.14}$$

$$P S_o = \delta \left\{ E_o + \left(E_o {}^* G_o \right) \right\} \tag{2.15}$$

Construction (or infrastructure) costs are intrinsically linked with the cost of various resources. A scarcity of resources due to limited supply means that unit resource costs are likely to increase leading to an overall increase in construction (or infrastructure development) costs. In the UK, this is reflected in the development of various indices such as materials, labour, plant and equipment to show changes in resource input costs over time. Construction costs are therefore affected by the demand and supply situation in the resource markets. Traditional cost structure of construction projects is normally presented in the form of a static elemental cost plan. Whilst this approach provides estimates of likely construction cost on an element basis, there are obvious limitations to its use for resource and production management. The alternative resource-based cost planning approach provides not only estimates of cost requirements, but more importantly provides a better understanding of the resource mix and the implication for changes in the resource markets which can be better accounted for in the cost plans. This will enable cost changes as a result of the availability of labour, material and plant resources to be carefully and accurately managed during the cost planning process.

References

Ashworth, A. (1999) *Cost Studies of Buildings*, 3rd edn, Longman Scientific & Technical, Harlow.

Ashworth, A. and Hogg, K. (2000) *Added Value in Design and Construction*, Pearson Education Ltd, Harlow.

Atwood, S. (2008), Sustainability, Alumni Event, London South Bank University, Keyworth Centre, 8 February, 2008, London.

Bathurst, P.E. and Butler, D.A. (1973) *Building Cost Control Techniques and Economics*, Heinemann, London.

BWA Associates (2006) Benchmarking of Construction Efficiency in the EU Member States – Scoping Study. Report commissioned by the Enterprise and Industry Directorate-General of the European Commission.

Cabinet Office (2011) *Government Construction. Construction Cost Benchmarks, Cost Reduction Trajectories & Indicative Cost Reduction*. April 2011, Addendum July 2012.

Carter, E. (2006) *Making Money from Sustainable Homes: A Developer's Guide*, CIOB, Ascot.

Cartlidge, D.P. (1976) *Construction Design Economics*, Hutchinson & Co., London.

Cartlidge, D. (2006) *New Aspects of Quantity Surveying Practice*, 2nd edn, Elsevier/Butterworth-Heinemann, Oxford.

Construction Excellence (2004) *Whole Life Costing*, 1 April 2004.

Cooper, I. (1999) Which focus for building assessment methods – environmental performance or sustainability. *Building Research and Information*, 27(4/5), 321–331.

Crotty, R. (2012) *The Impact of Building Information Modelling-Transforming Construction*, Spon Press, Oxford.

Dell'Isola, A. (1982) *Value Engineering in the Construction Industry*, Van Nostrand Reinhold Company Inc., New York.

Department for Communities and Local Government (2013) *Community Infrastructure Levy, the (Amendment) Regulations.* 2013 (Statutory Instruments). 2013 No. 982. Community Infrastructure Levy, England and Wales.

Egan, J. (1998) Rethinking Construction: Report of the Construction Task Force on the Scope for Improving the Quality and Efficiency of the UK Construction Industry, Department of the Environment, Transport and the Regions, London.

Ferry, D., Brandon, P. and Ferry, J. (1999) *Cost Planning of Buildings*, Blackwell Science, Oxford.

Gilchrist, A and Allouche, E.N. (2005) Quantification of social costs associated with construction projects: state of the art review. *Tunnelling and Underground Space Technology*, **20**, 89–104.

International Energy Agency (2007) *Climate Policy Uncertainty and Investment Risk*, OECD Energy, Vol. 2007(1), OECD Publications, Paris, pp. i–144.

Kaatz, E., Root, D. and Bowen, P. (2005) Broadening project participation through modified building sustainability assessment. *Building Research and Information*, **33**(5), 441–454.

Kelly, J. and Male, S. (2002) Value management, in *Best Value in Construction* (eds J. Kelly, R. Morledge and S. Wilkinson), Blackwell Publishing, Oxford, pp. 77–99.

Kelly, J., Male, S. and Drummond, D. (2004) *Value Management of Construction Projects*, Blackwell Publishing, London.

Latham, M. (1994) *Constructing the Team*, HMSO, London.

Lee, J.-S., Lee, H.-S. and Park, M.-S. (2011) Schematic cost estimating model for super tall buildings using a high-rise premium ratio. *Canadian Journal of Civil Engineering*, **38**(5), 530–545.

Martin, J. (2012) *Cost Analysis and Benchmarking*. Presentation at QS Seminar No. 9, Council of Heads of Built Environment (CHOBE), November 2012, Birmingham City University.

Nesbit, J. (1989) *Called to Account: Quantity Surveying, 1936–1986*, Stoke Publications, London.

OECD (2000) *Frameworks to Measure Sustainable Development.* An OECD Expert Workshop. OECD Proceedings for Initiative on Sustainable Development, Paris, France.

Perera, S., Hayles, C. and Kerlin, S. (2011) An analysis of value management in practice: the case of Northern Ireland's construction industry. *Journal of Financial Management of Property and Construction*, **16**(2), 94–110.

Pickard, J. (2007) Green Credentials under Scrutiny. Financial Times, 22 May 2007.

Robinson, H.S. (2000) A critical systems approach to infrastructure investment and resource management in developing countries: the InfORMED approach. Unpublished PhD thesis, South Bank University, London.

Seeley, I.H. (1972) *Building Economics: Appraisal and Control of Building Design Cost and Efficiency*, 4th edn, Macmillan, London.

Shen, Q. and Liu, G. (2004) Application of value management in the construction industry in China. *Engineering, Construction and Architectural Management*, **11**(1), 9–19.

Slaughter, E.S. (2004) Design Quality Indicators (DQI): the dynamics of design values and assessment, Building Research and Information, 32(3), 245–246

Sorrell, S. (2003) Making the link: climate policy and the reform of the UK construction industry. *Energy Policy*, **31**, 865–878.

Worpole, K. (2000) *The Value of Architecture: Design, Economy and the Architectural Imagination*, RIBA Future Studies, Royal Institute of British Architects, London.

Chapter 3
New Approaches and Rules of Measurement for Cost Estimating and Planning

Barry Symonds, Peter Barnes and Herbert Robinson

3.1   Introduction

In the UK, cost estimating and cost planning, have for the past 50 years been used by Quantity Surveyors and Cost Consultants to convey to the building client, the predicted cost of a project. The basis of the preparation of these estimates and cost plans originates from a system of 'elemental cost planning' (Seeley, 1972) and owes its origins from the construction economist pioneers who created the Building Cost Information Service (BCIS) of the Royal Institution of Chartered Surveyors (RICS), which provided the first rules for the measurement of the elements of a building. The rules were largely created, to enable historic cost data from Bills of Quantities to be archived in a standard format, to allow the UK surveyors, architects and engineers, to not only access the information, but use it to 'model' the costs of future projects. Remarkably the cost data was freely provided by members of the RICS, to allow fellow members to access what had hitherto been only available to an individual practice. The development of cost analyses and cost planning has been extensively documented in textbooks (e.g. Ferry *et al.*, 1999) together with many published conference papers and guidance rules, on cost planning methods.

However, in reality, whilst the cost consultant might have adopted 'elemental format', and accessed cost information from the BCIS, as well as those from their own sources, the reporting methods to building clients, by the building team, have varied widely. Partly this was in response, in earlier years (and even today on small projects), to clients, whom by and large, did not believe that extra fees for carrying out extensive cost planning exercises, was a necessary requirement. Later, as clients realised the potential of the Cost Plan, cost consultants responded by developing and producing individual cost reports to the client. However whilst loosely based on 'elemental cost planning' the quality of the reporting process depended much on the capability, expertise and innovation of the cost consultant.

Design Economics for the Built Environment: Impact of Sustainability on Project Evaluation, First Edition.
Edited by Herbert Robinson, Barry Symonds, Barry Gilbertson and Benedict Ilozor.
© 2015 John Wiley & Sons, Ltd. Published 2015 by John Wiley & Sons, Ltd.

Cost plans from cost consultants, might be produced on a totally different basis from one another. Submitted cost plans could easily report costs as either current, predicted to tender date or completion date, contain or not contain allowances for named risks and could even be based on different assumptions regarding measurement rules. From a client perspective, the lack of standardisation in cost reporting has produced unacceptable risk and confusion among the building team.

Following this introduction, Section 3.2 examines the impetus of change that might be brought about by the standardisation of cost estimating and the purpose and use of New Rules of Measurement (NRM) for cost planning (NRM 1) in conjunction with other similar documentation (BCIS) and the new Government guidelines on benchmarking (cost limits). This chapter also discusses how the production of NRM 1 provides the opportunity to map the RIBA (Royal Institute of British Architects) Plan of Work stages together with the OGC Gateways applicable to projects, against defined stages of estimating and cost planning. With the production of NRM 1, this chapter also explains how the BCIS elemental standard form of cost analysis has been revised to ensure that cost data will be stored appropriately. Finally it considers the impact of Building Information Modelling (BIM) on the process.

3.2   The standardisation of cost estimating

The RICS Quantity Surveying and Construction Professional Group, in recognition of the difficulties faced by the building client, established a Steering Group with the remit 'to research the problems associated with the measurement of building works at all stages of the design and construction process' (RICS, 2012c).

The Steering Group discovered that one of the root causes of inconsistency between cost consultants lay effectively in the lack of clear measurement rules and guidance for estimates and cost plans to the Construction Industry. Whilst various Standard Method of Measurements (e.g. RICS, 1989) had been produced by the RICS since 1922, these rules were largely created to provide consistency on the measurement of building work to enable relative accuracy of providing cost estimates for unit rates and builder's overheads and profit. These were embodied in Bills of Quantities that became the traditional method of cost management/control for the most part of the twentieth century.

The Steering Group discovered that 'the lack of consistency in the measurement and description…for estimates and cost plans…makes it extremely difficult for the employer and project team to understand what is included in the cost estimate, cost limit or cost target advised by the quantity surveyor; often resulting in doubts about cost advice provided. Moreover, the lack of uniformity afforded a just ground of complaint on the part of the employer who was often left in doubt as to what was really included in a cost estimate or a cost plan' (based upon the Foreword to the first edition of NRM 1; RICS, 2011).

Thus, the concept of 'RICS new rules of measurement. Order of cost estimating and cost planning, for capital building works (NRM 1)' (RICS, 2012c) was born, culminating in its first publication in 2009, being later revised in 2012. All RICS surveyors were requested to implement these rules in January 2013. However it is too early to establish if this has proved effective.

Interestingly, the reader should note, that in the UK, standard methods of cost measurement are not, like their counterparts elsewhere in Europe, enforceable in law. For example, in Germany, DIN 276 [DIN (Deutsches Institut für Normung), 2006] and DIN 277 (DIN, 2005) are standards that *must* be adopted, as are all DIN standards (Symonds, 1996). Only recently these standards were complemented by DIN 18960 which translates as '"the determination of costs in the Construction Industry' (DIN, 2008). These documents are not dissimilar to NRM 1, and the Code of Measuring Practice (RICS, 2007). However as DIN standards they must be used by everyone operating in the Public Sector. Arguably, the UK professional body RICS is only able to recommend the use of NRM 1 to ensure best practice to enable high standards of professional competence to be achieved. Should the UK be more like Germany, where the various stages of cost management are described and attached to the equivalent stages of the RIBA Plan of Work (HOAI) with each attracting a different percentage fee, then cost consultants might be more eager to follow measurement standards. However as fee scales were abandoned in the UK in the 1980s this is an unlikely scenario either now or in the future.

The effective 'standardisation'" of the cost estimating and cost planning process and production of documents, in principle, embodies the long established systems of 'initial cost estimating' and 'elemental cost planning' but now provides clear guidance in terms of definitions and measurement rules.

Clearly the new 'standard', if widely adopted by the construction industry, should undoubtedly improve the quality and standard of cost information provided to the client. However the nature of the UK Construction Industry with its many actors (i.e. engineers, architects, contractors, sub-contractors, etc.), may initially result in slow 'take-up'. Nevertheless the introduction of such a 'standard' is to be applauded and will hopefully create an opportunity for the construction economist to report and capture cost data that will bring added value to construction projects.

Almost simultaneously to the introduction of the 'standardisation' of the cost planning process, a relatively new phenomenon, BIM, has entered the construction market place. BIM's impetus is in part enhanced by UK Government Strategy (Cabinet Office, 2011) which has linked the various professional bodies, contracting organisations and a host of other bodies, to drive the construction industry to take up BIM techniques, in an effort to improve UK construction efficiency. The effect of BIM on cost information provides huge scope for change in cost estimating at an early phase and enabling this information to be updated automatically as the building model evolves. However, as the working group for the Government Construction Client Group's strategy paper discovered, BIM is being used, mostly by contractors to produce schedules of quantities (normally into some form of Excel spreadsheet) to allow pricing of the model. These quantities however are different from those derived for traditional cost estimating (e.g. based on a standard method of measurement and bills of quantities). This then provides a major challenge for the UK construction sector.

3.3   The RICS NRM 1

The authors of NRM 1, clearly state that NRM 1 is not a text that explains estimating methods or cost planning techniques. Such techniques that have evolved over the last century by cost consultants and constructors are, as earlier noted, the skill and expertise gained by construction economists and taught by academics and industry to graduates of the industry. As noted by one of the lead authors of NRM 1:

> NRM 1 does not intend to re-define estimating and cost planning-it captures best/common practice and documents it as a single reference source for everyone (Earl, 2012).

To this end, knowledge of formulating unit rates, the use of wall/floor ratios and other various cost modelling techniques (Seeley, 1972; Cartlidge, 2006) and the development and use of cost indices (Myers, 2004) together with evaluation of shape, plan and height (Morton and Jaggar, 1995) are techniques and innovations not specified by NRM 1.

The major aim and purpose of NRM 1 is in the words of the document:

> ...to provide a standard set of measurement rules that are understandable by all those involved in a construction project...and assist the QS/Cost Manager in providing effective and accurate cost advice... (RICS, 2012c).

These rules are specifically created to enable the preparation of:

- Order of Cost Estimates (Preliminary Estimates)
- Cost Plans
- Elemental Cost Plans.

And within the areas of cost analyses and benchmarking (RICS Practice Standards, UK) preparation of:

- Cost Analyses
- Benchmarking Analyses.

This is based upon the structured and consistent basis for measuring building work. To underpin this approach the rules are backed up by a series of definitions, for example cost limits, cost targets, gross internal floor area (GIFA), and so on. This is extremely important and provides the industry with a standard set of definitions that should create less confusion. However, it should be noted that whilst the RICS Standard form of Cost Analysis (SFCA) shares elemental definitions and data structures, they have effectively different objectives, namely, the SFCA provides rules for allocating cost to their functional elements, whereas the detailed tabulated rules (NRM 1: Part 4) are rules for measuring 'designed' elements of future buildings (Martin, 2012a). As many of these standards are hugely different from those used hitherto then academics, students and consultants will need to take extreme care when using traditional texts relating to measurement, estimating and cost planning. In addition historic cost data bases will need to be aligned to NRM 1 and NRM 2.

3.4   RIBA plan of work, RICS estimating, cost planning and NRM 1

The production of NRM 1 has provided the opportunity for the authors to map the RIBA Plan of Work Stages (RIBA, 2008), together with the OGC Gateways (Office of Government Commerce, 2007) applicable to projects, against defined stages of estimating and cost planning. This should provide a clear understanding for the construction team when estimates, cost plans, pre- and post-tender estimates and Bills of Quantities are to be produced within a sequential time line. The *RIBA Plan of Work* (RIBA, 2013) has only recently been updated to include BIM and guidance is provided by Sinclair (2013), on the use of the new documentation. As BIM is still in its infancy it is perhaps too early to predict the stages for production of cost estimates and cost plans that will fit building models. This effectively demands a different approach. However, some BIM software companies are indicating that NRM 1 is reasonably compatible, and that model objects can be quantified to match elements. However the fact remains that design models are not created (or should be created) to fit rules of measurement. To that end, the conflict between measurements derived from a model, for example floor areas, and areas defined by The Code of Measuring Practice (RICS, 2007), will inevitably be different from model quantities and compatibility can only be achieved by the software companies adapting their software to fit rules of measurement.

However, it is evident that some cost consultants in the UK, are already overcoming the problems of compatible information (Patchell, 2012) where full working elemental cost plans may be created in a BIM file incorporating NRM 1 and National Building Specifications. Thus, it is expected that quantitative data will increasingly be derived from BIM.

3.5   Cost estimating and cost planning

NRM 1 effectively divides provision of cost information relative to the Outline RIBA Stages of Work and OGC Gateways.

Stage A (Appraisal) and stage B (Design Brief): Order of cost estimate (NRM 1 Part 2)

Many cost consultants would identify this stage with the terminology 'Preliminary Estimates'. However these are identified by NRM 1 as Cost Estimates. A standard template of the constituents of a Cost Estimate (Figure 3.1) is recommended, and detailed rules and formulae for deriving quantities for floor areas (cost/m² GIFA) and functional units [e.g. cost/m² of net internal area (NIA) for offices/factories, cost per bedroom for hotels, cost per student for schools/universities, cost per bed space for hospitals and nursing homes] which might be used at this early stage to create basic estimates, are provided.

Perhaps the most significant standard referred to by NRM 1 is that of GIFA which has been used by cost consultants for many years (since the 1960s) and based, most likely, upon definitions provided by the BCIS of the RICS. However, it should be noted that GIFA is defined as the Gross Internal Area (GIA) as defined by the *RICS Code of Measuring Practice* (RICS, 2007). Care should be taken by cost consultants with the GIFA definition when working on projects as unfortunately GIFA and GIA together with NIA are hugely confused by construction clients. This is due to the following reasons:

- Definition by building cost estimator
- Definition by estate agent, valuer and property developer
- Definition for property management (Agency)
- Definition for rating purposes.

This is further complicated both in the UK where definitions for different building types may differ, and at the international level where many countries have differing definitions, making cross comparisons of costs and values somewhat hazardous. Thus extreme care should be taken by cost consultants, to make clear to clients the meaning of GIFA/GIA for the purposes of reporting estimates, especially to global clients operating in the UK.

Research by Kippes (2005) on residential property in Germany and Australia indicated that reporting floor areas to buyers could differ hugely from defined standards. In the case of offices, the problem of up to 24% variance is now addressed by the International Property Measurement Standards Coalition (IPMSC) of 57 countries, with the publication of the first international measurement standard (IPMSC 2014).

If sufficient information is available, then the cost estimate could be derived using an elemental method. Most surveyors will be familiar with both the rules and the formulae for the calculations which are now standardised.

COST ESTIMATE   001

RAPID 5D

PROJECT:	Offices Chelmsford			Project ref	RSD.5.15				
Dated.	01.11.2012.								
GIFA		38000	m2	£ Sub-totals	£ Total Cost	£ cost/m2 GIFA	£ % of total	£ % of total	NRM1 Ref
Ref Code	**Constituent Part of Estimate**								
1	Facilitating Works Estimate	sum			400,000	10.53	0.33		3.1
2	Building Work Estimate	sum			90,000,000	2,368.42	75.27		3.11
3	Main Contractor's preliminaries estimate	sum			10,000,000	263.16	8.36		3.14
4	Sub-Total	sum	100,400,000			-	-	83.96	
5	Main Contractor's overheads and profit estimate	sum			5,000,000	131.58	4.18		3.15
6	WORKS COST ESTIMATE	sum	105,400,000			-	-	88.14	
7	Project/Design team fees estimate	sum or %			3,000,000	78.95	2.51		3.16
8	Sub-Total		108,400,000			-	-	90.65	
9	Other development/project costs estimate	sum			500,000	13.16	0.42		3.17
10	BASE COST ESTIMATE		108,900,000			-	-	91.07	
11	RISK ALLOWANCE ESTIMATE	sum			2,750,000	72.37	2.30		3.18
11.1	Design development risk estimate	sum	250,000			-			
11.2	Construction risk estimate	sum	500,000			-			
11.3	Employer change risk estimate	sum	1,500,000			-			
11.4	Employer other risk estimate	sum	500,000			-			
12	COST LIMIT (excluding inflation)	sum	111,650,000			-	-	93.37	(CL1) 3.18.9
13	TENDER inflation estimate		2%			2,233,000	58.76	1.87	3.19
14	COST LIMIT (excluding construction inflation)	sum	113,883,000			-	-	95.24	
15	CONSTRUCTION inflation estimate		5%			5,694,150	149.85	4.76	3.19
16	COST LIMIT (including inflation)	sum	119,577,150	119,577,150	3,146.77	100.00	1.87 100.00		(CL2).3.19.7
17	VAT Assessment								

Figure 3.1   Constituents of a Cost Estimate. Source: based on NRM1 and Rapid5D cost reports.

Consultants however, may be less familiar with the rules governing the production of items such as risk. To this end, in the example given (Figure 3.1) the risks were derived as a percentage whereas in practice only exact computations of risk should be included and most likely reported to the building client separately. However rules detail all constituent parts of the Cost Estimate and these should provide a uniform approach that will enable all members of the team to more easily understand what is included in the various forms of Estimate.

Perhaps the most significant outcomes of such a standardised approach are that all the building team will be able to easily identify the following:

- Works Cost Estimate
  (Facilitating Works + Building Works Estimate + Preliminaries + Overheads and Profit)
- Base Cost Estimate
  (Works Cost Estimate + Project/Design Team Fees)
- Cost Limit (Base Cost Estimate + Risk Allowances).

Rules for the measurement of all these items are rigorously explained in NRM 1. Guidelines for the reporting of 'order of cost estimates' (OCE) is provided with a reminder that the cost consultant should take considerable care with 'inclusions and exclusions' from the OCE.

It is recommended that value-added tax (VAT) and other forms of taxation are excluded from all estimates. This can effectively only be assessed by the client organisation. The cost limit may be expressed either with or without construction inflation and there are separate calculations for the provision of construction inflation. In addition provision is made for tender inflation to be calculated. Allowances for risk, in a formalised way, is perhaps the most significant addition to the process of cost reporting. However as Mann (1992) succinctly established this is an area of 'what we must know but cannot control' and therefore cost consultants should ensure that the construction team is fully aware of risk allowances and what they do and do not cover. Forecasting and forecasting techniques are now disciplines in their own right and cost consultants need to become more conversant with the science of risk management. As NRM 1 states, risk allowances are not standard percentages. Risks in the cost management process are given as:

- Design development risks
- Construction risks
- Employer change risks
- Employer other risks.

Whilst definitions for each type of risk are provided, NRM 1 clearly advises that the definitions are not meant to be definitive or exhaustive, but simply a guide. In reality risk assessment is a specialist skill and needs quantitative analysis.

Cost planning phase: RIBA outline stages of work stages C-E (NRM 1 Part 3)

In accordance with past good practice, NRM 1 recommends that after the completion of Cost Estimates ('preliminary estimates'), and when more design information is available then, 'formal' cost plans should be prepared. NRM perceives that separate Cost Plans should be submitted at each of the stages of the RIBA Outline Plan of Work:

- RIBA: Stage C: Concept   Cost Plan 1   (OGC Gateway 3A)
- RIBA: Stage D: Design Development   Cost Plan 2
- RIBA: Stage E: Technical Design   Cost Plan 3   (OGC Gateway 3B)

However, for experienced cost consultants, this is likely to prove a difficult hurdle in practice, as seldom are the RIBA Stages of Work, as clearly sequential as that envisaged by NRM 1. The first Cost Plan (see example Figure 3.2) is based purely on GIFA calculations. When more information is available it is then possible to produce full elemental cost plans. Figure 3.3 provides an example of this but restricts reporting to condensed elements.

Nevertheless, the cost planning phase of the 'pre tender' cost advice stage is formalised within NRM 1 Part 3, the purpose of which is to provide advice to employers and designers of:

- Value for money
- Cost consequences (i.e. alternative design/specification/layout, etc.)
- Practical and balanced design
- Expenditure within budget (cost limits)
- Cost information to allow informed decisions.

Whilst not explicitly stated, these objectives are underpinned by another RICS publication, *RICS Cost Analysis and Benchmarking* (RICS, 2012a), which perceives cost consultants using the cost analyses of other projects to benchmark costs and elements, for new projects. This publication indicates that considerable care should be taken in the use of existing data to benchmark future projects. The prospect of Government Departments linking future costs of building projects for schools, social housing, hospitals, or infrastructure such as roads, looms large in a cash deficient public sector. Similarly, the commercial sector may well adopt a similar stance with offices, factories and speculative housing. Whether this predicates a return to the 'yardstick' era (Seeley, 1972) of the 1960–1980 period remains to be seen. However as the Thatcher government of the late 1970s was soon to discover, Government *cost yardsticks* were also wasteful (especially for housing) of resources. These were also backed up by high standard specifications (e.g. Parker Morris Standards (Parker Morris Committee, 1961) in housing design) that the commercial sector found vied with profits. To that end, according to the Greater London Authority (2006) today's housing provides less space per square metre per person than during the cost yardstick period. Developers profit and value in use make for poor bed-partners and the UK lays claim to the dubious honour of providing the smallest dwelling space per person than any other country in comparable European states. Although this is in part due to UK real estate surveyors and buyers, focusing value on the number of bedrooms per dwelling, rather than evaluating the square metre cost/value.

NRM 1 Part 3 perceives the widespread adoption of 'elemental cost planning' (defined as 'an iterative process, which is performed in steps of increasing detail as more design information becomes available') (RICS, 2012c) and provides detailed rules for measurement of the 'Constituents of a Cost Plan' which is effectively an update of the 'Constituents of an Order of Cost Estimate'.

These rules are stated as 'measurement rules for cost planning'. They not only refer to measurement, but also define *unit rates (EUR), that is the total cost of an*

COST PLAN   001

RAPID 5D

| PROJECT: | Offices Chelmsford | | | Project ref | RSD.5.15 | | | | |

| Dated. | 01.11.2012. | | | | | | | | |

GIFA		38000	m2	£ Sub-totals	£ Total Cost	£ cost/m2 GIFA	£ % of total	£ % of total	NRM1 Ref

Ref Code	Constituent Element of Cost Plan								
1	Facilitating Works Estimate	sum			400,000	10.53	0.34		3.1
2	Building Work Estimate	sum			85,000,000	2,236.84	73.18		3.11
3	Main Contractor's preliminaries estimate	sum			8,000,000	210.53	6.89		3.14
4	Sub-Total	sum		93,400,000		-	-	80.41	
5	Main Contractor's overheads and profit estimate	sum			5,000,000	131.58	4.30		3.15
6	WORKS COST ESTIMATE	sum		98,400,000		-	-	84.72	
7	Project/Design team fees estimate	sum or %			5,750,000	151.32	4.95		3.16
7.1	Consultants Fees			4,000,000		-			
7.2	Main Contractor's pre-construction fee estimate			250,000					
7.3	Main Contractor's Design Fee Estimate			1,500,000		-			
8	Sub-Total			104,150,000		-	-	89.67	
9	Other development/project costs estimate	sum			500,000	13.16	0.43		3.17
10	BASE COST ESTIMATE			104,650,000		-	-	90.10	
11	RISK ALLOWANCE ESTIMATE	sum			2,750,000	72.37	2.37		3.18
11.1	Design development risk estimate	sum		250,000			-		
11.2	Construction risk estimate	sum		500,000			-		
11.3	Employer change risk estimate	sum		1,500,000			-		
11.4	Employer other risk estimate	sum		500,000			-		
12	COST LIMIT (excluding inflation)	sum		107,400,000		-	-	92.46	(CL1) 3.18.9
13	TENDER inflation estimate		2%		3,222,000	84.79	2.77		3.19
14	COST LIMIT (excluding construction inflation)	sum		110,622,000		-	-	95.24	
15	CONSTRUCTION inflation estimate		5%		5,531,100	145.56	4.76		3.19
16	COST LIMIT (including inflation)	sum		116,153,100	116,153,100	3,056.66	100.00	100.00	(CL2).3.19.7
17	VAT Assessment								

Figure 3.2   Cost Plan. Source: based on standard template from NRM 1 and Rapid5D cost reports.

element divided by the element unit quantity (EUQ), their use and methods of updating. The rules of measurement for elemental cost planning are tabulated in about 300 pages of detailed documentation.

Any cost consultant conversant with many years of providing cost advice via 'elemental cost plans' will understand with relative ease the requirements of the RICS self-regulatory standard. However, the rules governing the process, that is the submission of Formal Cost Plans 1, 2 and 3, might best have been written as recommendations rather than the self-imposition of a 'straitjacket' which prevents innovation. Nevertheless, no doubt the writers of NRM 1 have only best practice in mind, and should cost consultants adopt the principles outlined, few could doubt that the standardisation of cost planning should result in a better understanding of costs than hitherto.

Recommended templates are produced both for 'Constituents of a Cost Plan' and both condensed and expanded 'Formal Elemental Cost Plans'. In addition,

ELEMENTAL COST PLAN 001								RAPID 5D
PROJECT: Offices Chelmsford				Project ref	RSD.5.15			
Dated.	DATE: 28.12.2012.		£	£	£	£	£	
Cost Centre	GROUP ELEMENT/ELEMENT	GIFA 38000 m2	Sub-totals	Total Cost of ELEMENT (TARGET COST)	cost/m2 GIFA	% of total element	% of total sub-total	NRM1 Ref
	Facilitating Works and Building Works							
0	Facilitating Works Estimate			400,000	10.53	0.34		
1	Substructure			12,600,000	331.58	10.77		
2	Superstructure			38,000,000	1,000.00	32.49		3.1
3	Internal Finishes			8,350,000	219.74	7.14		
4	Fittings, furnishings and equipment			3,000,000	78.95	2.56		3.11
5	Services			19,000,000	500.00	16.24		
6	Prefabricated buildings and building units			1,450,000	38.16	1.24		3.14
7	Work to Existing Buildings			150,000	3.95	0.13		
8	External Works			4,500,000	118.42	3.85		
	SUB-TOTAL: FACILITATING WORKS AND BUILDING WORKS (A)		87,450,000		-	-	74.76	
9	Main Contractor's preliminaries (B)			8,745,000	230.13	7.48		3.15
	SUB-TOTAL: FACILITATING WORKS AND BUILDING WORKS (A) (including Main Contractors Preliminaries)(C) when (C=A+B)		96,195,000		-	-	82.24	
10	Main Contractor's overheads and profit (D)			5,771,700	151.89	4.93		
	TOTAL: BUILDING WORKS ESTIMATE (E) when(E=C+D)		101,966,700		2,683.33	87.17	87.17	
	PROJECT/DESIGN FEES and and other DEVELOPMENT/PROJECT COSTS		-		-	-	-	3.16
11	Project /Design team Fees (F)		5,750,000		-	-	-	
12	Other Development / Projectcosts (G)		1,500,000		-	-	-	
	TOTAL: PROJECT/DESIGN FEES AND OTHER DEVELOPMENT/PROJECT COSTS ESTIMATE(H) when (H=F+G)			7,250,000	190.79	6.20	-	
	BASE COST ESTIMATE (I) when (I=E+H)		109,216,700		-	-	93.37	
13	TOTAL: RISK ALLOWANCE ESTIMATE(J)			2,184,334	57.48	1.87		3.17
	COST LIMIT (excluding inflation) (K) when (K= I+J)		111,401,034		2,931.61	95.24	95.24	
14	TOTAL INFLATION ALLOWANCE (L)			5,570,052	146.58	4.76	-	
	COST LIMIT (excluding VAT assessment) (M) (M=K+L)		116,971,086		3,078.19	100.00	100.00	3.18
16	VAT Assessment					-	-	

Figure 3.3 Elemental Cost Plan. Source: based on standard template from NRM 1, Appendix G: based upon level 1 codes and Rapid5D cost reports.

the authors of NRM 1 have mercifully recommended methods of codifying Elemental Cost Plans but also for work packages which it is recognised may be the process by which the project is managed.

3.6 Elemental Standard Form of Cost Analysis (SFCA)

With the production of NRM 1 the BCIS of the RICS has completely revised the elemental SFCA (RCIS, 2012b) to ensure that cost data will be stored appropriately. As stated in the preface to this document, the new edition of the SFCA has been produced to meet the Government's construction strategy for 'implementation of cost-led procurement, benchmarking, life cycle costing, BIM, which requires cost information to be presented consistently in a standard format'.

Whilst the SFCA is a radical change in terms of the standardisation related to NRM, the basic rules of cost analysis remain the same. However it is recognised that the Government's preoccupation with cost reduction (benchmarking) and BIM, will need to be accommodated. 'The development of BIM calls for information to be supplied from the BIM model at various stages along the project timeline so that costs can be produced or validated' (RICS, 2012b). SFCA envisages that the employer and the project's team will need to clearly adopt rules for measuring the building and its elements. However, it should be recognised perhaps, that BIM does not automatically produce elemental quantities or costs. These need to be imposed upon the model. Also cost data derived from contracting organisations involved in Design Build projects is not elemental in format. Nevertheless, the new SFCA is another step in the direction of standardising cost information.

3.7   Benchmarking (cost limits)

The UK Government has a long post war history of 'benchmarking' costs of construction, dating back to the post war period of stringent budgets. The then Ministry of Education created the first 'cost limits' for the construction of schools to enable the greatest expansion of school building since 1870. The Ministry of Education formed the Architects and Building branch in 1949 and together with pioneers such as Herefordshire County Council, innovated new construction techniques such as prefabricated units and flexible spaces (i.e. open plan) within schools. Recent commentators such as the Institute of Education (2007) have referred to this innovation as 'rat trad' or 'rationalised traditional style'. It could be said that this innovation, was the foundation stone of 'cost planning' within the UK, and led by the 1960–70 period to 'cost yardsticks' and 'costs per functional unit' for all types of public sector construction, most notably perhaps the 'housing cost yardstick' by which vast numbers of the UK housing stock were built in 20 years.

By the 1980s the Thatcher Government, had by and large divested itself of 'yardsticks' and embraced the methods of the private sector, in the belief that 'cost yardsticks' and the huge bureaucracy that implemented them was an encumbrance to speculative development, which it was believed could drive down costs by market forces of supply and demand. However, towards the end of the twentieth century there was some move to return to cost control but by and large the Governments of the day were more engaged in attempting to gain value for money via the Private Finance Initiative (PFI) which effectively took many construction costs off the Government balance sheet. More recently, and mostly as a result of the 2008–9 subprime debt revelations and the subsequent banking crisis, Government budgets are being hugely reduced to cut the UK deficit, and Government Departments are returning to benchmarking (i.e. yardsticks) of construction costs.

The Cabinet Office (Government Construction) published guidelines for benchmarking and cost reduction in 2011 and 2012 (Cabinet Office, 2011, 2012a, b). In brief, the intention of the Government is to 'to produce, a sustainable reduction of construction costs of between 15–20% by May 2015', which is effectively the end of the current parliament. A visual example of this can be seen in Figure 3.4 where it is envisaged prices per square metre of GIFA may be driven down by approximately 30% over 4–5 years. It is clear that it is not the intention to reduce costs by cutting on quality and it is stated that reductions are 'to be achieved without impacting either on the whole life value or the long term health of the construction industry'. As in 1949, the Government appears to be stating that it is not intending to cut construction budgets but obtain more building for the same budget. So no doubt the Government is looking to the construction sector to innovate with for example prefabrication, and procurement techniques, such as employed by

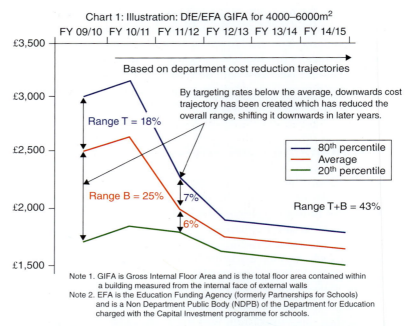

Figure 3.4   Benchmarking illustration. Source: Martin, 2012b. Taken from Cabinet Office, 2012b.

European constructors in Belgium, Holland, Germany and Scandinavia, where many global clients believe buildings are 20–30 % cheaper than in the UK. To this end, the UK Government's drive with BIM could be considered as one of the innovations they believe can bring about change and cost savings.

Not surprisingly perhaps, the BCIS of the RICS has worked closely with the Government to assist with the implementation of the policy. Martin (2012b) has from the BCIS defined benchmarking as 'the continuous process of measuring products, services and practices against the toughest competitors or those recognised as industry leaders'. The intention being to 'learn from the best in class'. It is not the objective of this chapter to inform the reader of the detail involved in benchmarking exercises. However, in many ways the new Government guidelines call upon construction economists and cost managers to use the 'Order of Cost Estimate' as defined in NRM 1 in the knowledge that this will normally be created from cost data related to a specific building rather than the cost of a building of a specific design. Thus we should be aware that by and large we know what buildings should cost rather than what they will cost.

The Cabinet Office's 2013 publication (Cabinet Office, 2013) claims that since 2012, Government Departments have made reductions in cost of £447m and that the sustainable reduction in construction cost of 15–20% is achievable circa 2015.

According to the Government's publication *Cost Benchmarking Principles and Expectations* (Cabinet Office, 2012a) cost benchmarks are described as follows:

- **Type 1 Benchmarks (Spatial Measures)** encompass the most common formats used by clients and industry to benchmark total construction costs, for example: £/m, £/m², £/m³. They are related to *throughput* (quantity) in the sense, for example, of square metres of accommodation delivered by a project.

- **Type 2 Benchmarks (Functional Measures)** encompass a range of more department-specific benchmarks, which address *business outcomes* per £ for example: £/Place; Flood Damage Avoided £/Investment £.
- **Type 3 Benchmarks** address a range of more department-specific benchmarks but where *business outcomes* are related only indirectly to the benchmark, for example: ratio of product cost (or alternatively development cost) to total construction cost.
- **Type 4 Benchmarks** are similar to Type 1 benchmarks but applied at an *elemental throughput* (quantity) level, for example: foundation costs £/m, £/m^2 or £/m^3. They are only applied within this document, when elements taken together represent majority of spend.

3.8   Building information modelling

BIM is already in use and will become a common feature on construction projects over the coming years. BIM will revolutionise the way the building industry thinks and works. The basis of BIM is a single multidimensional collaborative project model which will see a project through from its initial conception to its eventual demolition. In other words, the model will deal with the entire life cycle of a project. There are many perceived benefits of BIM relating to design, including full co-ordination of the various consultants' design elements; remodelling of alternative layouts, elements and construction techniques; and the modelling of 'as built' design on completion for maintenance, facilities management and life-cycle replacements. However, one of the key elements of the full BIM model when in use is the ability to integrate scheduling of quantities and/or materials (referred to as 4D of the model) and estimating and pricing of the works (referred to as 5D of the model).

The designed model can be measured and priced by a cost manager using an automated system with ad-hoc adjustments being incorporated for site specifics, abnormal issues and specification requirements. By using a BIM model, the take-offs and measurements can be generated direct from the underlying model, therefore the information is always consistent with the design; and where a change is made to the design, the take-offs and measurements are also automatically altered. (See Figure 3.5 for VICO example.) The adoption of this process reduces the time that is spent on taking off quantities and eliminates the potential for human error. Time can then be more usefully spent on ensuring that the pricing levels for elements of the work are consistent with the nature of the works.

In addition, the model can be continually updated as work is completed so that the valuation of works executed can be compared with the budget allowance. Further, savings, extras and value engineering possibilities can easily be tested and/or incorporated into the budget through an entire or a partial remodelling exercise. For the 4D and 5D models to be successful, the annotation of the various design elements will be critical (with regards to the level of specification and coding) in order to enable each individual element to be accurately priced. Therefore, although standard components and allowances can be incorporated, there will be a need for the component descriptions/specifications to be an accurate description of the ad-hoc nature of each construction project, and the schedule of rates will need to be both comprehensive and capable of adjustment for ad-hoc specifications and particular site circumstances.

Figure 3.5   Screen shot: BIM showing 3D model and 4D and 5D attributes of time, measurement and cost. Source: VICO and RAPID5D.

The automated pricing system would usually need to be refined and aligned with the designer's specification level and range of products, and may also need to be designed to take account of or otherwise allow for the impact of inflationary influences (e.g. economic climate, supply and demand, technological changes, etc.). Given that virtually all construction projects are unique, the requirement for ad-hoc adjustments for project specific and/or abnormal elements is a challenge that the 4D and 5D elements of the BIM model still need to address.

3.9   Concluding remarks

New approaches in cost estimating and cost planning in the UK are largely related to the drive for efficiency not only from the public but also the private sector. Whilst in part this is due to the austerity of Government budgets it is also due to Global Clients identifying that construction costs in the UK are often higher than those in comparable economies.

It is difficult perhaps in the above to identify the enormity of the changes that will come into being as a result of BIM and standardisation of construction cost documentation. The introduction of NRM 1 (in addition to NRM 2 and 3), plus new standard forms of cost analysis and benchmarking, provide not only the greatest challenge to the construction economist and cost manager, but the best opportunity in a generation, of improving the prediction and control of construction costs. This in turn will drive innovation in construction management and techniques of construction to new models of production.

As outlined above, the rapid standardisation of cost documentation is not in itself a radical innovation. However, the implications of standardisation together with BIM will hugely change not only the methods by which we build but the way we procure construction and work together. Integrated working is undoubtedly the keyword. However as Ray Crotty (2012) has noted, standardisation will only assist if it fits the need and that standards can, if not carefully thought out, impose difficulties:

> the idea of a shared language, of uniformity and consistency of meaning across the disciplines of project management, is stymied from the beginning. Home-made applications, spreadsheets, and baseless but impressive looking planning graphics proliferate-all presenting mutually contradictory views of the project.

It has to be accepted that in many respects the massive standardisation of the RICS of New Rules of Measurement, was commenced long before the full implications of BIM were understood. In addition BIM is only at the inception of its development and it will take some time to reach its full potential. However, already software companies are writing standard libraries and creating cost databases to fit, and construction companies are beginning to see the advantages that standardisation can bring.

There can be little argument that the standardisation of the cost estimating process will lead to better efficiency and greater understanding. However, we are only at the beginning of the process. Standardisation is a positive but cost prediction will only remain as good as the sum of intelligent standardised systems, integrated information, cost data and the ability of the cost consultant. Thus the skills of tomorrow's construction economists and managers will need to encompass much more than now and this will require a massive level of investment in re-education and training. This then is the challenge for the construction sector.

References

Cabinet Office (2011) A report for the Government Construction Client Group Building Information Modelling (BIM) Working Party Strategy Paper. March 2011.

Cabinet Office (2012a) Government Construction. *Cost Benchmarking Principles and Expectations.* **10** February 2012.

Cabinet Office (2012b) Government Construction. *Construction Cost Benchmarks, Cost Reduction Trajectories & Indicative Cost Reduction.* April 2011, Addendum July 2012.

Cabinet Office (2013) Government Construction. *Construction Cost benchmarks, Cost Reduction Trajectories & Cost Reduction Trajectories to March 2013.* **2** July 2013.

Cartlidge, D. (2006) *New Aspects of Quantity Surveying Practice,* 2nd edn, Elsevier/Butterworth-Heinemann, Oxford.

Crotty, R. (2012) *The Impact of Building Information Modelling-Transforming Construction.* Spon Press, Oxford.

DIN (Deutsches Institut für Normung) (2005) *DIN 277:2005-02, Grundflächen und Rauminhalte von Bauwerken im Hochbau,* Beuth-Verlag, Berlin.

DIN (2006) *DIN 276-1:2006-11, Kosten im Bauwesen – Teil 1: Hochbau,* Beuth-Verlag, Berlin.

DIN (2008) *DIN 18960:2008-02, Nutzungskosten im Hochbau,* Beuth-Verlag, Berlin.

Earl, S. (2012) Managing costs consistently. *RICS Construction Journal,* **April–May,** 6–7.

Ferry, J.F., Brandon, S.B. and Ferry, J.D. (1999) *Cost Planning of Buildings,* 7th edn, Blackwell Science, Oxford.

Greater London Authority (2006). Housing Space Standards. A report by HATC Ltd for the Greater Lonondon Authority. August 2006.

Institute of Education (2007) Institute of Education Archives. http://www.ioe.ac.uk/services/4389.html.

IPMSC (2014). International Property Measurement Standards. IPMS (2014).

Kippes, S. (2005) *Problems Concerning Calculation of Habitable Dwelling Surfaces– an Empirical Analysis*. The 21st ARES annual conference, 13 April 2005, Santa Fe, NM, USA.

Mann, T. (1992). *Building Economics for Architects*, Van Nostrand Reinhold, New York.

Martin, J. (2012a) Changing requirements. *RICS Construction Journal*, **April–May**, 12–13.

Martin, J. (2012b) Cost Analysis and Benchmarking. Presentation at QS Seminar No. 9, Council of Heads of Built Environment (CHOBE), November 2012, Birmingham City University.

Morton, R. and Jaggar, D. (1995) *Design and the Economics of Building*, E & FN Spon, London.

Myers, D. (2004) *Construction Economics, a New Approach*, Spon Press, London.

Office of Government Commerce (2007) OGC Gateway, OGC ITIL.

Patchell, B. (2012) *A Full Working Elemental Cost Plan in a BIM File*. Proceedings of the RICS BIM Conference, 9 February 2012, London.

Parker Morris Committee (1961) *Homes for Today and Tomorrow*, HMSO, London.

RIBA (Royal Institute of British Architects) (2008) *RIBA Outline Plan of Work* 2007. Amended 2008, RIBA, London.

RIBA (2013) *RIBA Plan of Work 2013*, RIBA, London.

RICS (Royal Institution of Chartered Surveyors) (1989) *Standard Method of Measurement*, 7th edn, RICS, London.

RICS (2007) *RICS Code of Measuring Practice*, 6th edn, RICS, London.

RICS (2011) *RICS New Rules of Measurement. Order of Cost Estimating and Cost Planning for Capital Works (NRM1)*, 1st edn, RICS, London.

RICS (2012a) *RICS Cost Analysis and Benchmarking*, 1st edn. Guidance Note: RICS QS and Construction Standards: GN 86/2011, RICS, London.

RICS (2012b) *RICS Elemental Standard for of Cost Analysis. Principles, Instructions, Elements and Definitions*, 4th (NRM) edn, RICS, London.

RICS (2012c) *RICS New Rules of Measurement. Order of Cost Estimating and Cost Planning for Capital Building Works (NRM 1)*, 2nd edn, RICS, London.

Seeley, I.H. (1972) *Building Economics*, Macmillan, London.

Sinclair, D. (2013) *Guide to Using the RIBA Plan of Work*, RIBA, London.

Symonds, B.C. (1996). The German Construction Industry 1994: Some aspects of economics, procurement, design, costing and supervision of projects. *The Surveyor,* The Professional Journal of the Institution of Surveyors Malaysia, **31.**1, 30–39.

Chapter 4
The Relationship between Building Height and Construction Costs

David Picken and Benedict Ilozor

4.1   Introduction

Conventional wisdom in the construction industry suggests that for the same areas of accommodation, tall buildings are more expensive to construct than low rise buildings. In the literature which includes books on construction economics, design economics and building cost planning, statements and definitions can be found which demonstrate this widely held view on the relationship between height and cost. The following shows a sample of the various views with references relating to the effect of height on cost:

Two-storey building performs cheapest (Nisbet, 1961). (Although this statement was included in an early study of height and cost, and only referred to buildings in the low-rise range.)

Prices per square foot tended to rise as the number of storeys increased in Britain. Housing in tall multi-storey blocks is around 50% more expensive than those in two-storey dwellings (Stone, 1967).

Multi-storey buildings/high-rise buildings would be a design choice only if they could make savings from the tremendous land cost by building upwards (Cartlidge, 1973).

Generally the cost of building per square of floor area can be expected to increase with the addition of extra storeys (Bathurst and Butler, 1980).

Constructional costs of buildings rise with increases in height (Seeley, 1995).

Here it is possible, and desirable to be dogmatic. Tall buildings are invariably more expensive to build than two- or three-storey buildings offering the same accommodation, and the taller the building the greater the comparative cost (Ferry *et al.*, 1999).

The constructional costs of tall structures are greater than low-rise buildings offering a similar amount of accommodation (Ashworth, 2004).

Design Economics for the Built Environment: Impact of Sustainability on Project Evaluation, First Edition.
Edited by Herbert Robinson, Barry Symonds, Barry Gilbertson and Benedict Ilozor.
© 2015 John Wiley & Sons, Ltd. Published 2015 by John Wiley & Sons, Ltd.

All this seems to be rather persuasive and most authors describe the key issues: increased cost of mechanical plant, tower cranes and so on, for constructing high rise buildings; the cost of vertical transportation (lifts, staircases) increases with height; the need for the lower sections of the building to be able to carry the weight of the structure above. A fairly obvious one, at least in many countries, relates to the cost of labour where site operatives are paid enhanced rates for carrying out work at higher levels. Typically, these supplements are incrementally based. It is possible to imagine how perspectives on height and costs have been handed down over the generations to become almost *nostrums*, of sorts.

This chapter examines these conventional views by focusing on the research that has been undertaken to analyse them and then propose new ideas and alternatives on how the cost–height relationship might work. The chapter starts which a discussion of earlier research work in the 1970s and 1980s revealing a more complex relationship than the simple linear notion that as height increases so does cost. This is followed by an analysis of more recent work which has examined this further including other factors such as geographical location which can be influential. Data are explored from Shanghai and Hong Kong to establish the nature of the cost–height relationships and to generate a better understanding of cost profiles and the factors influencing building costs.

4.2   Research in the 1970s and 1980s

Flanagan and Norman (1978) focused on the issue in a short article, returning to the widely held theories and wondering whether this aspect of design economics was not well understood. They highlighted various studies which perhaps might not have been easily noticed in the mainstream – Tregenza in 1972 in the UK, separate studies by Jarle and Pöyhönen in 1969 in Finland, and Steyert in 1972 in the USA. Essentially, Flanagan and Norman (1978) summarised the conventional wisdom noted above by arguing that what those views were suggesting is that there is a simplistic linear relationship between height and cost (cost rising as height increases). Following the work of Steyert (1972), Flanagan and Norman (1978) questioned the linear relationship theory. They argued that the design variables likely to be affected by height were many and various, and that changes brought about by increases in height were not discrete and independent, but were interrelated, an observation also made by Proverbs *et al.* (1999a,1999b,1999c) in their research which examined the movement of materials on construction sites involving multi-storey buildings. Flanagan and Norman (1978) put forward four categories of cost in connection with height which are also identified by Ashworth (2004, and in earlier editions). They are as follows:

(i)   Costs which fall as height increases (an example is a roof). This can be demonstrated fairly easily by looking at the simple example of a house – single storey and then, with exactly the same footprint with two and three storeys. The total cost of the roof will be pretty much the same in each case, but much less in terms of cost per square metre of floor area in the two- and three-storey buildings.

(ii)   Costs which rise as height increases (an example would be lifts). This would include the need for greater capacity in the lifting machinery as the building height increases. In addition, design standards require an increase in the number of lifts as height increases.

(iii)   Costs which do not change with height (examples are floor finishes and internal doors).

(iv)   Costs which fall initially and then rise as the height increases (the example offered here was external cladding). This is an example of the effect of an increase in pay rates for working at higher levels. There is also the cost of scaffolding (on low to medium rise buildings) and more sophisticated access equipment at higher levels.

An observation on item (iii) is that, taken on face value, it is difficult to imagine any costs which should not be affected by height – presumably the floor finishing material has to be transported to the higher floors. In addition, there is the situation noted earlier where workers are paid more for working at higher levels. So, for example, the joiner is paid more for fixing a door on the thirtieth floor than for fixing precisely the same door on the first floor. There is a need to mention the issue of the learning curve increasing productivity where work repeats floor after floor, but eventually a point will be reached where optimum labour constants are achieved. Flanagan and Norman (1978) postulated a U-shaped total cost curve as a result of adding their four categories together. That is to say, overall costs will fall initially and then, at some point, bottom out and start to rise as the costs that are sharply affected by height come to bear. They went on to test their theory with a study of UK data, and the curve plotted is shown diagrammatically in Figure 4.1. Their U-shaped curve became something of a classic in construction economics to describe the height–cost relationship. It is worth noting that they made it clear that their study provided a theoretical basis and the precise nature of the impact of height on construction cost needed further investigation.

Work by Newton (1982) confirmed the U-shaped curve and also added some interesting observations. First, the trials carried out using a simulation model held the gross floor area constant whilst increasing the height. Here the low point of the U-shaped curve was slightly less (around three storeys) than that shown in Figure 4.1, but thereafter the rise was much steeper. When the plan area (or building footprint) was held constant, Newton suggested that, for smaller footprint areas, increasing height will not cause costs to increase so quickly (not until the 12-storey mark is reached for a 600 m² footprint). The flatter curves for constant footprint areas, as the height increases, are to be expected. For each additional

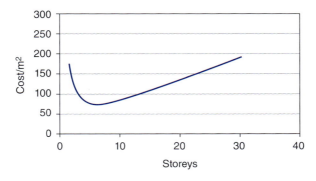

Figure 4.1   Cost (*in £*)/m² of gross floor area *versus* number of storeys.
Source: Flanagan and Norman (1978).

storey one is adding to the gross floor space for, presumably, a relatively small increase in total cost. These effects were also noted to a certain extent by Bathurst and Butler (1980).

4.3   More recent research in Hong Kong and Shanghai

The Hong Kong perspective

Picken and Ilozor (2003) explored the relationship between height and cost further using Hong Kong as a case study. The built environment in Hong Kong is characterised by tall buildings and was perceived as something of a 'laboratory' for tall buildings to study various issues relating to construction economics.

The Hong Kong study was not concerned with those cost–height phenomena associated with low rise buildings, namely those which cause costs per square metre to fall initially with height changes from a single storey to four or five storeys. An example of such would be the substructure for which statutory regulations related to structural stability would be required as a minimum for a single storey structure. The same substructure can also carry a two-storey building of the same plan (footprint) area. The conventional wisdom referred to earlier then assumes that there is a point where this fall in cost would 'bottom out' as the cost of substructures starts to reflect the need for stronger and more complex foundations in higher buildings.

Initially, a broad view was taken that, for all intents and purposes, all Hong Kong buildings are multi-storey and are, so to speak, 'beyond' the bottom of any U-shaped curve which may exist such as that described by Flanagan and Norman (1978). Based on this view, taken together with the so-called conventional theories described earlier, the expectation before the study was that whilst the relationship between height and cost might not be linear, it would be represented by an upward sloping curve. When data were collected for housing development projects, it transpired that that there were some data related to low rise projects and these were included in the analysis. They collected data for 24 buildings ranging from three storeys at an overall height of 9.45 m to 39 storeys at an overall height of 107.35 m. Figure 4.2 shows a diagrammatic representation of

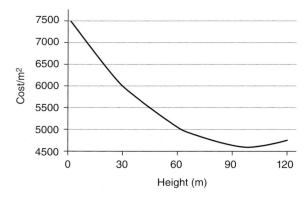

Figure 4.2   Cost (in $)/m² of gross floor area versus height (Hong Kong data). Source: Derived from Picken and Ilozor (2003).

the regression curve plotted from the Hong Kong data. The upward turn of the curve after 100 m was added and is speculative; it is not actually supported by these research data. What can be seen is that the curve is beginning to flatten as it approaches 100 m. As Picken and Ilozor (2003) suggest, to explore precisely the continuation of the curve, it would be necessary to obtain more data for buildings over, say, 115 m.

Based on the data and analysis for buildings in Hong Kong, Picken and Ilozor (2003) suggested that it appears that increasing height does not seem to cause increasing cost until a height of around 100 m (or just over 30 storeys) is reached. This was a surprising result and displayed a quite different bottoming out point than that shown earlier in the study by Flanagan and Norman (1978). An inspection of the plan (or footprint) areas revealed quite small sizes in some samples which may echo the flatter curves observed by Newton (1982) when plan areas were held constant. Indeed, the term 'pencil blocks' is used for some buildings in Hong Kong, with plan areas often constrained, and working within tight plot ratios and maximising site area usage is a critical task for designers.

Picken and Ilozor (2003) explained that they still regarded the theory of Flanagan and Norman (1978) as the guideline for studying the relationship between height and cost of high-rise buildings. However, they indicated that if there is something in the Hong Kong data, then perhaps a different set of criteria should be applied in the judgement of how height affects cost depending on the context and commonality of buildings in the location under consideration. A further observation is that, perhaps, buildings could be grouped for analysis in terms of cost–height relationships; for example a group for buildings up to 10 storeys, then groups for 11–20 storeys, 21–30 storeys, and so on. Each group would display its own downward curve in the same manner shown by the analysis of the Hong Kong data. However, there would be an element of the bottoming out referred to earlier at the top of each height range. Each succeeding range, though, would start at a higher cost per floor area with the influence of the same special considerations for low rise buildings at work as shown in the U-shaped curve of Flanagan and Norman (1978). A graph expressing this would have a saw tooth shape as shown diagrammatically in Figure 4.3. What this notion is saying is that, in effect, the saw tooth shape is what we might see if we were to zoom into the graph in Figure 4.2 which is merely a generalised expression of this reality.

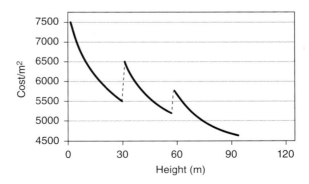

Figure 4.3 Cost (in $)/m² of gross floor area versus height – grouped by height ranges.

In summarising the Hong Kong study, various questions can be posed:

- Do Hong Kong contractors have some special expertise in multi-storey construction which accounts for the results observed?

  Certainly it can be assumed that they have substantial practical experience. The phenomenon of contractors performing best in bidding for work in which they are most experienced can be observed here. The situation is quite different from Indonesia where it was observed in a study of Jakarta and Yogyakarta that inflationary increases in material cost, inaccurate material estimating and project complexity play major roles in raising building construction cost relative to height (Kaming *et al.*, 1997).

- Assuming the results are right for Hong Kong, perhaps one might assume that different countries would show significantly different results, and these results can be equated to the amount of, and therefore expertise in, multi-storey building work.

  This point was supported by many authors (Chan and Kumaraswamy, 1995; Proverbs *et al.*, 1999a,1999b,1999c). Perhaps the issue of the method of choosing contractors affecting the cost–height relationship (Holt *et al.*, 1994; Holt, 1995) does not apply in Hong Kong.

- Is there a cost–height relationship factor which Hong Kong contractors fail to recognise and therefore do not allow for it in their bids?

  Perhaps, they are making a profit of some sort and are not inclined to look more closely at their costing. The risk of making any height allowances is passed onto the multitude of sub-contractors who actually carry out the work on site. Five layers of sub-contractors (sub- sub-, etc.) is not uncommon, and these firms often use very unsophisticated pricing procedures. In addition, they are operating in a very competitive market place.

- Is the disparity as a result of this study being based on public housing projects?

  There can be differences in cost depending on whether the building is privately or publicly owned. There are more rigorous controls on government financing of building projects. Extensive standardisation and the use of large prefabricated assemblies for public premises can also effectively reduce the design deviations, construction time variations, rework, and hence cost (Chan and Kumaraswamy, 1995; Love *et al.*, 1997).

- There are no formally agreed wage rates in Hong Kong. It is very much a *laissez faire* approach and there are no provisions in Hong Kong for paying enhanced wages for working at higher levels. Workers associations, notionally trade unions, do exist. However, they do not wield the power, or have the same potential, to achieve a regime of pay and conditions that would reward working at height in the same way as that described in other countries.

Picken and Ilozor (2003) concluded by making suggestions for further study. Assuming that their analysis is at least partially representative of cost–height relationships in Hong Kong, it is possible that locations like Hong Kong, where there are intensive concentrations of tall buildings – they suggested, for example, New York and Chicago – might also exhibit these different cost–height relationships. In addition, data sets for locations where there are less intensive concentrations of tall buildings would assist in testing the theories described by Flanagan and Norman (1978).

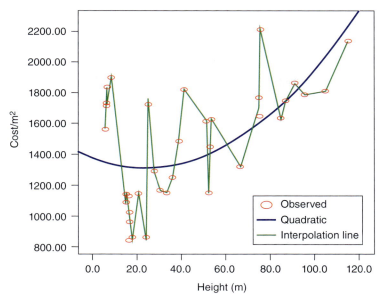

Figure 4.4   Cost (in ¥)/m² of gross floor area versus height. Source: Blackman and Picken (2010)

Studies in Shanghai

Steps to carry out the extended research suggested by the Hong Kong study were reported by Blackman and Picken (2010), who analysed cost data related to buildings in Shanghai. This is a location that has become well known as a high-density city and is regarded as one of the fastest growing cities in the world. The Emporis Corporation (2008) produced a ranking of cities presented in terms of the 'most compelling skyline'. Essentially, it identifies those cities with a significant number of tall buildings and the visual impact that this presents. Shanghai was ranked 7th amongst cities with the most compelling skylines in the world, and at the time was experiencing a major high-rise building boom. The distinguishing aspect of the Shanghai study was that it examined more detailed costs than merely total costs. Costs for various elements were also obtained. Data for 36 buildings were assembled – ranging from two storeys (with an overall height of 6 m) to 37 storeys (with an overall height of 115.3 m).

The analysis of the Shanghai data for total cost versus height is shown in Figure 4.4.

The curve bottoms out at around 24 m – less than the 100 m in Hong Kong, but still considerably higher than the UK analysis at five to six storeys. This starts to show some support for the idea of a different set of criteria being applied in the judgement of how height affects cost depending on the context and commonality of tall buildings in the location under consideration. There are some factors related to buildings in Shanghai as compared with those in Hong Kong which are worthy of mention at this point. In Shanghai, any building over 100 m is referred to as a super-high tower, which requires a different set of fire prevention measures for design, including a refuge storey for fire escape purposes. These limitations could be the reasons why Shanghai does not have as many buildings over 100 m as Hong Kong, especially in the residential high-rise category. There are no special regulations set up for buildings over 100 m in Hong Kong.

The number of six-storey buildings in the Shanghai sample is relatively large. This is because six-storey buildings are widely built in Shanghai. There is no regulation requiring the installation of lifts in residential buildings up to and including six storeys. Once a high-rise residential building is over 11 storeys, a minimum of two lifts should be installed. In Hong Kong, a more structured, mathematical approach is applied for the calculation of the lift installations as required by the building authorities.

The incidence of high-rise residential building developments is therefore different between Shanghai and Hong Kong. Hong Kong is a city with an overwhelming proportion of such buildings with most of the public sector housing being high-rise due to the need for high densities. In contrast, high-rise living is still a relatively new and, to an extent, fashionable concept in Shanghai.

The intensity of using smaller plots and building higher is greater in Hong Kong than Shanghai due to the varying densities of these two cities. High-rise residential buildings, typically constructed in compact groups in Shanghai usually contain 10, 20, 30 or more similar buildings with a height of less than 35 storeys. Residential buildings in Hong Kong commonly reach heights of 60, 70 or even 80 storeys.

It can, therefore, be suggested that in comparison with Hong Kong there is less experience in the construction of tall buildings in Shanghai.

Turning to the analysis of the elemental cost data related to the Shanghai buildings, it is important to recall the work of Steyert in the USA in 1972. Steyert (1972) concluded that different elements of a building would have different responses in terms of cost when the height changes. He suggested that the cost of some elements might decrease with height. Steyert summarised two reasons for cost reduction. First, there would be a learning curve effect. Secondly, the total cost of some items would increase less than proportionately with gross floor area, such as roof and substructure.

More recently de Jong et al. (2007) studied the economic context of high-rise office buildings in the Netherlands by interviewing experts working on high-rise projects and cost modelling techniques. They analysed seven elemental costs and total construction cost against increases in height. Structure, installations (building services), and elevator costs were the main factors contributing to the total cost increase with an average of 16, 25 and 3%, respectively. Site costs were also heavily influenced by height. Their research discovered that eight storeys is the height of the lowest cost per square metre of the façade structure of buildings categorised as 'high-rise' building in the Netherlands. This being, in effect, in the context of the classic U-shaped curve, the lowest point of their curve for this element. They noted that eight storeys is regarded as the starting point for the experts studying such buildings. They introduced a term, 'high-rise-ability', and suggested that building cost is one of the important factors influencing this aspect. They expressed the view that a good understanding of these issues is important in seeking cost reductions during the design process.

From the Shanghai data set, the costs of several elements were analysed. For the fifth, services, this was also broken down into plumbing and electrical.

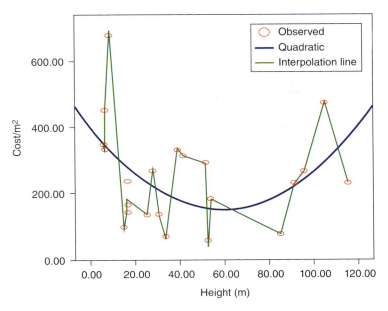

Figure 4.5   Substructure cost (in ¥)/m² of gross floor area versus height. Source: Blackman and Picken (2010).

Substructure

For the analysis in respect of the substructure costs, 22 of the buildings in the sample were selected. Figure 4.5 shows the resulting graph. The curve is beginning to flatten as it approaches 60 m with a U-shaped profile that is somewhat similar to that of the total cost and height relationship.

Roof

To analyse the roof element, 27 of the buildings were used. Figure 4.6 shows the resulting graph. It can be observed that the curve is again beginning to flatten as it approaches 65 m, and exhibits a similar profile to the curve of the total cost and height relationship.

Upper floors

Data for 22 buildings were assembled from the 36 to analyse the upper floors element. Figure 4.7 shows the graph plotted for this element. It can be observed that the curve is beginning to flatten as it approaches 40 m, with a similar profile to the curve of the total cost and height relationship.

Doors and windows

Data for 27 buildings were assembled for the windows and doors element. Figure 4.8 shows the graph plotted from the doors and windows data. It can be observed that the curve is beginning to flatten as it approaches 50 m, with a similar profile to the curve of the total cost and height relationship.

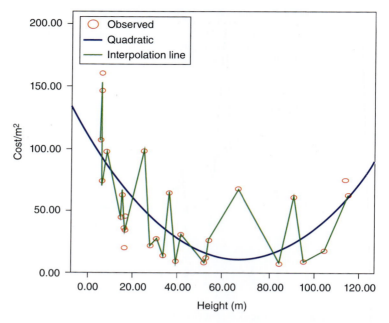

Figure 4.6 Roof cost (in ¥)/m² of gross floor area versus height. Source: Blackman and Picken (2010).

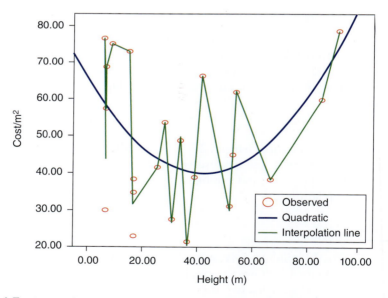

Figure 4.7 Upper floors cost (in ¥)/m² of gross floor area versus height. Source: Blackman and Picken (2010).

Services

Services is a group element including plumbing, mechanical, fire, electrical, transportation, and special services (to provide services or installations not covered by other elements). Data for 35 buildings were selected for this group element.

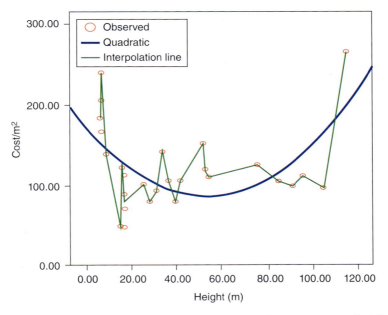

Figure 4.8   Doors and windows cost (in ¥)/m² of gross floor area versus height. Source: Blackman and Picken (2010).

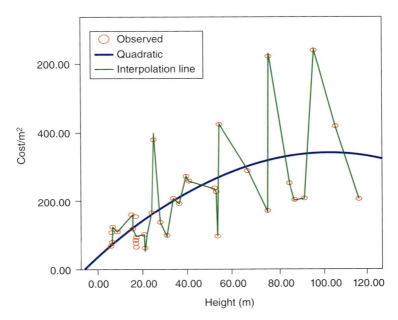

Figure 4.9   Services cost (in ¥)/m² of gross floor area versus height. Source: Blackman and Picken (2010).

Figure 4.9 shows the graph plotted from the services data. It can be seen that there are some quite different results with a profile almost the opposite approach of the total cost and height relationship. The highest cost per square metre was reached at 106 m. This result prompted the idea of looking at this element more

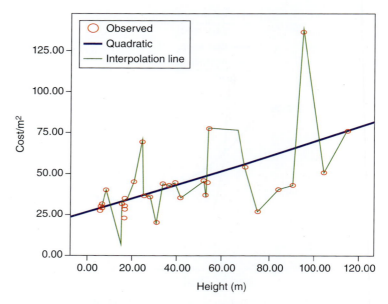

Figure 4.10   Plumbing cost (in ¥)/m² of gross floor area versus height. Source: Blackman and Picken (2010).

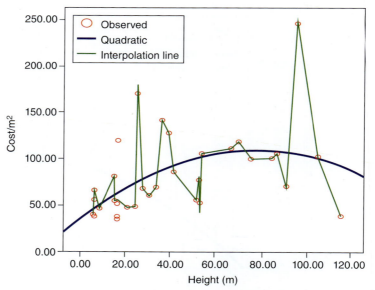

Figure 4.11   Electrical cost (in ¥)/m² of gross floor area versus height. Source: Blackman and Picken (2010).

closely and the available data allowed for further analysis of plumbing and electrical work.

Plumbing

There was sufficient detail in the data for the plumbing costs of 33 of the buildings to be analysed. Figure 4.10 shows the graph plotted from the plumbing data. For the first time, we see a linear relationship between cost and height.

Electrical

Lastly, 34 of the buildings in the sample were analysed for electrical costs. Figure 4.11 shows the graph plotted from the electrical data. The profile of the curve is convex – more akin to the overall services element with the height being around 78 m.

4.4   Conclusions

The various studies described in the chapter show that the interrelationships between construction cost and building height are complex. Clearly, it would appear that a statement which claims 'construction costs become more expensive as building height increases' is too simplistic, and perhaps, patently incorrect. Certainly there are no universal rules on the matter. The way in which height affects cost will be different in different locations, and this will be due to the commonality of tall buildings as well as other factors such as building standards, regulations and the expertise of the local construction industry. Analysis of local data to determine the particular cost profiles relevant to the location in question is needed to generate a better understanding. Shanghai and Hong Kong exhibit similar distinctive cost–height relationships. However, the profiles of the curves are different. This shows that different sets of criteria should be applied in making a judgement of how height affects cost in these two locations.

Taking the analysis of the Shanghai data as an example, it can be suggested that the results help to identify targets where cost reduction efforts can be made. For example, substructure construction costs reduce as the height increases from 0 to 60 m. The lowest cost range falls into 61–90 m with a dramatic decrease. The highest cost range in terms of height is from 91 to 120 m. Therefore, it could be concluded that 0–60 m and 91–120 m should be the main focus of designing and building the substructure in Shanghai. This could involve identifying design considerations, as well as construction techniques. The curves for the upper floors element showed a dramatic down and up movement. In addition, after the bottoming out point (around 43 m), as the height increases, the faster the cost increases. This would lead to the conclusion that upper floor systems would be a worthwhile target for structural designers and cost planners for identifying opportunities for cost reduction and checking cost escalation. The curves of electrical and services costs appear to be convex when compared with the other elements above. Initially, they increase with an increase in height, and then level out. The relationship between plumbing costs and height appears to increase with cost increasing as height increases consistently. Here is another element where finding a cost effective method of undertaking the task is essential.

The usefulness of analysis of cost–height relationships can, perhaps, be summarised in the cost management techniques of value engineering and value management. Research can assist in the process when high-rise buildings are under consideration by identifying the elemental costs more influenced by height. The considerations could assist in the creation of value for money plans, as well as more energy and labour efficient methods to undertake certain tasks to reduce the total construction cost. If the minimum cost point of every element could be moved, as it were, from the current position, it is easy to imagine that the total lowest cost per square metre would move to a higher metre range.

References

Ashworth, A. (2004) *Cost Studies of Buildings*, 4th edn, Pearson/Prentice Hall, Harlow.

Bathurst, P.E. and Butler, D.A. (1980) *Buildings Cost Control Techniques and Economics*, 2nd edn, Heinemann, London.

Blackman, I.Q. and Picken, D.H. (2010) Height and construction costs of residential high-rise buildings in Shanghai. *Journal of Construction Engineering and Management*, **136**(11), 1169–1180.

Cartlidge, D.P. (1973) *Cost Planning and Building Economics*, Hutchinson, London.

Chan, D.W.M. and Kumaraswamy, M.M. (1995) A study of the factors affecting construction durations in Hong Kong. *Construction Management and Economics*, **13**(4), 319–333.

de Jong, P., van Oss, S.C.F. and Wamelink, J.W.F. (2007) High rise ability, in *Earthquake Resistant Engineering Structures VI* (eds C.A. Brebbia *et al.*), WIT Press, Southampton, pp. 1–11.

Emporis Corporation (2008) Skyline ranking. http://www.emporis.com/en/bu/sk/st/sr/ (accessed 10 April 2008).

Ferry, D.J., Brandon, P.S. and Ferry, J.D. (1999) *Cost Planning of Buildings*, 7th edn, Blackwell Science, Oxford.

Flanagan, R. and Norman, G. (1978) The relationship between construction price and height. *Chartered Surveyor Building and Quantity Surveying Quarterly*, **5**(4), 68–71.

Holt, G.D. (1995) *A methodology for predicting the performance of construction contractors*. PhD thesis, University of Wolverhampton.

Holt, G.D., Olomolaiye, P.O. and Harris, F.C. (1994) Factors influencing UK clients' choice of contractor. *Building and Environment*, **29**(2), 241–248.

Kaming, P.F., Olomolaiye, P.O., Holt, G.D. and Harris, F.C. (1997) Factors influencing construction time and cost overruns on high-rise projects in Indonesia. *Construction Management and Economics*, **15**(1), 83–94.

Love, P.E.D., Mandal, P. and Li, H. (1997) *A Systematic Approach to Modelling the Causes and Effects of Rework in Construction*. The First International Conference on Construction Industry Development: Building the Future Together, 9–11 December, Singapore, pp. 347–353.

Newton, S. (1982) Cost modelling: a tentative specification, in *Building Cost Techniques: New Directions*, E & FN Spon, London, pp. 192–209.

Nisbet, J. (1961) *Estimating and Cost Control*. Batsford, London.

Picken, D. and Ilozor, B.D. (2003) Height and construction costs of buildings in Hong Kong. *Construction Management and Economics*, **21**, 107–111.

Proverbs, D.G., Holt, G.D. and Love, P.E.D. (1999a) Logistics of materials handling methods in high rise in-situ construction. *International Journal of Physical Distribution & Logistics Management*, **29**(9), 659–675.

Proverbs, D.G., Holt, G.D. and Olomolaiye, P.O. (1999b) Productivity rates and construction methods for high rise concrete construction: a comparative evaluation of UK, German and French contractors. *Construction Management and Economics*, **17**(1), 45–52.

Proverbs, D.G., Holt, G.D. and Olomolaiye, P.O. (1999c) Construction resource/method factors influencing productivity for high rise concrete construction. *Construction Management and Economics*, **17**(5), 577–587.

Seeley, I.H. (1995) *Building Economics*, 4th edn, Palgrave, London.

Steyert, R.S. (1972) *The Economics of High Rise Apartment Buildings of Alternate Design Construction*, American Society of Civil Engineers, Reston, VA.

Stone, P.A. (1967) *Building Design Evaluation: Costs-in-use*, E & FN Spon, London.

Chapter 5
Appraisal of Design to Determine Viability of Development Schemes

Herbert Robinson

5.1   Introduction

Appraisal is essential to ensure that the economic, social or environmental implications of a design or alternative design solution is known at the beginning to establish whether to build or not, to modify a design or simply to progress to the next stage of the development process. This chapter focuses on the appraisal of design based on the analysis of key variables and their impact on costs and benefits at different time periods during the life cycle of a project. Key variables include land cost, yield, building cost, planning, professional fees, marketing costs and interest rates and their effects on profitability and value of a scheme.

There are various methods used for the appraisal of design. This chapter examines the principles of design appraisal using examples of a discounted cash flow (DCF) technique and non-discounted method called the residual valuation method to explore the implications of design variables on project costs and development value.

5.2   Assessing costs and benefits of design alternatives

The factors affecting design costs and benefits, and their implications in terms of capital costs have been discussed extensively in Chapter 2. Using the design costs and benefits matrix (reproduced in Table 2.1), various design alternatives and associated design variables can be explored to determine the best solution.

Some costs such as land, planning, design and construction costs and benefits such as rental income, sales income and asset value are easily assessed during the development process and/or when a development is completed. Other costs can be determined indirectly or directly through regulations such as charges, taxes, fees for planning permission and building regulations. It is also recognised in Chapter 2 that there are categories of costs and benefits that are too difficult to price or

Design Economics for the Built Environment: Impact of Sustainability on Project Evaluation, First Edition.
Edited by Herbert Robinson, Barry Symonds, Barry Gilbertson and Benedict Ilozor.
© 2015 John Wiley & Sons, Ltd. Published 2015 by John Wiley & Sons, Ltd.

quantify (e.g. savings as result of reduced carbon emissions, flooding damage, and productivity), social and environment costs (e.g. noise and air pollution, and traffic congestion). Appraisal involves the identification and analysis of various costs and benefits occurring at different time periods during the life cycle of a project.

In the appraisal of design, specific priorities are established by a client, for example to reduce waste, savings in energy, reducing insurance costs associated with flooding, conserve resources by using recycled, recyclable or energy efficient technologies to minimise energy consumption.

The climate change and sustainability agenda reflects a move from the traditional approach of appraisal focusing on the economic dimension to an inclusive approach including social and environmental costs. The total cost associated with any project is the sum of the project's economic, social and environmental costs and the same approach is adopted for benefit assessment. These developments have provided the momentum to consider environmental costs of alternative design using assessment tools such as BREEAM and LEED (discussed in Chapter 12) and social costs using various valuation methods to assess social impact (Gilchrist and Allouche, 2005).

The BREEAM rating/score is a trade-off between economic and environmental costs of a design. According to Nick Hayes, Head of Sustainability at EC Harris Built Environment Consultancy, failure to consider the impact of achieving the desired rating in the early design stages (RIBA Stages A–C) can lead to key credits becoming unobtainable, which can have a negative impact on design, and subsequent economic and environmental costs. Hayes further argued that after outline design is completed, the credits available reduce by up to 30%. Various studies have shown that to achieve an increase in rating, an uplift in cost is required which varies from one standard to another.

A BRE information paper (BRE and Sweett, 2005) indicated that to obtain a BREEAM Excellent rating in offices, construction costs are expected to increase by 3.3% for a good location or 7.0% for a typical location (Figure 5.1).

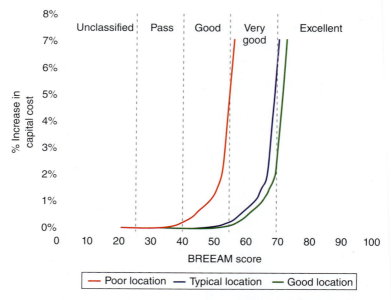

Figure 5.1 Cost of achieving various BREEAM standards. Source: BRE and Cyril Sweett (2005).

Another case study of three office buildings carried out by Halcrow supported by Davis Langdon and Cyril Sweet shows that the additional cost for achieving BREEAM Excellent range from 4.6 to 7.4% (Halcrow *et al.*, 2010). CBRE (2009) also noted that development of a greener building designed to achieve one of the higher standards of accreditation is likely to add between 5% and 7.5% to construction costs. The study concluded that 'even the ambition of producing a zero-carbon development – which is more demanding than even the highest levels of BREEAM or LEED accreditation – would potentially add less than 15% to development costs'. However, BRE reported that the development of Campus M Business Park, a high profile office and technology park in the eastern district of Munich, built by AIG/Lincoln, achieved a BREEAM Excellent rating at no extra capital cost.

The variations in the cost of compliance suggest that considerable care is required in determining the additional environmental costs in the appraisal of design. Whether there are additional costs or not for BREEAM compliance, the outcome could result in reduced operational cost due to lower energy usage, enhanced capital allowances and other tax advantages. There are also reduced costs associated with void as well as the opportunities for significantly higher rental returns or sale income. For example, BRE reported a case study of Bletchley Leisure Centre, Milton Keynes in 2010 which "achieved BREEAM Excellent" and provided a significant reduction in running costs helping the Council to save over £3 million in revenue funding during a 15 year period.

The social cost of construction projects can also be assessed. In the UK some social costs are incorporated by statute within the Department for Communities and Local Government (2013). For example, the planning obligations (s106 T&CPA 1990 *as amended*) and payments arising from Community Infrastructure Levy (s206 Planning Act 2008) are used to ensure that project owners contribute to the additional social costs (and environmental costs) arising as a result of the pressure for extra social and economic infrastructure created by new development projects.

5.3   Appraisal of design using discounting methods

For any design, some costs (whether economic, social or environmental) are incurred at the beginning whilst others including benefits such as rental and sales income are realised at a later stage when a development is completed. Discounting methods address the problem relating to costs and benefit streams of a project occurring at different time periods. Examples of discounting methods are Net Present Value (NPV), Internal Rate of Return (IRR) and Whole Life Costing (discussed in Chapter 8). Discounting technique examines how much an investment today will be worth in the future. This will depend on one's perception of demand, inflation and uncertainty. The fundamental question is illustrated by asking what will be today's value of £1,000,000 received in 10 years' time. The reverse question is how much you would need to invest today to realise £1,000,000 in 10 years which can be determined by applying compound interest.

The value £100 today is not the same as £100 in 5 years' time. Time value allows for costs and benefits at different time periods to be translated to a single equivalent monetary value and interest rates allow for comparisons to be made between design alternatives. Time value reflects the cost of capital or the opportunity cost of capital and it is not the same for everyone. For the bank lending money to the developer, interest rate is the compensation for loss of earnings. For example, if a developer borrowed £10 million with an agreed interest of 10%, the amount to be paid after 12 months will be £11 million. In the example above, £10 million is the principal and £1 million is the interest. There are different types of interest. In simple interest, only the original capital earns interest. In compound interest, interest accrues on the original capital as well as the interest in the previous time period. Interest can also be compounded several or many times within a year such as monthly, weekly or daily. For example, if a developer borrowed £100 million from the bank with an agreement to pay 5%, the payments due under the two different interest regime are as shown in Table 5.1.

Compounding allows for the way a present sum of money will grow over time. For example, £100 million will grow to £110.25 in 2 years' time given an interest rate of 5%. The compound formula is $(1+r)^n$ where r is the cost of capital or interest rate and n is the number of years.

Discounting is the reverse of compounding – how much a future sum of money would be worth today. As shown in Table 5.2, you need to invest £90.70 today to generate £100 in 2 years' time. Similarly, £95.24 can be invested today to generate £100 in 1 year's time. Thus £1,000,000 received in 10 years' time, assuming a discount rate of 10%, will be the equivalent value of £385,544 received in cash today. The relationship between present value and future value is:

Present value of cash flow $(PV) = 1/(1+r)^n \times$ Future value of cash flow (FV)
or
$FV = PV (1+r)^n$,
where r is the interest rate and n is the number of years.
Hence $£1\,000\,000 = PV (1+r)^n$.
Where $r = 10\%$ and $n = 10$ years
Present value (PV) of £1,000,000 = £385,544.

However, it is highly unlikely that interest rates will remain the same, due primarily to uncertainties. Thus the selected interest rate reflects a view of the

Table 5.1   Simple and compound interest.

Year	Cost	Simple interest factor	Value	Compound interest factor	Value
0	100				
After 1 year	100	(1.05)	£105 million	(1.05)	£105
After 2 years	100	(1.05)	£105 million	(1.1025)	£110.25

Table 5.2   Discount factor.

Year	Cost	Discount factor (5%)	Value
0	100		
1	100	0.9524	£95.24
2	100	0.9070	£90.70

future and the uncertainties associated with a particular project. This principle can be applied by estimating annual cash flow for each year of a project, comprising Capital Expenses (CE), Operating Expenses (OE) and Revenue Income (RI). Since resources expended and income gained will be accruing annually, each year's cash flow will need to be adjusted according to the discount factor for that year. Assuming a project to have estimated CE of £98.5 million over a design and construct period of 5 years, this expenditure can be expressed as a series of annual budgets. Similarly the operating period of 9 years can be expressed in the same manner using OE and RI.

It is essential that correct decisions are taken at the design stage when the cost of making changes is relatively small and the potential to make improvements is high. Therefore all possible options should be considered and the best possible design solution selected to meet the client requirements. Table 5.3 illustrates a comparison between two design alternatives with annual discounted cash flow taken through the period of the whole project. The calculation does not allow for the deduction of tax on revenue. The discount rate is set at 10% which reflects the cost of capital or the minimum acceptable rate of return. Design A generates a cumulative NPV of £96.25 million, thus the project returns a shortfall of £2.25 million compared with the capital expense of £98.5 million. Design B incurs additional capital cost, however operating expenditure is less and there is the added benefit of the design and construction period being reduced to 4 years, thereby providing an additional year's income. Taking these factors into account the cumulative NPV increases to £122.165 million which exceeds the £105 million capital expenditure by £17.17 million. Investment in Design B generates a greater return. Assuming that there are no other factors to be taken into account then it is reasonable to conclude that Design B offers a better proposition, given that it produces a further £17.165 million above the NPV discounted at 10%. With NPV which measures all cash flows over the duration of the project, and then discounted to the present, projects are selected *if* NPV > 0. The IRR is the discount rate that results in a NPV of zero. In general projects are selected if IRR is greater than the cost of capital. The Net Terminal Value (NTV, cash flow) approach uses a compound factor (instead of a discount factor). Using the NTV and NPV approaches result in the same decisions in terms of the choice between Designs A and B. The figures are only different because they are located at different points in time.

5.4   Appraisal of design using residual technique

Residual valuation technique is an example of non-discounting method for considering investment proposals and establishing project feasibility. It depends mainly on two factors: development value and development costs. The main purpose is to ensure that the cost of development is reasonable and there is a satisfactory return or benefits. The method is used to determine Residual Profit and Residual Land Value:

Residual Profit = Development Value – Development Cost
Residual Profit = Development Value – (Land Cost + Cons. Cost)
Residual Land Value = Development Value – (Cons. Cost + Profit)

Table 5.3 Infrastructure Designs A and B.

DESIGN A

Year	0	1	2	3	4	5	6	7	8	9	10	11	12	13	14
	Design and Build						Operate								
Capital Expense (CE) - Cash Flow	(−98.5)	(−5)	(−4.5)	(−26)	(−35)	(−28)									
Operating Expense (OE)							−15	−15	−15	−15	−15	−15	−15	−15	−15
Revenue Income (RI)							30	40	45	45	45	45	45	45	45
Gross Revenue*							15	25	30	30	30	30	30	30	30
Discount Rate 10%	1	0.9091	0.8264	0.7513	0.683	0.6209	0.5645	0.5132	0.4665	0.4241	0.3856	0.3505	0.3186	0.2897	0.2633
NPV							8.4675	12.83	13.995	12.723	11.568	10.515	9.558	8.691	7.899
CUM NPV							8.4675	21.2975	35.2925	48.0155	59.5835	70.0985	79.6565	88.3475	96.2465

Project Capital Expense (CE) £98.5 million
Total Operating Expenses (OE) £135 million
Total Revenue Income (RI) £385 million
Discount Rate 10%
Project NPV shortfall Shortfall £2.2535 million

DESIGN B

Year	0	1	2	3	4	5	6	7	8	9	10	11	12	13	14
	Design and Build					Operate									
Capital Expense (CE) - Cash Flow	(−105)	(−7)		(−45)	(−35)										
Operating Expense (OE)						−13	−13	−13	−13	−13	−13	−13	−13	−13	−13
Revenue Income (RI)						30	40	45	45	45	45	45	45	45	45
Gross Revenue*						17	27	32	32	32	32	32	32	32	32
Discount Rate 10%	1	0.9091	0.8264	0.7513	0.683	0.6209	0.5645	0.5132	0.4665	0.4241	0.3856	0.3505	0.3186	0.2897	0.2633
NPV						10.5553	15.2415	16.4224	14.928	13.5712	12.3392	11.216	10.1952	9.2704	8.4256
CUM NPV						10.5553	25.7968	42.2192	57.1472	70.7184	83.0576	94.2736	104.4688	113.7392	122.1648

Tax liability not deducted
Project Capital Expense (CE) £105 million
Total Operating Expenses (OE) £130 million
Total Revenue Income (RI) £430 million
Discount Rate 10%
Project exceeds NPV by £17.1648 million

where Cons. Cost is the construction and associated costs which include professional fees, planning, and interest charges on building costs and professional fees.

According to the RICS valuation information paper (VIP) 12 (RICS, 2012), the residual method recognises that the value of a development scheme is a function of a number of elements:

- Value of the completed development (gross development value, GDV)
- Direct costs of developing the property (gross development cost, GDC)
- Return to the developer for taking the development risk
- Cost of any planning obligations
- Cost or value of the site.

The development cost establishes all expenditures incurred including financing costs and interest charges. The development value establishes the benefit to the developer in financial terms but a key question is: how to quantify the value? There are two key variables required to establish a scheme's value which are: (1) income or the rent – amount of money a tenant is likely to pay to occupy the proposed development; and (2) investment yield used to discount future income stream to calculate the capital value of the scheme today.

$$\text{Capital Value (or Development Value)} = \text{Rent} \times 100/\text{Yield (\%)}$$

Establishing rental values

The amount of rent (income) is determined by the relationship between demand and supply of different types of property. The key variables affecting rental income are as follows:

> Demand – investment and occupation demand
> Market segmentation – location, types or uses
> Labour markets – services, industry, retail, and so on
> Population – demographic trend, growth rate, and so on
> Regulations and laws – land use, taxation, and so on.

Establishing rental income is based on rules for measuring areas (e.g. RICS Code of Measuring Practice) using rates per square foot or metre. This is usually expressed as net lettable area (e.g. office), gross internal area (e.g. industrial) and zones (e.g. retail units). The rent can also be determined by comparable evidence of recent lettings of similar schemes in the location and making adjustments for differences in age, quality, specification and market trends such as the growth of green buildings. There are other factors to consider such as the balance of current supply between new or refurbished versus second-hand space, current climate and surrounding markets. Strong demand and modest completion of speculative space can combine to force down the vacancy rate for quality space. Judgements therefore have to be made in difficult and unpredictable economic climates whether to build in anticipation that there will be a significant reward in future when economic conditions improve or become favourable. Surrounding markets have a huge influence on rental levels. For example, average central London prime rents continued to rise as rents in both the City and West End markets reached their highest levels since 2008.

In terms of the rental value of green or sustainable buildings, there is still a considerable debate. According to Geoffrey Steward of SGBA Limited, the benefits of green buildings include a reduction in operating costs (7–9%), increase in occupancy by 3.5%, increased building value by 7.5% and an increase in rental by 3%. A CBRE (2009) study concluded that "there are a number of unresolved issues in assessing the scale and source of payback for incurring these additional costs, particularly in terms of investment value and pricing". They further stated that evidence on rental transactions indicates that "green buildings achieve a rental premium similar in proportion to the scale of additional development costs for mid-range levels of certification". The authors argued that the "future accumulation of evidence on the relative rent levels, running costs and, in due course, investment prices, of green over conventional buildings, will reinforce these market differentials".

Establishing yield

Yield is a term widely used in stock market transactions to compare totally different stocks with one another. An interest in property is no different from an interest in any other form of investment. Establishing property yield involves analysing the sales of comparable properties to the proposed development and making adjustments to reflect investors' perception of future rental growth against the risk of future uncertainty. Freehold properties are deemed to produce a perpetual income, for the purposes of valuation. As with shares or any other form of investment, this income may vary considerably in time. Currently acceptable yields for various property types are determined by the market. Higher yields are used to reflect greater perceived risks or are associated with additional risks in participating in a development.

HIGH YIELDS = GREATER perceived risks/problems
LOW YIELDS = SMALLER perceived risks/problems.

100/Yield (%) = Years Purchase (YP) Factor

The YP value is simply the expression of an investor's expected return. Thus, should an investor expect a return of 15% on an investment in a house to rent, then the YP is equal to $100/15 = 6.67$.

HIGH YIELDS = LOW Capital Values
10% Yield means the multiplier is 10 YP

LOW YIELDS = HIGH Capital Values
5% Yield means the multiplier is 20 YP

Yields are affected by risk of loss of capital, loss or irregularity of income and are influenced by liquidity of investment, cost of transfers (sales and purchase) and management of property. For example within each category, it is possible to detect a wide range of yields according to the factors affecting specific property categories. However, location, both in terms of geographical and locality, provide the most important determinant of yields. Property yield has a relationship with general investment yields (other non-property assets, e.g. money, bonds, shares). The key variables affecting yield are as follows:

> Asset preferences – portfolio of assets depends on balance between liquidity and yield
> Cost of switching – ease and cost associated with changing from one asset to another (e.g. cost of property purchases and sales)
> Risks – balance between income certainty and capital gains.

5.5   Case study of the blackfriars development project

A developer is interested in purchasing land for the Blackfriars Road Development Project, in the South Bank Area in London SE1 and intends to submit a planning application for a proposed development. The company requires a feasibility/profitability assessment of the proposed development. The proposed development will be of a high specification design to maximise the potential rental income, and aims to achieve a Lettable Floor Area of 85% of Gross Internal Floor Area.

Assessing the viability of a development scheme

For any development, it is important to do a thorough site survey and viability assessment with all assumptions clearly stated to complete the appraisal of the proposed development. According to the RICS Guidance Note on financial viability in planning (RICS, 2012), a proper assessment is required to ensure the following:

1. land is appropriate and can be released for development;
2. developers are capable of obtaining an appropriate market risk adjusted return for delivering the proposed development;
3. proposed development is capable of securing funding.

It is crucial to explore the planning obligation liabilities to assess how they will adversely affect the site value to the landowner and return to the developer. If these conditions are not favourable, the land is not likely to be available and the development will not take place as it will not be profitable.

Assessing the development location and site

Southwark will benefit significantly due to The Shard and other high profile developments in the surrounding areas. There is a current plan over the next 15 years to create an extraordinary site containing successful business districts, sustainable residential neighborhoods and world class services (Southwark Council, 2010). See Figure 5.2 for the South Bank Development Plan showing Southwark Council's approach to change in Bankside, Borough and London Bridge.

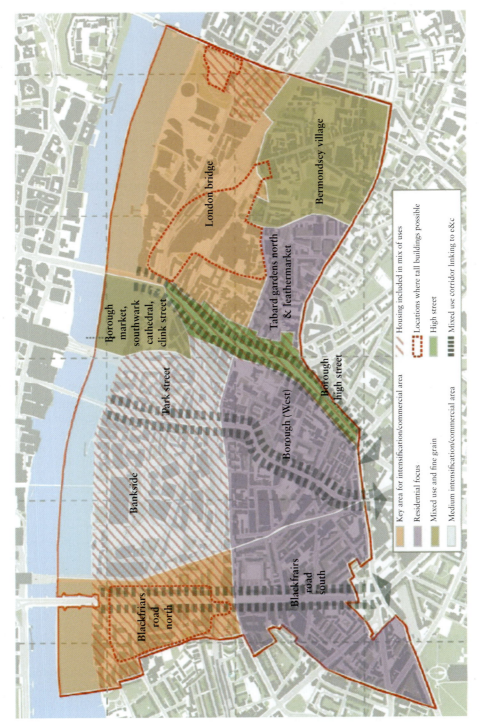

Figure 5.2 South Bank Development Plan. Source: Southwark Council.

Bankside

Park street

Borough market, southwark cathedral, clink street

London bridge

Bermondsey village

Tabard gardens north & feathermarket

Borough high street

Borough (West)

Blackfriars road north

Blackfriars road south

Key area for intensification/commercial area

Residential focus

Mixed use and fine grain

Medium intensification/commercial area

Housing included in mix of uses

Locations where tall buildings possible

High street

Mixed use corridor linking to e&cc

Regeneration around Southwark consisted of a £1.5 billion Elephant and Castle regeneration programme, major redevelopment projects at Canada Water, Bermondsey Spa and Bermondsey Square, the Peckham Programme, with recent regeneration at Borough and Bankside.

Demand for investment is growing because the Southwark market is characterised by a lack of modern, good quality developments. The demand for commercial space is relatively good, with demand for higher quality space significantly outweighing supply. There has been a considerable pressure for good quality land for development in London due to increasing competition, particularly for larger high quality floor spaces in the traditional 'core' locations. This has prompted occupiers and tenants to search further for good value. The business case for companies to remain within the 'Square Mile' is strong creating a shift towards Southwark and good quality fringe locations (see Figure 5.3 showing the area where the Blackfriars development project is located). Blackfriars Station's new south entrance will improve transport links.

The development site should be thoroughly appraised (see Figure 5.4 showing the land identified for the Blackfriars development project). The RICS Guidance Note (RICS, 2012) stated that "viability appraisals may be used in connection with a number of issues in respect of both planning policy and development control". Some of the issues include:

assessing the nature and level of planning obligation contributions/ requirements;

the timing of planning obligations contributions;

applications for enabling development;

reviewing land uses;

dealing with heritage assets and conservation;

formulating planning policy through core strategies (local development frameworks and plans.

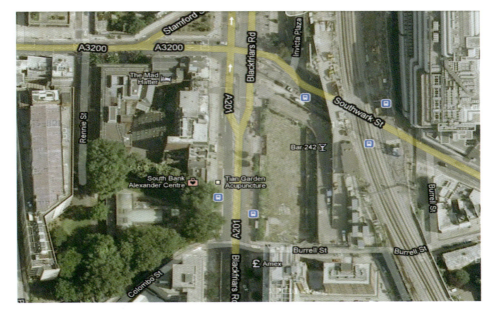

Figure 5.3   Development site.

Figure 5.4   Land for development.

Table 5.4   The key variables in establishing development costs.

- Land costs – price of land, stamp duty, legal fees, agents fees (e.g. 1–2%)
- Site costs – ground investigation and land survey fees
- Building costs – Based on the gross area of the building and price per square metre (e.g. single rate methods such as superficial area method for initial appraisals and other multiple rate methods, e.g. elemental cost plan and approximate estimating as appropriate)
- Professional/management fees – usually based on a percentage of the building costs or a scale of charges, negotiated or fixed fee for each profession involved
- Planning fees – costs involved in making planning applications and securing consent
- Building regulation fees – scale of charges depending on the building cost or size (Building Control Department)
- Funding fees – incurred for arranging finance and usually reflects the size of the loan
- Finance/interest charges – cost of borrowing money or opportunity cost (interest on land costs, professional fees and building costs)
- Letting fees – usually varies as a percentage of rental value
- Sales costs – include agent's and solicitors fee (usually a percentage of Net Development Value)
- Other development costs (e.g. relocation, planning agreements such as Section 106 planning agreement, commissioning, taxation, etc.)

The RICS Guidance Note also acknowledged that the variables used in the appraisal (Table 5.4) may change over time and will reflect the movement in the property market generally.

Establishing development costs

The direct costs for development includes construction and associated costs such as professional fees, planning and building regulation fees, interest charges and other costs associated with using or disposing of a building (Table 5.4).

Table 5.5   Examples of sources of building cost.

Source A	BCIS Online
Source B	Building.co.uk; High Spec; 11-storey office building
Source C	Company database, e.g. Davis Langdon Cost Model
Source D	*The Architects' Journal*; City of London office developments
Source E	Price books, e.g. Spons, Wessex, Griffith

There are well established (and sometimes sophisticated) methods to determine variables such as Lettable Floor Area, building costs, construction duration, professional fees and other costs.

Establishing building costs

To estimate the building cost, analysis of similar and comparable development is required. An Elemental Analysis of similar development can be taken from a variety of sources (Table 5.5) such as from company project databases, external sources such as BCIS, *Building Magazine* and *The Architects' Journal*. Other sources could be used such as price books (e.g. Spons, Wessex and Griffiths), and sub-contractors' prices to determine and adjust building costs depending on the estimating technique used.

The choice of design in the form of materials selected to reduce embodied energy, orientation of building, geometry and height of building, use of renewable systems, minimising waste, use of recycling materials and the application of environmental assessment techniques such as BREEAM will affect the cost. The examples selected should therefore be the best matched with the proposed project in terms of specification, scope and scale.

First, from the analysis of each comparable project, indices for time, location and quality can then be applied to adjust the building costs. Time and location indices are published by BCIS. Due to the high specification expected in the development, quality ratio will require an assumption based on published information on comparable developments. Secondly, the client may require a BREEAM Excellent Score. So the question is how much adjustment or uplift is required in building cost to achieve BREEAM Excellent?

Determining the cost of planning obligations

Difficulties can arise in estimating the values of planning obligations and/or certain variables so assumptions have to be made in certain circumstances. For planning obligations such as Section 106 agreement, Southwark Council has clearly defined procedures including suggestions from the public on how to mitigate the impact of a new development in the surrounding area as part of the planning application process. Section 106 agreement is designed to ensure that new developments enhance local communities and to reduce the impact of developments on local communities. As part of the Southwark approach, they have the Section 106 Supplementary Planning Document (SPD) to guide the negotiations for Section 106 planning contributions and to ensure greater transparency and openness. They have also set up a community infrastructure project list (CIPL) to identify priority local infrastructure projects for physical improvements supported either by Section 106 contributions or CIL funds. Examples of contributions arising from planning obligations range from the provision of affordable homes, open space including landscape and tree planting, facilities for sports development, and transport enhancement to funding of school places or employment and training schemes.

Southwark Council has developed a Section 106 Planning Obligations Workbook as a tool to calculate the planning contributions required for a proposed development project.

Extracts from the site specific transport section of the Workbook is shown in Table 5.6

The planning obligations, often the social and environmental costs, relating to particular development projects can be implemented in two ways – either the developer can implement it or pay the council for the works to be carried out. There is also a separate element for strategic transport initiatives applied to major residential, commercial and retail development and cross rail charge based on a strategic transport investment rate. The total contribution from a proposed development is based on the number of residents and/or employees.

Similarly, the planning obligation contribution costs for other elements relating to open space, sports development, children's play equipment, archeology priority zone, affordable housing, employment and training and school places based on a number of factors and ratios such as contribution per occupant.

Determining rental income

Several factors determine rental income such as rental growth, availability and take-up rates, vacancy rates, development projects in the pipeline (under construction and

Table 5.6 Examples of planning contribution costs for site specific transport.

Examples of specific transport initiative	Unit of measurement	Number required	Cost per item (£)	Contribution required (£)
Pedestrian crossing	Item		40,000	
1. Pelican			50,000	
2. Toucan			30,000	
3. Zebra				
Traffic calming	Item		12,000	
1. Flat top hump			3,000	
2. Round hump			1,000	
3. 3 cushions				
Widening footway over road	Linear metre		60	
Widening highway over footway	Linear metre		90	
Install cycleway on road	Linear metre		25	
Securing land footway/cycleway				
Cycle stands	Item		300	
Travel plan monitoring			3,000	
Other initiatives				
Engineering fees (15%)				

planned development), and proximity to other areas. In a difficult economic climate characterised by redundancies and business failures, there will be increasing vacancy rate and oversupply leading to falling rental values. It is prudent to use the average rental price for a property in a similar location and then adjust for differences. The Shard is reported to have received rentals starting from £50 per square foot, with the highest rate nearer £70. In terms of the value of sustainability, a recent development (192,252 ft² of office space) awarded BREEAM Excellent rating receives an annual rent of £5.8 million per year which works out at £30 per square foot (£323 per square metre). Based on research of available information, a data and assumptions sheet is created (Table 5.7) which will then feed into the financial model.

Determining the value of other variables

Other variables, for example, finance rate depends on several factors such as capital put towards loan (deposit) - term of loan, whether long term or short term and ccompany credit history and a comparison with other projects of the similar size and duration.

The residual appraisal method is traditionally used in two ways as discussed earlier. First, to assess the level of profitability (return generated) for a proposed project where land value is an input into the appraisal; and, secondly, to establish a residual land value where a predetermined level of return is expected by the developer. In practice, the technique can also be used to determine other variables in the development process other than the residual land value and profit such as the level of planning obligations that can be afforded (social cost), direct costs for property development, building cost (including environmental cost for BREEAM

Table 5.7   Example of data and assumptions sheet.

Variable Name	Quantity	Unit	Notes
Gross Internal Floor Area	27,600	m²	
Lettable Floor Area (85% of Gross Internal Floor Area)	23,460	m²	
Construction Cost	2,644	£/m²	
Yield	5.5	%	
Contributions from Planning Obligations			SPD[a]
Annual Rent	350.00	£/m²	
Finance (Interest Rate)	0.6%	per month	
Land Purchase to Start of Construction	6	months	
Construction Duration	28	months	
Void	3	months	
Total Development Period	37	months	
Profit on development value	20%		
Professional Fees (Architects, Quantity Surveyors and Engineers)	15%		
Contingency Sum	3%		
Letting Fees as a % of annual rental	15%		
Sale Fees as a % of capital value	1.5%		
Land Acquisition Fees as a % of land value	4%		

[a] To comply with Section 106 Supplementary Planning Document (SPD).

Table 5.8   Residual calculations.

Development Value			
Annual rental income	8,211,000		
Capitalise (@ 5.5%)	18.1818		
Capital value			149,290,909
Less profit at 20% of capital value			29,858,182
			119,432,727
Development Costs[a]			
Construction Costs			
Building	72,974,400		
Professional fees	10,946,160		
Contingencies and risks	2,189,232		
		86,109,792	
Finance			
Finance on half construction costs for construction period	7,850,630		
Finance on construction costs for void period	1,428,561		
		9,279,191	
Letting and Sale costs			
Letting fees 15% of annual rental	1,231,650		
Promotion, say	430,000		
Sale Fees 1.5% of capital value	2,239,364		
		3,901,014	
Total Costs			99,289,997
Land value including finance & fees			20,142,730
Less finance over development period			0.8014
			16,143,190
Less acquisition costs 4%			645,728
SITE VALUE			15,497,462
		say	**£15,500,000**

[a] Section 106 and other planning obligation/contribution costs not included.

compliance) to generate a desired level of profitability. However, it should be noted that the value of development is not directly related to its cost.

The RICS Guidance Note recognises that the residual approach can be applied with differing levels of information and sophistication and it is up to the appraisal team to decide on the most appropriate application of any financial model, bespoke or otherwise. An example of a financial model is shown in Table 5.8 but in practice different approaches are used.

Costs relating to Section 106 and other planning obligation liabilities are crucial and could have a significant impact on the site value to the landowner and return to the developer. These costs should therefore be carefully determined in accordance with Section 106 SPD of Southwark Council.

Table 5.8 provides a summary of the key variables to determine the site value. Using data available and assumptions made, the estimated value of the site is approximately £15,500 as detailed in the Residual Land Valuation. The value of £15.5 m is the

Table 5.9   Sensitivity of property yield and rent on capital value of Blackfriars development scheme (in £ Millions).

		Yield (%)					
		4.5	5.0	5.5	6.0	6.5	7.0
	300		140.76	127.94	117.30		
	325		152.49	138.63	127.08		
Rent (£/m²)	350	182.47	164.22	149.29	136.85	126.32	117.30
	375		175.95	159.95	146.63		
	400		187.68	170.62	156.40		

maximum price which should be offered, and it is advisable to start off negotiations much lower to try and maximise potential profits taking into account the influence of planning obligation costs. From a purchase of £15.5 million for the site a profit of 20% can be realised from the development if the site value is reduced by an amount equal to the cost of planning obligations.

Purchasing the land over this price will reduce the profit margin which could be further reduced if planning obligation costs are factored in. It is therefore appropriate for sensitivity (scenario) analysis to be undertaken. First, this would help to examine the effect of changes in the value of key variables on the residual land or site value (or developer's return). Secondly, it would help to test the main assumptions to ensure that they are robust, before a final decision is made. Certain scenarios can be considered such as:

- If rental values drop by 20%
- If building costs increase by 20%
- If costs relating to section 106 and other planning obligations are included
- If the yield changes.

It is useful to explore, for example, the sensitivity of changes in rental income and property yield combined on the value of the development value. See examples of impact on capital/development value as shown in Table 5.9.

Changes in the rent and the investment yield will therefore significantly affect the capital value of any development scheme. A key challenge for any developer is to arrive at a design solution that will maximise rental income for a given development cost. Another scenario can be explored such as the sensitivity of property yield on profitability (see example shown in Figure 5.5).

5.6   Concluding remarks

Uncertainty and risks are usually taken into account by examining a range of outcomes such as pessimistic, most likely and optimistic to generate a range of values sometimes called the sensitive range. The objective of any developer is to maximise return or profitability determined by two major variables – development value and development costs. The risks associated with the key variables for any development should therefore be reduced in terms of exploring the factors that affect development value (e.g. prices, rents, yields) and development costs [e.g. land, project (including build) cost, finance cost, void periods].

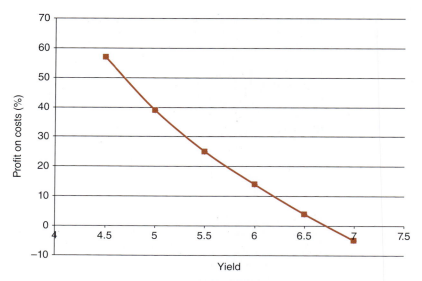

Figure 5.5 Sensitivity of property yield with respect to profit.

For example, to minimise risk, a house builder could sell units "off-plan" and a developer could arrange a "fixed price" construction contract and/or borrow at a fixed rate. Reducing void and uncertainty in rental income could be achieved through for example pre-letting and good market research to attract appropriate tenants. To reduce the risks associated with building cost due to unusual design with innovative sustainability features, excellent design co-ordination, good communication, good cost control, good construction team and specialists on sustainability including an experienced architect can be appointed.

It is recommended by the RICS Guidance Note that the appraisal team should check residual development appraisals with market evidence. The checks should focus on the following:

- Comparison with the sale price of land for similar development
- Calculation of the ratio of the residual site value to the capital value of the scheme
- Assessing how this ratio compares to other evidence of similar transactions.

References

BRE and Cyril Sweett. (2005) Costing Sustainability: How much does it cost to achieve BREEAM and EcoHomes ratings? BRE information paper 4/05, IP4/05, March 2005.

CBRE (2009) Who pays for green? The economics of sustainable buildings, CB Richard Ellis and EMEA Research. www.cbre.eu/environment (accessed 20 October 2013).

Department for Communities and Local Government (2013) Community Infrastructure Levy, the (Amendment) Regulations. 2013 (Statutory Instruments). 2013 No. 982. Community Infrastructure Levy, England and Wales.

Halcrow, Davis Langdon and Cyril Sweett (2010) Capital Cost of Sustainable Offices. Full Report of South West of England Regional Development Agency (SWRDA).

Gilchrist, A. and Allouche, E.N. (2005) Quantification of social costs associated with construction projects: state of the art review. *Tunnelling and Underground Space Technology*, **20**, 89–104.

RICS (2012) *RICS Guidance Note: Financial Viability in Planning*, 1st edn. GN 94/2012, RICS, London.

Southwark Council (2010) Draft Bankside, Borough and London Bridge Supplementary Planning Document dated 9 February 2010. http://moderngov.southwarksites.com/mgConvert2PDF.aspx?ID=7818 (accessed 29 July 2012).

Chapter 6
Eco-cost Associated with Tall Buildings

Peter de Jong and J.W.F. Hans Wamelink

6.1 Introduction

This chapter examines the eco-costs/value ratio (EVR) concept and a tool based on life cycle analysis (LCA) from production, operating and end-of-life phases to evaluate the ecological impact of alternative design interventions. Following this introduction, a brief overview of the Dutch housing market heavily influenced by the subprime and Euro crisis and the role of land use in the development of tall buildings in the Netherlands is provided. The EVR concept is then discussed. Based on practical experience in several redevelopment and renovation projects, the applicability of the eco-cost model is tested in a complex of approximately 200 apartments. Three cases of tall buildings are also measured according to the EVR system and tested against four environmental ranking methods. The need for embedding EVR in other sustainable ranking methods is also discussed. The chapter concludes with the major challenges to overcome in the EVR method such as lack of standardisation and practical data to make the eco-cost model fully operational.

6.2 Overview of the Dutch housing market and land use planning

The financial (subprime) crisis of 2007–9, pinpointed by Lehman's bankruptcy, and the Euro crisis had a severe impact on Dutch real estate. It is probably not the worst situation in Europe, but after a rather comfortable period of real estate development, followed by real boom in cities and regions, the Dutch office market was characterised by an unhealthy structural vacancy. Before the crisis, the market could deal with the oversupply of speculative office development, but it turned

Design Economics for the Built Environment: Impact of Sustainability on Project Evaluation, First Edition.
Edited by Herbert Robinson, Barry Symonds, Barry Gilbertson and Benedict Ilozor.
© 2015 John Wiley & Sons, Ltd. Published 2015 by John Wiley & Sons, Ltd.

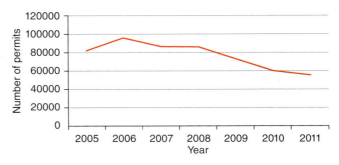

Figure 6.1   Number of permits in the Dutch housing market (Statistics Netherlands, 2013).

into an unmanageable surplus. The crisis made this failure in the market more visible. Another hurdle was the liquidation of open-ended German funds with massive involvement in the Dutch office market.

Figure 6.1 showing the number of permits confirmed that there were a few difficult years characterised by a reduction in building production activities after the crisis of 2007–9. Although the period of unrestricted growth in housing prices was considerable, buying a house was always considered by house owners as a safe investment, Dutch politicians are still too reluctant to take the required measures.

A second relevant aspect is the increased focus on brown field development instead of the traditional green fields. For the Rotterdam area, there is another dimension to this aspect relating to the enlargement of harbour facilities, demanding a such as Maasvlakte 2, the port of the future. With the position of the city, this port development is causing a massive supply of brown fields, arising from previous uses of all kinds of harbour related activities. Whereas many other cities are searching for decent areas to upgrade their business districts, Rotterdam is confronted with an oversupply of reusable areas. The area development strategies for the city, drawn only a few years ago during the optimistic decade before 2008, have to be adjusted to the new circumstances.

This element of land use is particularly important for the development of tall buildings in the Netherlands. The land policy is characterised by the local government as dominant actor with substantial public ownership (80%), land supply and interference in price setting by the municipality, combined with a restrictive planning system (Van der Post, 2007). Furthermore the price of land is almost entirely determined by the gross floor area and the density/size of the development. For example, land on which you can build 50 floors costs 10 times more than land for 5 floors. Given these circumstances, it is not surprising that only a few tall buildings are under construction at the moment. All developments which started before the crisis particularly tall buildings in preconstruction phase are facing significant difficulties.

In the context of Dutch land use planning, Needham (2007, p. 98) stated that it is not politically acceptable unless it is labelled 'sustainable'. The environmental consequence is considered, but only if it would be socially and economically acceptable in the short term. 'Sustainability' using the Brundtland (1987) definition, focuses also on areas to be preserved for future generations. Densification of cities is rational on the more abstract level but impossible to take into account on all buildings.

In order to emphasise the sustainable aspect of tall buildings, focus is needed on more physical aspects as well such as the use of material.

6.3   Eco-costs/value ratio and the EVR model

In the search for a proper tool to evaluate the ecological impact of alternatives to interventions in the housing stock, De Jonge (2005) compared many options before selecting the EVR (eco-costs/value ratio). The original design of the method by Joost Vogtländer (2001), an industrial designer, focused on general product development. Tim de Jonge (2005) adopted the method for (renovation of) social housing, while Joost Hoffman (2008) elaborated the application of the method for tall buildings in comparison with other alternative methods.

EVR is an assessment tool based on LCA (life cycle analysis) that expresses the ecological burden of a product or service in terms of 'eco-costs'. The ratio compares eco-costs with value. A low EVR indicates that the product is fit for use in a future sustainable society. A high EVR suggests there is no market for such a product (Vogtländer, 2001). Eco-costs are defined as the cost of technical measures intended to prevent pollution and resource depletion to a level that is sufficient to make society sustainable. The model is based on 'virtual eco-costs '99', which is the sum of the marginal prevention costs of:

- Depletion of materials: equal to the market value of raw materials when these are not recycled. When (a fraction of) the material is recycled, a correction is made.
- Energy consumption: energy eco-costs are based on the assumption that fossil fuels have to be replaced by sustainable energy sources. The eco-costs of energy are thus set to the price of renewable energy. This includes the whole life cycle, so that energy consumption during the production phase, the operating phase and the end-of-life (demolition and waste separation) phase is taken into account.
- Toxic emissions: the cost of the measures is determined according to the virtual pollution prevention costs '99, completed with the marginal prevention measures to reach the (Dutch) target for reduction of landfill.
- Environmental burden of labour: labour itself hardly causes any environmental burden, but the conditions related to labour do have an impact (heating, lighting, commuting and equipment).
- Environmental burden of the use of equipment, buildings, and so forth: the eco-costs related to the fact that fixed assets are used to produce a building are also taken into account. The calculations relating to the indirect eco-costs of the use of fixed assets have the same characteristics as cost estimates for investments.

The EVR compares the 'eco-costs' to the value of the product or service. Like many other ranking methods, the EVR of a building is rather meaningless, unless compared with the EVR of an alternative solution. On that basis, it allows comparing new construction to renovation or maintenance.

Production phase

Building projects consist of a combination of semi-finished products, which are assembled at the building site. Therefore, the environmental burden of a building in the production phase can be considered as consisting of the eco-costs of those semi-finished products plus the eco-costs of the assembling activities (including all additional works like preparation, building site facilities and management). It is possible to estimate the eco-costs of a building applying 'eco-cost unit prices' of building elements. As in a traditional cost estimate based on unit prices, the composition of the concerned elements is determined in terms of quantities of semi-finished products and assembling activities. The emission and depletion data, which serve as a basis for eco-costs assessments, can be found in all kind of databases although there are significant differences in the way such data can be presented.

Hence, the eco-costs per unit of element can be determined by inserting the eco-costs of the semi-finished products and the assembling activities into the various components of the elements. Finally, the elemental bills of quantities (for estimating traditional economic costs) can be transformed into eco-costs estimates by substituting the traditional economic unit prices with the eco-cost unit prices. For the relation to the value assessment, it is even preferable to keep eco-costs and building costs side-by-side. In this way, eco-costs have been implemented by Tim de Jonge (2005) in a materials database of an estimating system used to produce elemental bills of quantities for determining the construction costs of new construction and renovation projects.

Operating phase

In the operating phase, the most important factors for determining the ecological burden are the energy demand and the maintenance of the building in use, based upon existing energy demand models. Maintenance models seem to be too complicated for use in (early) design stages. At this stage, elaborated calculations of maintenance efforts are very unusual. At Delft University of Technology, an estimating model was developed for investigating the impact of design decisions on the maintenance costs of residential buildings. This estimating model was deemed suitable for integration in the EVR assessment approach because of its basic structure. In the housing sector, management and administration costs are usually treated as independent from the specific building design elements. For estimating the related eco-costs, these costs can be considered as mainly related to office jobs.

End-of-life phase

The costs of demolition and the separation of waste are covered by traditional economic costing. The pollution prevention costs of these activities can be estimated without considerable problems. The eco-costs of recycling or upgrading are assigned to the new products emerging from these processes. So, all eco-costs in the end-of-life phase after the separation of waste are related to the waste component or fraction that is not fit for upgrading or recycling. This fraction is related to the 'eco-costs of land fill'. Table 6.1 provide examples of the eco-costs of selected materials with their EVR.

The value of houses

For commodity goods, of which many items are sold on a day-to-day basis, the value of products can be determined by observing sales prices. In real estate and housing markets, however, it is more challenging to establish the value of products by observing sales prices, as this is influenced by quality, time and location aspects.

Table 6.1   Eco-costs of materials (www.winket.nl).

			Traditional costs		Eco-costs	EVR
Code	Description	Unit	material/unit	labour/unit	material/unit	material/unit
21.32.10	Traditional formwork of footings	m²	3.04	0.70	4.10	135%
	Traditional formwork of walls	m²	12.15	1.75	16.41	135%
21.32.32	Wall formwork (1 dwelling/day)	m²	2.25	0.25	11.43	509%
	Tunnel formwork (1 dwelling/day)	m²	2.25	0.25	11.43	509%
21.33.10	Dovetail sheet	m²	18.83	0.20	3.74	20%
21.40.10	Reinforcement steel FeB 500HK	kg	0.89	0.03	0.38	43%

The value of a dwelling as a real estate object equals the (discounted cash flow of the) net future profits of that object. The net future profits are estimated, considering the reduction of the quality of the housing services, which are provided by the dwelling. It should be kept in mind that after a certain period (e.g. 30 years), the quality of the dwelling will be perceived (by the customers) as being insufficient, and a reinvestment will probably be required for further operation. The (residual) value of the dwellings should be estimated based on the expected reduction of various quality dimensions of the housing services provided and the possibilities of recovering quality and value, by refurbishment, extensive renovation or new construction. So, the residual value at the end of the operating term is produced by the difference of the value of the dwelling after an intervention and the (all-in) costs of the intervention.

$$V_e = V_n - C,$$

where V_e = (residual) value of the existing dwelling
V_n = value of the new dwelling created by the intervention
C = all-in construction costs of the intervention.

Requirements for an estimating model

The environmental burden and, by consequence, eco-costs relate to all phases of the life cycle of houses. So, eco-cost estimating should have the scope of a Life Cycle Costing approach. In order to fit in with normal practices in housing projects, the applied technique should be an operating estimate, in which, for example maintenance and energy costs can be adjusted to varying design specifications. Many architects prefer to relate building cost data to their own experiences from previous design commissions of similar buildings. They do so mainly because in the early stages of the process no better alternative is usually available. However, using your own cost data for the early stages has several drawbacks:

1. They are unable to communicate relevant eco-cost information, since the raw data are not readily available.

2. The (greater part of) project documents in architectural firms are not structured in such a way that the cost data can be modelled according to the (main) dimensions of preliminary design.

3. In general, the preliminary cost information from the reference projects are poorly connected to the information in later development stages.

In addition, the need for more specific cost data become evident during the design development process. Design deals with alternative building forms and several combinations of functional and/or spatial entities may be considered. Technical specification of building elements, however, may be far away. At this stage of preliminary design, information is needed that relates costs to alternative combinations of (functional) project sections and varying dimensions of buildings.

At the early stages of the process of design and specifications, detailed cost information referring to more specific elements (i.e. technical solutions) is neither required nor applicable. It is only in the final stages that the cost effects of applying different materials and semi-finished products are considered on a more extensive scale. Cost analysis should be closely related to the requirements from the design process. That means being specific if required but global when the decisions involved have a global character; and, moreover, the model should be able to follow the designer 'up and down the design ladder'.

Filling in the missing link

At this point, the existing tools for cost estimating appear to have a missing link. At the top end of the composition hierarchy, a general idea of building costs may be available, based on square metre prices of previously designed and completed projects. At the bottom end, unit prices of technical solutions may be available from a database of cost analyses, which links specified elements (i.e. technical solutions) to the costs of materials, labour, and so on through various elemental categories. In between, however, the existing estimating tools do not provide information about which combination of technical solutions is characteristic for the actual type of building in a particular development project. To fill in this missing link, the Reference Projects Model has been developed. It provides the required data, based on the idea that (within a building market region, e.g. the Netherlands) a building is a unique product, not so much because of the unique technical solutions it consists of, but much more because of the unique combination of similar technical solutions.

The reference projects model

The idea behind the Reference Projects Model is that an architect deduces the construction costs of a new design from the construction costs of a project already completed usually referred to as: the reference project. Evidently, projects that contain the architect's own designed buildings are the reference projects most suitable for him/her. So, in general, an architect should relate the new project, in which he/she is actually involved as a designer, to other projects from his/her own portfolio.

In estimating, two exceptions can be discerned:

1. The architect is confronted with a commission referring to a category of buildings he/she is not acquainted with.
2. There is not a database with well-structured cost data referring to the architect's portfolio.

In these situations, a public database of reference projects could provide 'second best' cost data for early design development stages. The Reference Projects Model has been designed as such a database (www.winket.nl). By using the model, architects (and clients) are able to estimate the costs of housing projects on an appropriate scale and level in all stages of the development process. From the point of view of the estimating technique, there is only one difference between traditional construction costs and eco-costs in the model: eco-costs cannot be verified on the basis of realised tender prices.

Estimating tools for the EVR in housing projects

At this stage of the research, for calculations referring to the production phase and the end-of-life phase, the Reference Projects Model is operational. For calculations referring to the operating phase the spreadsheet facility for Estimating Energy Demand (www.dgmr.nl) and the Delft Maintenance Calculating Model can be combined and connected to the input interface of the Reference Projects Model. Some engineering is still needed to make this combination of tools for the operating phase available for architects in real life projects.

In order to illustrate the type of results that can be obtained by means of the developed models, two case studies have been conducted. First, the results of eco-cost calculations in fourteen recently completed building projects are presented. The emphasis in these projects is on housing, that is new construction as well as renovation. However, some non-residential projects are added to get a (preliminary) indication of the position of the housing sector as related to other building categories.

The results of the calculations show that new construction of houses and offices have EVRs on similar levels. Renovation, however, shows significantly lower EVRs than new construction. Analysis of the results indicate that the difference between new construction and renovation is mainly related to the combination of the relatively high ecological burden of Substructure, Structure and Skin elements of buildings in the production phase, and the fact that the elements have different implications for new construction and renovation projects. Analysis of the results also indicates that the greater part of the eco-costs of buildings in the production phase can be traced back to a relatively small group of (raw) materials: if the eco-costs of wood and paint are known, the eco-costs of a wooden window frame can be calculated.

6.4   Applying the EVR model to housing

Based on practical experience in several redevelopment and renovation projects, a case has been constructed in order to test the applicability of the developed model. In a complex of approximately 200 apartments, built in the 1960s, the landlord,

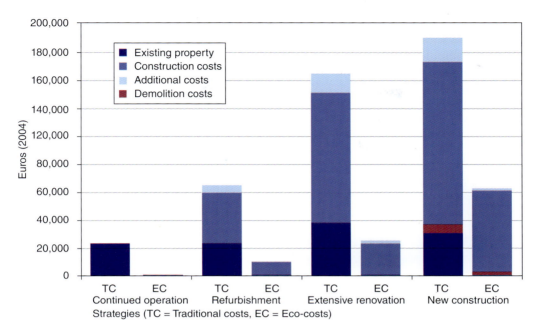

Figure 6.2   Traditional costs and eco-costs of investments of several strategies (De Jonge, 2005).

a Dutch housing association, is planning to start an intervention project. The characteristic approach of such a project would be to conduct a feasibility study concerning various options to support a final project definition.

Apart from selling the apartments, in principle, four strategies – that is four types of interventions – are possible: (1) unchanged continued operation; (2) refurbishment aimed at improving one or more quality dimensions of the apartments as they are; (3) extensive renovation aimed at creating (virtually) new apartments within the structure of the existing block; and (4) redevelopment aimed at the construction of completely new houses. For all of these strategies investment costs (traditional and eco-costs) have been estimated.

On an investment level, the EVR of new construction clearly is the highest. Figure 6.2 is a comparison of traditional costs and eco-costs of investments based on several strategies. Moreover, the all-in construction costs and eco-costs per apartment are the highest in new construction.

The allocation of eco-costs takes place in line with economic principles (based on the Present Value). This means that the eco-costs of the indirect yield value (i.e. the present value of the operation) equal the eco-costs of the investment. Hence, the eco-costs of renting houses can be deduced from the eco-costs of the investment and the eco-costs of the operating expenses.

Apart from rent, housing expenses also include energy costs. The levels of energy costs following the varying interventions are assessed with the help of an energy demand estimating tool. In a feasibility study, the final evaluation of the various strategies could be conducted by comparing the results of the estimates with the findings of a customer value assessment. In this case study, however, all strategies are assumed to result in acceptable levels of housing expenses (for different target groups), in the traditional economic sense. In other words: the housing expenses of the varying apartments can be considered to represent the values of the housing services provided. So, in line with the model of the EVR, the environmental burden of the discerned strategies for interventions in the housing stock can be compared with their value by comparing them with the (traditional economic) housing expenses.

The EVR of refurbishment calculated this way, turns out to be lower than the EVR of an unchanged continued operation. The EVR of new construction turns out to be lower than the EVR of continued operation but it is higher than the EVR of renovation. In the cases of refurbishment and renovation, a relatively larger part of the expenses consists of energy costs than in the case of new construction. The energy costs raise the EVRs of refurbishment and renovation. However, they remain clearly below the EVR of new construction.

6.5   EVR and tall buildings

The previous section is almost entirely based on the research of Tim de Jonge (2005), and Joost Hoffman (2008) who developed the use of the method specifically for tall buildings. This section is based on selected cases. The selected cases (shown in Table 6.2 and Figure 6.3) are not very tall, but have various complexities which are characteristic of tall buildings.

Table 6.2   Selected cases (Hoffman, 2008).

	Location	Height(m)	Floors	Structure	Offices(m²)	GFA(m²)
Grotius tower	The Hague	86	22	Concrete	36,670	39,929
KJ-square (KJP)	The Hague	97	27	Concrete	43,344	73,439
Hoog aan de Maas (HadM)	Rotterdam	72	20	Steel	12,736	12,736

GFA, Gross Floor Area.

Figure 6.3   Selected cases: Grotius tower, KJ-square and Hoog aan de Maas (from left to right) (Hoffman, 2008).

Table 6.3   Results of the cases based on different ranking methods (Hoffman, 2008).

	Energy label		GreenCalc+		BREEAM		EVR	
Standard	E (1,5) >>B (1,1)		163–197 (2007)		Pass		39%	
Grotius tower	A	1	176	2	Very good	1	44%	3
KJP	A	1	176	2	Good	2	39%	2
HadM	D	3	195	1	Good	2	35%	1

In the period of the research all three designs were planned. Unfortunately all three cases are not yet realised and will probably not be realised at all. The Grotius tower, designed by Kees van Dongen of the 'De ArchitektenCie', consists of a plinth and two towers of, 16 and 22 floors always seemed a realistic option but the clients have withdrawn. The KJ-square (KJP) was much more ambitious. Designed by OMA, it should have been the icon for the central station area, but did not fit anymore, basically due to a collapsing office market. Finally, the Hoog aan de Maas (HadM) was a very creative solution, an office block hanging over a monumental building, with an impressive structural backbone. In the end, a solution which should be driven by high land prices in order to justify high construction costs. Land is not the problem in Rotterdam. A common characteristic is they are not designed for sustainability although the Grotius tower had high ambitions for energy performance.

The three cases are all measured according to the same system and tested against four environmental ranking methods. The results of the cases and as well as a standard low rise office building are given in Table 6.3.

The reference buildings performed well in comparison with the standard low rise. An average level of sustainability can be achieved with significant gain from installations and materialisation. These items are the main drivers for building costs and environmental costs. Lowering these costs will have a positive influence on the feasibility of tall buildings. High rise in itself is already more expensive than low rise. Dutch research (Van Oss, 2007; De Jong and Wamelink, 2008) reveals an increase of the total building costs of 0.8% for every additional floor. In an integrated sustainable design approach certified buildings do not necessarily have higher building costs per unit (Morris and Matthiessen, 2007) but in the reflected cases there was already a design which had to be upgraded to a higher level.

There are no positive outliers in the cost items of high rise in the different methods. Negative outliers are the structure, the foundation and the installation. Stepping up in the ranking, within a given design, seems expensive. In the test on BREEAM a step upwards from "Pass" to "Good" increase costs by about €100/m² GFA, while the next step will increase cost by €270/m² GFA (elaboration on all three cases). Figure 6.4 shows the BREEAM score of the Grotius tower which averages about 57 based on the eight categories assessed. Also the research on the EVR method expressed high cost increases where more sustainable materials were applied.

6.6   Embedding EVR in other sustainable ranking methods

There is no universal integrated system for ranking sustainability yet. Different tools such as BREEAM, EVR, GreenCalc and EPC serve different goals for different target groups. The ranking tools using a checklist and reporting based on classification

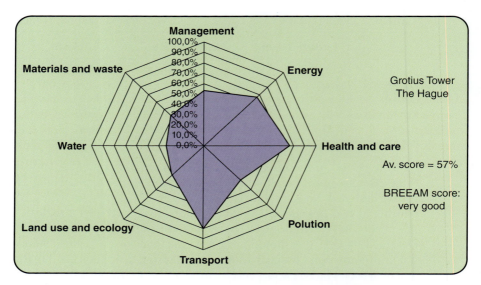

Figure 6.4   BREEAM score of the Grotius tower (Hoffman, 2008).

categories such as BREEAM and LEED are very useful for setting sustainability ambitions in the building process. It is not surprising that these tools are adapted by developers for attracting clients and satisfying their demands. The calculation based tools such as EVR, GreenCalc and energy performance tools are much more detailed at the design stage. However, there is a real need for assessing actual performance during the operation period to bring life-cycle cost (LCC) calculations to a higher level. There is a need for an integrated system for assessing sustainability during the design, construction and operation stages that is valuable to developers, architects, investors and facility managers.

6.7   Conclusion

A criticism of the EVR method is that there is a lot of data needed before it reaches an operational level. The extension of the method by Tim de Jonge[1] for standard housing technical solutions and Joost Hoffman[2] are good examples demonstrating its applicability and usefulness. Unfortunately, the lack of practical and reliable data as part of the discourse on whole-life cost (WLC) and LCC relating to key components in terms of construction, operation, maintenance and end-of-life (as shown in Figure 6.5) is a major constraint. Given the discussions on operating and maintenance cost in relation to LCC (Evans et al., 1998; Hughes et al., 2004; Ive, 2007) the standardising and collection of data will be one of the main challenges in the coming years to make the EVR model fully operational. Standards on

[1]www.winket.nl (Dutch) shows numerous examples of estimates for both building costs and eco-costs in project analysis.
[2]repository.tudelft.nl is a database of (doctoral) thesis and reports of TU Delft, in which the thesis of Joost Hoffman (2008) can be found (in Dutch), supplying the data of the tall building cases. Also, the thesis of De Jonge (2005) can be retrieved here.

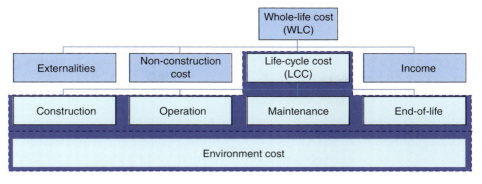

Figure 6.5   The elements of whole-life cost (NEN-ISO, 2008).

building cost will be critical, given the significance of cost data, but an open mind and a huge effort is needed.

EVR is mirroring the elements of LCC as described in the international standards (NEN-ISO, 2008), including the End-of-life and the Environmental cost. It is therefore be promising for further development and application. For the assessment of tall buildings from a sustainable perspective, the element of 'externalities' as stated in the standard should be further elaborated. For example, the proper use of land, preserving rural land for the future can be rewarded in land policy and should appear by some means in the project balance sheet.

References

Brundtland, G.H. (1987) *Our Common Future. World Commission on Environment and Development*, Oxford University Press, Oxford.

De Jong, P. and Wamelink. H. (2008) Building cost and eco-cost aspects of tall buildings. 8th World Congress Tall & Green: Typology for a Sustainable Urban Future, 3–5 March 2008, Dubai. http://www.ctbuh.org (accessed 14 January 2015).

De Jonge, T. (2005) Cost effectiveness of sustainable housing investments. PhD thesis, Delft University of Technology. http://repository.tudelft.nl/ (accessed 14 January 2015).

Evans, R., Haryyott, R., Haste, N. and Jones, A. (1998) *The Long Term Costs of Owning and Using Buildings*, The Royal Academy of Engineering, London.

Hoffman, J. (2008) Duurzame Hoogbouw, Creating space for a sustainable future. MSc thesis, Delft University of Technology. http://repository.tudelft.nl/ (accessed 14 January 2015).

Hughes, W., Ancell, D., Gruneberg, S. and Hirst, L. (2004) Exposing the myth of the 1:5:200 ratio relating initial cost, maintenance and staffing costs of office buildings. 20th Annual ARCOM Conference, 1–3 September 2004, Edinburgh.

Ive, G. (2007. Re-examining the costs and value ratios of owning and occupying buildings. *Building Research & Information*, 34(3), 230–245.

Morris, P. and Matthiessen, L.F. (2007) *Cost of Green Revisited.*, Davis Langdon, London.

Needham, B. (2007) *Dutch Land Use Planning*, SDU Uitgeverij, The Hague.

NEN-ISO (2008) Buildings and constructed assets – Service-life planning – Part 5: Life-cycle costing (ISO 15686-5:2008,IDT). NEN, Delft.

Statistics Netherlands (2013) Building permits and newly built houses, core figures 1995–2011. http://statline.cbs.nl/StatWeb/publication/PrintView.aspx?DM=SLNL&PA=70009NED... (accessed 17 September 2013).

Van der Post, W. (2007) Vacant land and vacancy rates. 14th Annual European Real Estate Society Conference, 27–30 June 2007, London.

Van Oss, S. (2007) Hoe hoger hoe duurder. MSc thesis, Delft University of Technology. http://repository.tudelft.nl/ (accessed 14 January 2015).

Vogtländer, J. (2001) The model of the Eco-costs/Value Ratio. PhD thesis, Delft University of Technology. http://repository.tudelft.nl/ (accessed 14 January 2015).

Chapter 7
Productivity in Construction Projects

Shamil Naoum

7.1 Introduction

The construction industry has a significant role to play in the economic growth of a nation. The industry employs a large number of skilled, semi-skilled and unskilled workers, and its activities provide work for the economic sector. The nature of the design, size of project, choice of materials, equipment and technology employed are also factors that can affect productivity. The success or failure of the construction projects can therefore be seriously influenced by productivity. There have been a number of definitions of site productivity. For the purpose of this chapter, 'productivity' is defined as 'the number of units (output) to be produced within a span of time, utilising an optimum number of human and material resources (input) in a safe and efficient manner'. A high rate of productivity can be achieved by eliminating unnecessary wasteful resources from construction operations.

This chapter explores the factors affecting the productivity of construction projects utilising a socio-techno-managerial approach that was developed by Naoum (2011). This approach recognises that productivity can be influenced as follows:

- *technically* by an efficient planning and scheduling of the resources;
- *socially* by creating the work environment that can motivate and lead people effectively;
- *managerially* by designing an efficient management system to communicate, co-ordinate and control the work activities from design to construction.

A conceptual model is designed to demonstrate the interrelationship among the factors that can impair productivity on construction sites and affect the performance of construction projects.

Design Economics for the Built Environment: Impact of Sustainability on Project Evaluation, First Edition.
Edited by Herbert Robinson, Barry Symonds, Barry Gilbertson and Benedict Ilozor.
© 2015 John Wiley & Sons, Ltd. Published 2015 by John Wiley & Sons, Ltd.

7.2   Concept and measurement of productivity

Productivity is one of the most difficult factors to measure as its determinants can vary significantly depending on design, size of project, site characteristics and place of measurement. Moreover, productivity can range from industry-wide economic parameters to the measurement of crews and individuals. For instance, single-factor productivity measures such as Average Labour Productivity (ALP) looks at the impact of one factor input (labour), whereas total (multi-factor) productivity measures take into account the impact of all inputs and output. Crawford and Vogl (2006) provided an overview of methods used to measure productivity in the construction industry. It was concluded that most existing work provides a partial modelling of the production process, potentially resulting in biased productivity estimates. Furthermore, the simple-to-calculate output/labour input ratios used in most studies do not enable the establishment of robust cause and effect relationships, leaving the reader largely in the dark about drivers of performance or productivity and their relative importance in construction projects.

In a more technical approach to measuring labour productivity, Radosavljević and Horner (2002) examined the complex variability of 12 construction labour productivity data sets by analysing the central moments of tendency, and applying the Kolmogorov–Smirnov and Anderson–Darling tests of normality. The results consistently show that productivity is not normally distributed. In addition, undefined variance causes a failure of the central limit theorem, thus indicating that some basic statistical diagnostics like correlation coefficients and t-statistics may give misleading results and are not applicable. A brief comparison with volatility studies in econometrics has revealed surprising similarity with Pareto distributions, which can model undefined or infinite variance. Such distributions are typical of chaotic systems like the logistic equation, whose properties also are described briefly. Therefore, it is suggested that future research should be focused on studying the applicability of chaos theory to construction labour.

A study by Clarke and Herrmann (2004) into productivity in social housing construction in England, Scotland, Denmark and Germany was apposite in demonstrating structural differences in the organisation of the construction process, their implications for efficiency and productivity, and their impact on employment and contract relations, as well as innovation and skills. The effects of the overriding cost rationale of the British system are illustrated in terms of labour deployment and the efficiency and productivity of the site construction process. Clarke and Herrmann's (2004) study showed the high labour intensity in the British case, with 39% more labour needed to produce $1\,m^2$ compared with Germany and 50% compared with Denmark. At the same time, the nature of labour deployment is qualitatively different, being front-loaded in England, whilst in the other countries, it is end-loaded in the sense that there is a gradual build-up of labour on site. It was also noted that the Danish building industry is facilitated by extensive prefabrication processes.

7.3   Previous literature on factors affecting site productivity

Concern has been expressed for many years about the factors impairing site productivity. The view is generally held that factors at head office and site management level are the main constraints to productivity. These factors are discussed in the following subsections.

Project characteristics

The type and size of the project including design characteristics such as layout and complexity can have a great impact on site productivity. Naturally, a large construction site requiring a large number of workers would be relatively harder to manage than a smaller size. The difficulties in managing manpower on a large scale may result in productivity loss. A large proportion of high costs in construction works are as a result of excessive labour costs. These costs can be reduced if productivity on site is increased by improving labour efficiency. Thomas (1991) notes that work on a complex project such as the construction of a nuclear project becomes more difficult as the project advances. The construction method such as use of off-site pre-fabrication units will reduce the number of labour hours required. This was confirmed in recent research by Eastman and Sacks (2008) who compared the relative productivity of construction industry sectors with significant off-site fabrication with more traditional on-site sectors. In this research it was found that the off-site sectors, such as curtain wall, structural steel, and precast concrete fabrication, consistently show higher productivity growth than on-site sectors. Furthermore, the value-added content of the off-site sectors is increasing faster than that of the on-site sectors, indicating faster productivity growth.

Earlier, Proverbs et al. (1999) conducted research into construction resource and productivity level for high rise concrete construction contractors in France, Germany and the UK. For concrete placing productivity rates, none of the resource factors (material, plant or labour) when considered independently was found to be of significance. It was then concluded that international variations in concrete placing productivity rates were not connected directly to these individual factors. However, framework productivity rates were impacted by the type of framework utilised on column and beam work. The most unproductive rates were related to traditional timber solutions, while proprietary (for column work) and prefabricated (for beam work) solutions were associated with the most efficient (and hence most economic) productivity rates.

Labour Characteristics

Labour related factors affect the physical progress of any construction project. In order to improve labour productivity, site production should be measured on a regular basis, and then compared with acceptable standard benchmarks. The management of each contracting company should maintain its own record describing the baseline productivity in different previous projects with similar conditions. Enshassi et al. (2007) argue that such records can be used to help estimate labour productivity in future projects. For example, changes made to the original scope of work are costly and have an effect on labour productivity and should be recorded. Although some changes are inevitable, the impact on site productivity is nonetheless significant. The impact of changes to the original scope of work has been investigated by Thomas and Napolitan (1995). They studied the impact of changes in quantitative terms and discussed why change impacts on the labour force's efficiency. They also explored the relationship between changes and various types of disruption.

Moreover, improving labour productivity is very much linked with construction methods. For example, *in situ* reinforced concrete construction is necessary because of the importance of this material to the industry. Several factors influence labour productivity but buildability is among the most important. Despite the plethora of research into construction productivity reported over the years, a thorough examination of the literature revealed a dearth of research into the effects of buildability factors on the efficiency of the concreting operation. As concreting is an integral, labour intensive trade of *in situ* reinforced concrete construction, research by Jarkas (2012) explored the influence of primary buildability factors on concreting labour productivity. In achieving this aim, a sufficiently large volume of productivity data was collected and analysed by using the categorical-regression method. As a result, the effects and relative influence of: (1) concrete workability; (2) reinforcing steel congestion; (3) volume of pours; and (4) height relative to ground level, on labour productivity of skipped and pumped placement methods are determined and quantified. The findings show significant impacts of factors investigated on the efficiency of the concreting operation, which can provide designers with feedback on how well their designs consider the requirements of the buildability concept, and the tangible consequences of their decisions on the productivity of the operatives. Practical recommendations, moreover, are presented, which upon implementation may improve the buildability level of this trade, and thus translate into higher labour productivity and lower labour cost. On the other hand, the depicted patterns of results can provide guidance to construction managers for effective planning and efficient labour and plant utilisation

Management style and management systems

Previous studies have identified two main characteristics that have the greatest impact on the site manager's performance. These are: the site manager's involvement at the contract stage; and the delegation of responsibilities. Laufer and Jenkins (1982) examined various approaches to motivate construction workers and found that construction management would benefit from a general move toward a more participative decision-making style of leadership. Bresnen *et al.* (1987) identified site management involvement in planning as a very important factor because of their input at an early stage in developing an understanding of, and preparedness for, their job. It has been suggested that site managers should have a major say in setting the original targets, planning the process and organising the resources. In terms of good atmosphere to boost performance, a better atmosphere on site has been associated with increased decentralisation of decision-making authority within the firm, and a greater level of site manager influence over operations and decisions on site. Olomolaiye (1990) found that good supervision was the most significant variable influencing percentage productive time and that fluctuations in productivity are primarily the responsibility of on site management. This suggests that site managers have a powerful influence on the behaviour of site employees. At project level, management relations with the site can be a source of dissatisfaction or motivation.

Other researchers regarded the management information system (MIS) as the linking mechanism of the above factors, that is decision making, site supervision, communication and morale. For example, Reinschmidt (1976) argued that the productivity of firms depends on the management's access to accurate information to aid in faster decision making. Information which does not flow promptly from one group to another will cause delays, rework, low motivation and, hence, decrease productivity. Sanvido and Paulson (1992) empirically tested the possible utilisation of practical tools (from a productivity improvement viewpoint) that can support various theoretical decision-making phases. They demonstrated that jobs where the planning and control functions were performed at the right level in the site hierarchy were more profitable, finished sooner and were better constructed than those where the functions were performed at the wrong level.

Design variation orders

Key variables affecting efficiency are the nature, frequency of changes as well as the time of the change. Rework, disruption and presence of change work can lower labour performance (Thomas and Napolitan, 1995). Hanna et al. (1999) have identified the impact of changes on the construction site and described that disputes are common between the client and contractors when these changes occur. Their study used data from 43 projects and a linear regression model was developed that predicted the impact of changes on labour efficiency. The model allows labour efficiency loss to be calculated in a particular project enabling both the client and the contractor to understand the impact such changes will have on labour productivity. However, this study is limited to mechanical trade with some specific plumbing, fire protection, and process piping. From a study carried out by Thomas and Napolitan (1995) over a period of 4 years, based on three projects, an equation was derived to calculate the efficiency losses from the impact of change order. Efficiency was defined as the ratio of actual productivity to baseline productivity. Baseline productivity was also measured for this survey. Efficiency was determined by dividing the performance ratio equation value on a normal day by the performance ratio equation when change order had occurred. The survey result showed an average efficiency loss of 30%. Change order impact on a project lowers labour efficiency and productivity. The result of a survey by Hanna et al. (1999) indicated that labour efficiency on a job that is not impacted by change has a higher level of efficiency. Disruption which was also found to cause changes in the original plan of work increased the project cost through re-work and decreased labour efficiency for the main contractors and sub-contractors.

A survey on time performance of different types of construction projects in Saudi Arabia was conducted by Assaf and Al-Hajji (2006) to determine the causes of delay and their importance according to each of the project participants, that is the owner, consultant and the contractor. The field survey conducted included 23 contractors, 19 consultants and 15 owners. Seventy-three causes of delay were identified during the research. From the study, 76% of the contractors and 56% of the consultants indicated that average of time overrun is between 10% and 30% of the original duration. The most common cause of delay identified by all the three parties is "change order".

Resource management and motivational factors

Material

Ferguson *et al.* (1995) suggest that 50% of the waste deposited in disposal sites in the UK is construction waste. In order to reduce waste and increase productivity Just-In-Time (JIT) has been introduced on construction sites. Pheng and Tan (1998) investigated whether the introduction of JIT can reduce the level of wastage on site. Their investigation showed that wastage of materials could be kept to a minimum and consequently productivity improved. From the result of the survey, both project and site managers did not regard wastage on site as an important factor in improving construction productivity. Faniran and Caban (1998) suggested that wastage on site could be reduced if design changes were kept to a minimum during the construction work. The respondents also identified leftover material scraps, waste from packaging and unreclaimable non-consumables, design/detailing errors, and poor weather as being important sources of construction waste.

Material management is a worldwide problem and on-going research work has been conducted to highlight its effect on site productivity. Abdul Kadir *et al.* (2005) in their research into factors affecting construction labour productivity for Malaysian residential projects, found that material shortage at site as well as non-payment to suppliers causing the shortage of material delivery to site as highly important. Other factors that can cause time and cost overrun and subsequently affect productivity are change order by consultants; late issuance of construction drawings by consultants; and incapability of contractors' site management to organise site activities.

Leadership

The construction process is a collective effort involving a team of specialists from different organisations. The leader of the team may affect the productivity of the design and construction. The person who leads the team varies dependent upon the contractual arrangement adopted for the project. A number of studies have been conducted to investigate the relationship between leadership styles and productivity rate. For example, Cheung *et al.* (2001) carried out an empirical survey that aimed to establish the relationship between leadership behaviours of design team leaders and the satisfaction of the design team members. The results indicate that charismatic and participative leadership behaviours primarily determine the satisfaction of the team members.

Motivating factors

Motivation is a prime determinant of worker performance. As far as known, researchers have failed to develop a commonly agreed theory that addresses worker motivation that is valid and relevant to the construction industry. A key motivator for one worker compared with another worker in a certain situation may differ. Individuals tend to seek a job, which will satisfy personal needs. Key motivators may be one or a combination of the following: high achievement, recognition, the nature of the work itself, responsibility and personal advancement and growth. In the view of some researchers, it appears that there are differences in opinion whether workers' motivation contributes positively or at all to the level of productivity on construction sites.

A number of behavioural and psychological researchers have argued that the expenditure of effort by a worker is the physical manifestation of motivation. The greater worker motivation is, the greater the worker motivation becomes. For example, Kazaz and Ulubeyli (2007) showed that monetary factors in Turkey remain pre-eminent in influencing productivity but that socio-psychological factors such as giving responsibility and taking account of cultural differences appear to be increasingly importance in a developing economy.

On the other hand, ranking on questions related to motivation by Naoum and Hackman (1996) showed that there is little support for the suggested motivating factors on the construction site. The survey indicated that other factors such as 'ineffective project planning' and 'constraints on a worker's performance' as the most crucial factors influencing productivity. Other highly ranked factors are 'difficulties with material procurement', 'lack of integration of project information', 'disruption of site programme', 'lack of experience and training' and 'exclusion of site management from contract meetings'.

Other factors such as training are also considered to be influential factors in high productivity on construction sites. In recent years the growth in labour only sub-contracting on sub-employment bases in the UK construction sector and the level and quality of training within the industry is a source of concern. Productivity within the UK construction industry compared with West Germany and France is partly the result of low levels of training the labour force receives in the UK (Prais and Steedman, 1986). In this context, UK government policy has emphasised the role of skills development and training as a means of improving productivity performance across all sectors of the economy. According to Abdel-Wahab *et al.* (2008), there is inconsistency in the industry's productivity performance, despite the overall increase in qualification attainment levels and participation rates in training over the same period. However, the year-by-year change in the participation rate of training was not consistently associated with an improvement in productivity performance. Therefore, there is an urgent need to consider skills development and training within the context of construction businesses in relation to other factors in order to unpack how skills can bring about improvement in productivity performance.

Impact of external environment

For the past two decades or so, technology has had a great effect on productivity output within the construction industry. Technological advancement has resulted in considerable changes to most tasks on construction sites. Equipment and machinery have evolved significantly during the past century. During the same time, our ability to study and understand equipment economics and productivity has also gained significant ground. Kanan (2011) identified three areas that relate equipment and productivity: repair costs; residual values; and total cost of ownership (TCO).

This technological advancement requires new manpower skills to ensure proper and full use is made of the new developments. It is becoming increasingly problematic to separate the contribution made by management, labour and machinery, to productivity. Innovation within the construction industry faces many barriers such as diversity of standards, fragmentation, business cycles, risk aversion and other factors, which all produce a complex and unfavourable climate in which to work. A high labour cost within the construction industry is a strong reason for this industry to move towards new technology. One of the reasons that many construction firms are reluctant to move towards new technology is the risk it carries if the new technology proves to be ineffective. The cost of changes may prove to be too high for the firms and may put their future in jeopardy. Baldwin (1990) argued that information technology would revolutionise management information systems and help management obtain accurate information leading to faster and more accurate decisions on site. So far many productivity studies tend to concentrate on improving the technology related to the construction site. Rapid mechanisation within the industry has resulted in increasing productivity by the introduction of structural steel, system form work, pre-casting techniques, pre-fabrication and component manufacture but the construction industry requires more innovation to remain competitive among other sectors.

Under the issue of innovation, a 30-year retrospective analysis of resource use in a range of new construction and repair projects reveals insignificant productivity increase when applying traditional narrow measures, as shown in a case study by Bröchner (2012) of beam bridges. Changes in government regulations, in specifications, and the development of non-price criteria for contract awards emerge as important. Schemes for benchmarking the performance of construction projects as well as life-cycle analyses suggest that customer risk aversion and effects on customer productivity should be taken into account. The outcome is a set of measurements that can be applied to the selection of any type of proposed new construction or repair technology innovation according to the potential impact on industry productivity. Although some new technologies promise to improve construction productivity, their ability to deliver is not always realised. Building on a great deal of previous research, a four-stage predictive model was developed by Goodrum (2011), and validated to estimate the potential for a technology to have a positive impact on construction productivity. Statistical analysis confirmed that average performance scores produced by the model were significantly different across the categories of successful, inconclusive, and unsuccessful in the actual implementation experience of technologies.

7.4   Productivity survey

A major study was conducted in 2006 by Dejahang (2006) to investigate the factors that can impair site productivity. Exploratory interviews with 4 project managers and 4 site managers were conducted and a structured questionnaire containing a list of 55 productivity factors was designed. The questionnaire was then sent to 200 project and site managers working for construction firms. From the 200 questionnaires, 137 were analysed (87 from project managers and 50 from site managers). The participating firms were large sized companies and the nature of the projects that the project and site managers were involved in were housing development projects costing between £1.5 million and £10 million.

In the questionnaire, the respondents were asked to rate each factor by using a scale from 1 to 5 in order of importance. The range included 'not an important determinant' (given a value of 1) to 'very important determinant' (given a value of 5). The data gathered were than analysed using the descriptive and inferential method of analysis. This helped to distinguish the highly rated factors from those which were rated as low. The Spearman correlation (ρ) was used to analyse the data as well as the factor analysis technique to demonstrate the cluster of factors that can mostly affect productivity on site.

Analysis of the Spearman rank test shows that, overall, there are no significant differences in opinions between project and site managers about the factors that influence construction site productivity ($\rho = 0.91$). Both respondents identified, 'Project and Site managers experience and capabilities', as the most important factor likely to increase the level of productivity on site. The respondents then identified 'project planning' as an equally important factor likely to affect the level of productivity. Both samples also agreed on the 'delay caused by design error', 'communication system', 'Design and buildability', 'leadership style' and 'procurement method' as being high determinants. Interestingly these factors are mostly associated with activities that are set at the pre-construction stage, especially when they are considered with the second level of determinates such as client brief and specifications.

There was a difference in perception between the two samples regarding the motivational factors. Project managers gave a high weighting to 'supervision' but registered a moderate importance to "salary". The site managers thought the opposite giving salary a fairly high weighting. It would appear that there is a difference in views as to whether work force drive can affect site productivity. Several behavioural and physiological researches indicated that the expenditure of effort by a worker is the physical manifestation of motivation; the greater a worker's motivation, the greater his/her expenditure of effort.

To expand the research findings further, the factor analysis technique was applied on the 55 determinants to help identify clusters of factors that can impair site productivity and the relative importance of these clusters. The explanation of this technique is rather complex and beyond the scope of this book. In short however, the main purpose of the factor analysis technique is: (1) to *reduce* the number of variables; and (2) to *detect structure* in the relationships between variables, that is to *classify variables*. Therefore, factor analysis is applied as a data reduction or structure detection method. Table 7.1 shows the factor matrix.

Factor (1) represents the first dominant group of factors that can affect site productivity which includes variables relating to pre-construction activities, namely, design, estimating, scheduling, procurement method, contractual relationship with sub-contractors and subordinates. The name given to this group of factors is "pre-construction activities and contractual arrangements". Factor (2) includes variables mostly related to site planning, communication and control system. The name given to this group is "site management". Factor (3) includes variables mostly related to the waste, namely, waste on site following design change, waste on site following design error and waste on site following material mismanagement. The name given to this group is "waste management". Factor (4) includes variables related to teamwork and groups, group coordination and delay/disruption caused by contractor's action. The name given to this group is "work groups". Factor (5) includes variables related to clarity of tasks, safety management, plant control, job security and salary. The name given to this group is "resource management".

The result from the factor analysis showed an interesting outcome in that it reinforces the earlier study by Winch and Carr (2001) who made an attempt to benchmark the on-site productivity in France and the UK. They conducted a

Table 7.1   Group of factors affecting productivity on construction sites.

Factor	Groupings in order	Name given
Factor (1)	Determinants Contractual arrangement Scheduling Estimating Delay caused by design error Relationship among building team Relationship with sub-contractors Relationship with subordinate	Pre-construction activities and contractual arrangements
Factor (2)	Determinants Experience/capability Goal commitment Site planning Communication Controlling	Site management
Factor (3)	Determinants Material waste on site following variation orders Material waste on site following design error Material waste on site due to mismanagement	Waste management
Factor (4)	Determinants Teamwork Group coordination Delay and disruption caused by contractor's action	Work groups
Factor (5)	Determinants Clarity of task Safety management Plant control Job security Salary	Resource management

detailed analysis of structural concrete operations to compare the productivity performance of these two countries using the activity sampling approach called CALIBRE. Their results showed that the French productivity performance is better than that in the UK which was explained by some technical and economic variations between the two counties. However, they also argued that this difference in productivity rate cannot be understood without reference to the overall contracting system and the constraints upon action that it imposes.

7.5   Proposed framework for site productivity

Based on the literature and Naoum's (2011) research into people and organisational management in construction, a conceptual model was designed to demonstrate the interrelationship among the factors that can impair productivity on construction projects (Figure 7.1). The purpose of the model is to represent or explain the phenomenon of productivity. A key conclusion that can be drawn is

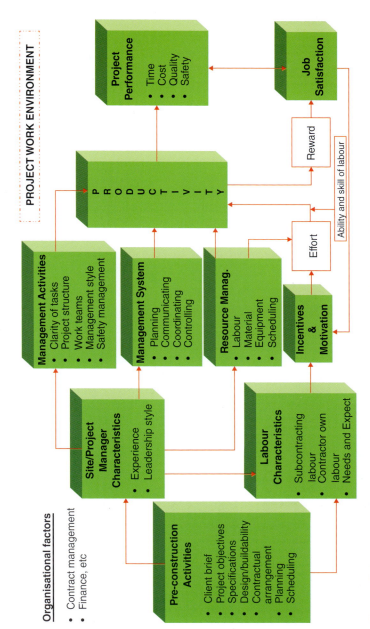

Figure 7.1 Proposed framework for site productivity. Source: Naoum (2011).

that the rate of productivity on site can mostly be affected by the *pre-construction activities*, namely, the selection of the personnel in terms of experience and ability, buildability of the design, delays caused by design error, pre-construction planning, decision on the relationship among the building team, clarity of the client brief and the communication system. Executive managers' responses or *site/project manager's leadership style and experience* can either reinforce and identify the productivity problem or create pressures to change it. Much depends upon how sophisticated a diagnostic model the manager uses. Other factors that can influence productivity are *labour characteristics* such as the needs and expectation of workers, and the appropriate balance between the contractor's own labour usage and the level of outsourcing or sub-contracting. The *motivation and incentives* of the workforce is also crucial. Achieving high motivation and incentivising workers depends on the working environment where financial or psychological rewards can be regarded as an incentive for the efforts needed to achieve high productivity. In order to assure a good working environment, the *management system* such as the planning, communication, co-ordination and controlling mechanisms should be effective and clearly understood by the project team. The *management of activities* reflected in a clear project structure, work teams with understanding of their role and clearly defined tasks, supported with strong management and safety management culture as well as the *resource management* needs to be highly effective, as demonstrated by previous research.

7.6   Conclusion and further research

This chapter discusses past and recent literature available on productivity in the construction industry as well as presenting the findings of a productivity survey that was conducted at London South Bank University. The proposed conceptual model is to explain the phenomenon of productivity and to understand the different factors that can influence the level of productivity from a socio-technical and managerial perspective. In this way, managers can predict what needs changing to improve the productivity level on a construction site and develop a strategy for achieving the change

References

Abdel-Wahab, M.S., Dainty, A.R J., Ison, S.G., Bowen, P. and Hazlehurst, G. (2008) Trends of skills and productivity in the UK construction Industry. *Engineering, Construction and Architectural Management*, 15(4), 372–382.

Abdul Kadir, M.R., Lee, W.P., Jaafar, M.S., Sapuan, S.M. and Ali, A.A.A. (2005) Factors affecting construction labour productivity for Malaysian residential projects. *Structural Survey*, 23(1), 42–54.

Assaf, S. and Al-Hijji, S. (2006) Causes of delays in large construction projects. *International Journal of Project Management*, 24, 349–357.

Baldwin, A. (1990) Why an IT strategy is important? *Chartered Builder*, July/August.

Bresnen, M., Bryman, A., Breadsworth, A. and Keil, T. (1987) Effectiveness of site managers. Chartered Institute of Building, Technical paper no. 85, pp. 1–7.

Bröchner, J. (2012) Construction Productivity Measures for Innovation Projects. *Journal of Construction Engineering and Management*, 137 No. (5), 670–677.

Cheung, S.O., Ng, S.T., Lam, K.C. and Yue, W.M. (2001) A satisfying leadership behaviour model for design consultants. *International Journal of Project Management*, **19**(7), 421–429.

Clarke, L. and Herrmann, G. (2004) Cost vs. production: labour deployment and productivity in social construction in England, Scotland, Denmark and Germany. *Journal of Construction Management and Economics*, **22**(10), 1057–1066.

Crawford, P. and Vogl, B. (2006) Measuring productivity in the construction industry. *Building Research and Information*, **34**(3), 208–219.

Dejahang, F. (2006) Determinants of productivity factors in construction sites. PhD thesis, London South Bank University.

Eastman, C. and Sacks, R. (2008) Relative productivity in the AEC industries in the United States for on-site and off-site activities. *Journal of Construction Engineering and Management*, **134**(7), 517–526.

Enshassi, A., Mohamed, S., Mayer, P. and Abed, K. (2007) Benchmarking masonry labour productivity. *Journal of Productivity and Performance Management*, **56**(4), 358–368.

Faniran, O.O. and Caban, G. (1998) Minimizing waste on construction project sites. *Engineering, Construction and Architectural Management*, **5**(2), 182–188.

Ferguson, J., Kermode, N., Nash, C.L., Sketch, W.A.J. and Huxford, R.P. (1995) Managing and Minimizing Construction Waste – A Practical Guide, Institute of Civil Engineers, London.

Goodrum, P. (2011) Model to predict the impact of a technology on construction productivity. *Journal of Construction Engineering and Management*, **137**(9), 678–688.

Hanna, A.S., Russell, J.S. and Vandenberg, P.J. (1999) The impact of change orders on mechanical construction labour efficiency. *Journal of Construction Management and Economics*, **17**(6), 711–720.

Jarkas, A. (2012) Influence of buildability factors on rebar installation labor productivity of columns. *Journal of Construction Engineering and Management*, **138**(2), 258–267.

Kannan, G. (2011) Field studies in construction equipment economics and productivity. *Journal of Construction Engineering and Management*, **137**(10), 823–828.

Kazaz, A. and Ulubeyli, S. (2007) Drivers of productivity among construction workers: a study in a developing country. *Building and Environment*, **42**(5), 2132–2140.

Laufer, A. and Jenkins, D. (1982) Motivating construction workers. *Journal of the Construction Division, ASCE*, **108**, 531–545.

Naoum, S. (2011) People and Organizational Management in Construction, ICE, London.

Naoum, S. and Hackman, J. (1996) Do site managers and the head office perceive productivity factors differently? *Journal of Engineering, Construction and Architectural Management*, **3**(1–2), 147–160.

Olomolaiye, P. (1990) An evaluation of the relationships between bricklayers' motivation and productivity. *Journal of Construction Management and Economics*, **8**(3), 301–313.

Pheng, L.S. and Tan, S.K.L. (1998) How 'just-in-time' wastages can be quantified: case study of a private condominium project. *Construction Management and Economics*, **16**(6), 621–635.

Prais, S.J. and Steedman, H. (1986) Vocation training in France and Britain: the building trades. *National Institute Economic Review*, **116**(1), 45–55.

Proverbs, D.G., Holt, G.D. and Olomolaiya, P.O. (1999) Construction resource/method factors influencing productivity for high rise concrete construction. *Journal of Construction Management and Economics*, **17**(5), 577–587.

Radosavljević, R. and Horner, R.M.W. (2002) The evidence of complex variability in construction labour productivity. *Journal of Construction Management and Economics*, **20**(1), 3–12.

Reinschmidt, K. (1976) Productivity in the construction industry. *Productivity in Engineering Design*. Proceedings of ASCE Conference, Lincoln, IL.

Sanvido, V.E. and Paulson, B.C. (1992) Site-level construction information system. *Journal of Construction Engineering and Management*, **118**(4), 701–715.

Thomas, H. R. (1991) Labor productivity and work sampling: the bottom line. *Journal of Construction Engineering and Management*, **117**(3), 423–444.

Thomas, H.R. and Napolitan, C.L. (1995) Quantitative effects of organizational changes on labour productivity. *Journal of Construction Engineering and Management*, **121**(3), 290–296.

Winch, G. and Carr, B. (2001) Benchmarking on-site productivity in France and the UK: a CALIBRE approach. *Journal Construction Management and Economics*, **19**(6), 577–590.

Chapter 8
Design Variables and Whole-Life Cost Modelling

Andrea Pelzeter

8.1   Introduction

Every variable in design may have an effect on costs, not only on initial construction cost but also on those costs that occur later, during the life cycle of the designed facility (Ellingham and Fawcett, 2006, p. 3). It is therefore important to ask fundamental questions to enable the designer to produce an efficient building structure in terms of life-cycle cost (LCC). Responses to questions such as (1) is it better to increase the initial investment, for example by investing in a geothermal installation to save consequential costs such as fuel, and (2) under what conditions might this be best achieved, need to be addressed. To assist in answering these questions, whole-life cost (WLC) should be calculated. According to ISO 15686-5 'Buildings and constructed assets. Service-life planning, Part 5: Life-cycle costing' (2008), WLC includes LCC, externalities, non-construction cost and income (Figure 8.1). LCC is defined as "cost of an asset or its parts throughout its life cycle, while fulfilling the performance requirements".

This chapter focuses specifically on the application of WLC theory which is central to design economics and essential in design development to minimise carbon emissions over the lifespan of buildings and built facilities. The key principles of LCC modelling methods will be examined together with design and cost variables in whole-life evaluation, steps required in the application of the technique, associated assumptions, limitations and critique of LCC. An attempt is made to deal with carbon emissions in whole-life analysis as a new approach by treating carbon emission as external costs and externalities in terms of the costs of emission certificates.

Design Economics for the Built Environment: Impact of Sustainability on Project Evaluation, First Edition.
Edited by Herbert Robinson, Barry Symonds, Barry Gilbertson and Benedict Ilozor.
© 2015 John Wiley & Sons, Ltd. Published 2015 by John Wiley & Sons, Ltd.

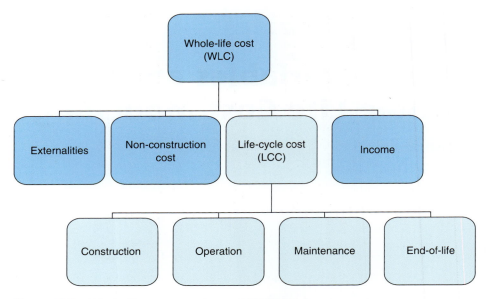

Figure 8.1   Whole-life cost according to ISO 15686-5.

8.2   Whole-life cost modelling

WLC can be calculated based on design opportunities with different outcome. In the case of judging different design variables, the choice would be the identification of that alternative which leads to lowest WLC or LCC. According to the elements given in Figure 8.1, only with the inclusion of external costs for carbon emissions, as externalities, will LCC be truly transformed into WLC. In whole life appraisal, other elements such as non-construction costs and income are included (as shown in Figure 8.1) as the focus is not only on construction cost. Figure 8.2 shows that an increase in the initial costs for construction does not necessarily lead to savings during the life cycle. The histogram in Figure 8.2 provides examples for the alternatives A–D (values given are only examples):

- Alternative 0: state-of-the-art design.
- Alternative A: initial costs rise for energy saving installations, less energy consumption in the usage or operational phase which reduces consequential costs.
- Alternative B: cheaper materials for surfaces reduce initial costs but raise consequential cost through a shorter life time expectancy (life cycle) of the materials that lead to higher expenses on maintenance. Thus the LCC may even be higher than in alternative 0.
- Alternative C, the ideal case: reduction of initial costs through low-tech measures, for example covering the north façade of a building with earth, saves construction costs and costs for heating and/or cooling, plus costs for maintenance.
- Alternative D, the worst case: installations that are not necessary or too big will cost more in the beginning and during operation and maintenance.

Figure 8.2 shows that relying on initial cost alone might reveal false savings in the case of alternative B with the lowest initial cost compared with all other alternatives. However, alternative B has the highest consequential costs making

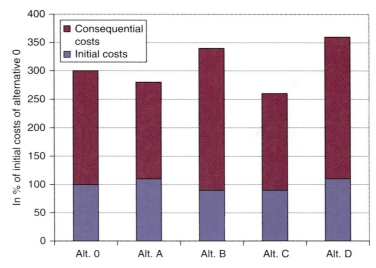

Figure 8.2   Relationship between initial costs and consequential costs.

it more expensive than alternative A when both initial and consequential costs are considered.

Understanding fully the purpose of LCC is essential to be able to address the concerns about the method of modelling. The previously mentioned ISO 15868-5 identified some helpful principles. For example, ISO 15686-5 (2008) integrates contents that derive from national standards, for example:

- ASTM E 917-02 (2002): "Standard practice for measuring life-cycle costs of buildings and building systems" from the USA
- NS 3454 (2000): "Life cycle costs for building and civil engineering work – Principles and classification" from Norway
- AS/NZS 4536 (1999): "Life cycle costing – An application guide" from Australia/New Zealand.

In addition there are standards with very detailed suggestions, of how to model LCC such as the German GEFMA 220 (2010) 'Lebenszykluskosten-Ermittlung im FM' (Life-cycle costing in Facility Management, only available in German), where GEFMA stands for German Facility Management Association. GEFMA 220 is replicated in Switzerland, and published by IFMA Switzerland (IFMA, International Facility Management Association). In IEC 60300-3-3 (2004) "Dependability management, Part 3-3: Application guide – Life-cycle costing" specific information on the relationship between reliability and LCC for electric facilities is given.

Certificates for sustainability in construction also may be regarded as a source for standardisation. In the UK, the BREEAM tool recognises life-cycle costing as an indicator for sustainable management of the building process. The German Sustainability Council defines LCC as one of about 60 criteria in the DGNB label. There the calculation method is given with all the details that will be required including benchmarks for some of the expenses in the operational or usage phase. LCC in DGNB is restricted to costs that are dependent on the physical and techni-cal configuration of the building but the method of modelling corresponds to GEFMA 220 as well as to ISO 15686-5.

8.3   Steps in LCC modelling

There are four steps in the modelling of LCC that have proven to be essential for the absolute as well as for the relative results of the whole-life calculation. These are:

Step 1: defining the period under consideration
Step 2: defining the system boundary
Step 3: defining the time value of money
Step 4: defining the sources of uncertainty, insecurity or risks for the calculation.

These are now considered in detail.

Period under consideration

The term "life-cycle cost" seems to indicate that it has to cover all phases of the life cycle of the facility in question, including construction, operation, maintenance and end-of-life. For whole buildings, the life cycle may continue over many decades or even centuries. This could result in an extremely long forecast. However, over long time periods, assumptions become less reliable due to uncertainty, for example prediction of oil prices in the future. Therefore the period under consideration should be chosen to be realistic and long enough in order to show the probable trade-off between initial capital costs and consequential/running costs but not transgressing phases of essential changes in the facility.

In GEFMA 220, a period of 25–30 years is suggested for buildings because after this period major repairs or rehabilitation of all technical installation is expected. The major repairs will lead to the installation of a new generation of technologies that cannot be predicted or forecasted with precision. In addition, a modernisation or revitalisation of a facility is often necessary to meet the expectations of future users. This may even lead to a general change of use that is not possible to forecast.

Other systems demand the use of a longer period under consideration, for example 50 years in DGNB (Table 8.1).

System boundary

It is important to recognise what should be integrated into the model of LCC and what ought to be excluded. The system boundary has to be stated clearly, usually by referring to national standards. The above mentioned DGNB standard restricts integrated costs to those that can be influenced by the design of the building such as costs for construction, water and energy consumption, cleaning, inspection and repair ("building related LCCs"). These are examples of the processes to be included. Also, the processes at the end of the period of analysis need some thought: is there a process of deconstruction and reuse/recycling/disposal to be

Table 8.1   Period under consideration according to different standards.

Named standards	GEFMA/IFMA 220	DGNB	ISO 15686-5
Period under consideration	10, 20, 30 years	50 years	To be chosen, no more than 100 years

Table 8.2   System boundary in LCC.

Aspects of system boundary to be defined	For example in LCC for DGNB
Process-related	Only what is influenced by the building, no process for end-of-life
Spatial	Building only, no outdoor facilities
Financial	Prices are defined in benchmarks
Stakeholder-related	Owner-occupier

modelled, or a complete renewal of the facility or no specific process at all, or just a cut or reduction in the cash flow? In general there are four aspects of system boundaries: in addition to the process-related aspects, you have to define the spatial system boundary (e.g. are outdoor facilities relevant, say, for rain water harvesting?), the financial system boundary (e.g. are in-house efforts priced?), and the system boundary with regard to the stakeholders (e.g. are the costs summed up from the point of view of the tenant, the landlord or the owner-occupier?) (Table 8.2).

Time value of money

Time is a key factor in LCC. Why is it relevant? At what time in the period under consideration do payments become due? Time is relevant because in most situations people prefer to get €100 today rather than in 10 years' time. This preference is normally expressed in terms of interest. Nominal interest rates reflect the price of owning or needing money, or the risk involved based on the inflation rate expected. In life-cycle costing, the time value of money is modelled as Net Present Value (NPV) using a defined interest rate for discounting future payments to its value in the present. Thus, the sum of all payments during the period under construction is made comparable even though singular payments may be due to very different phases of the life cycle.

Sources of uncertainty and risks

Due to the characteristics of LCC as a forecasting method, sources of uncertainty, insecurity or risks have to be addressed before using or applying the results of a comparison with LCC, for example before changing the design. Sources of risks in life-cycle costing are, among other things:

- choice of appropriate interest rate
- selection of inflation rate, especially for energy
- actual life-span of every part of the construction
- external factors that might shorten the life-span of parts of the construction, for example changes in use, in legal demands regarding fire prevention, access, and so on.

According to ISO 15686-5 and GEFMA/IFMA 220 the effect of uncertainty, insecurity or risks on relative results (comparisons with LCC) has to be analysed and if possible, quantified. GEFMA proposes a best-worst case analysis to show the effect of different risks and associated assumptions on the comparison. The result may be expressed in different forms, for example the span of costs for 1 kWh of power/fuel/heat that makes alternative *x* favourable compared with alternative *y*.

8.4   Design principles to optimise LCC

To enable LCCs to be reduced at the design stage it is necessary for the designer to establish an overview of the allocation of LCC and the different types of costs.

Cost variables

The allocation of LCC is strictly dependent on the assumptions made in modelling them. In Figure 8.3, LCC was calculated for an office building over a period of 50 years, using a discount rate of 5.5%, a general inflation rate of 2.0% and a specific inflation rate for energy of 4.0%. These values are given in DGNB (2011). Also, the types of integrated costs correspond to DGNB: costs for construction, cleaning, other maintenance, consumption of water and of energy are addressed. Renovation is not integrated.

With a discount rate of 5.5%, construction costs dominate LCC – as long as the building is used as an office or for living. The share of construction cost diminishes for buildings with very intensive use, for example in hospitals or schools where the costs for cleaning and other maintenance occupy a larger part of LCC. Optimising LCC is more than reducing energy consumption (for most kinds of building utilisation). Cleaning and technical maintenance may be of significant importance as well. Water consumption, until now, is still relatively cheap to

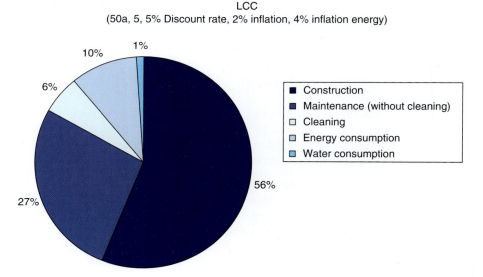

Figure 8.3   Example of a LCC calculation according to DGNB definitions for an office building in Berlin, Germany.

users, to justify expensive approaches to measurement to provide incentives for the reduction of water usage. Nevertheless some water saving principles will be identified in the design variables for optimising LCC, as costs of storage and supply are set to increase in the near future.

Design variables

In the planning phase design strategies may be applied to reduce the costs for construction, consumption, services, technical maintenance and renovation (Pelzeter and Sigg, 2011)

Designing for lower costs of construction

Cost efficient construction may reduce initial costs as well as costs for operation and maintenance if it uses space saving principles and synergies. Some examples are:

- Cost-saving construction: using prefabricated elements with standardised dimensions, easy to install constructions, simple geometry (rectangular, no inefficient bulges), bundle installations, no underground floors, and so on.
- Space-efficient planning: high ratio of space for primary use (e.g. office, medical treatment, etc.) to space for secondary use (movement/transport, technical installations, cleaning rooms, etc.), design rooms for standard furniture, and so on.
- Control comfort definitions: if you admit higher – or lower– indoor temperature for some days of the year, the size of technical equipment may be chosen smaller (and cheaper); is warm water necessary at every hand washbasin?
- Profit from synergies: integrate photovoltaic into elements for sun blinds, use waste heat from production or freezing (e.g. in commerce) to heat warm water; combine pile foundation with geothermic tubes.

Sometimes a reduction of costs for construction may be favourable even though this measure produces additional costs in the usage or operational phase. This may be the case if a technical device is only used very seldom. Depending on the individual circumstances it may be cheaper, for example. to pay for professional climbers who clean the glass façade than to construct facilities like a façade lift for glass cleaning by the usual staff.

Designing to lower costs of consumption

Costs for consumption belong to the group of costs for operation. The following three principles can reduce the costs for energy, associated with cooling, heating and power, water and sewage and for waste:

a) cheap procurement, for example by bundling and using discounts e.g. energy supplies
b) adaptation of consumption to actual needs with technical control and with the help of the end-user
c) Substitution of costs for consumption by costs for construction.

Some examples for (b) and (c) are shown in the following and illustrated in Figure 8.4.

- Energy for heating: insulation of external walls and roofs, installation of heat recovering systems in ventilation and air conditioning, place big windows on the south façade and small or no windows on the north façade (passive house principles), installation of solar heat panels on roof or façade, and so on.
- Energy for cooling: external sunblind elements reduce solar energy "gains" in summer (ideally broad-leafed trees grow in front of south façades), use natural cooling during nights (facilities for opening the windows needed), geothermal installations can be used for cooling or combined heat, cold and power generation, and so on.
- Energy for electric power: employ energy saving facilities for lighting, personal computer, printer, cooler, and so on, avoid stand-by losses, make use of daylight, control lamps according to daylight, and so on. Also the costs for consumed energy may be reduced by using chosen facilities in phases with cheap prices for power.
- Water: water saving taps and machines (i.e. dish washer), waterless urinals, collect rainwater for gardening and/or for flushing toilets (separate plumbing needed), and so on.
- Waste water: treat water from hand washbasins, showers and clothes washer for secondary use, analogous to rainwater harvesting. If rainwater cannot be implemented for secondary purposes it does not need to become waste water either: it may also be seeped away on the site. (In Germany this saves the rainwater charge in most towns.)
- Waste: waste becomes a resource if separated and recycled, but avoiding the development of waste is even more effective: procurement should pay attention to refill concepts (soap, cleaning liquids, printer cartridges, etc.)

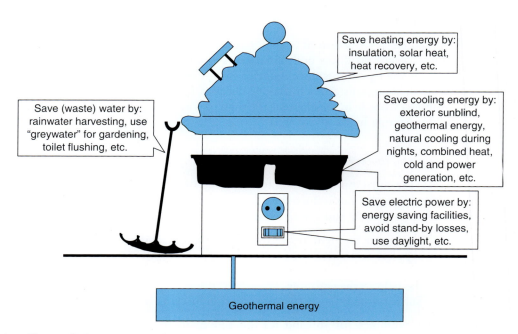

Figure 8.4   Measures to reduce consumption costs.

Designing to lower costs of services

Not every cost for services can be influenced by the building design. However, cleaning and security have building related aspects. Some ideas to reduce the effort required for these services are:

- Cleaning: minimise corners and facilities that are fixed on the floor, make the application of cleaning machines possible (no steps, broad doors, equipment of cleaning rooms), make glass surfaces accessible, employ surface treatment, for example lotus effect (reduces cleaning cycle) or antibacterial materials (with silver ions) in hospitals, and so on.
- Security: install alarm system, video surveying/surveillance, electronic door locks with access control; reduce number of entrances, and so on.

Designing to lower costs of technical maintenance

Principles to reduce the effort required for technical maintenance are:

- Accessibility: place technical facilities with short life cycles and/or with high demand for inspection and other maintenance at places that are accessible without – or with common – auxiliary devices (especially important in rooms with high ceilings), plan a secure access, if necessary (e.g. a façade lift), and so on.
- Longevity: surface materials or technical elements with a long life time expectancy reduce effort for replacement, for example natural stone on floors.
- Low tech: avoid the necessity of technical equipment, for example by passive cooling (no maintenance for air conditioning) or by intelligent planning of ventilation lines to avoid additional fire prevention installations (e.g. fire gates).

Designing to lower costs of renovation

The following principles may increase flexibility and reduce costs for renovation. Working in project teams creates an increasing demand for a flexible workplace or, if flexibility is not offered, moves and renovation may be needed (at additional costs). Also asset management expects high flexibility in floor layout to be adopted so that rented spaces in different size can be offered to any potential user. However, construction costs may rise with measures to improve flexibility. Using scenarios and the calculation of LCC will show the economic effectiveness for the individual cases.

- Universality: create rooms with high ceilings, no pillars, high bearing capacity in floors, narrow net of vertical access with stairs and installation lines, narrow grid of sanitary facilities (toilets, kitchenette), and so on.
- Adaptability: mobile walls and furniture may adapt the floor layout to changing needs, even the whole building may consist of flexible elements that can be put together again in different forms (modular system).
- Provision: for example, access to media (internet, power, etc.) can be provided by double flooring all over the facility, even in rooms that have no actual necessity for it at present.
- Options for retrofit: chambers for installations that are easily accessible (doors) and still have space for additional lines.

8.5   A worked example of an office façade

In order to show, how life-cycle costing can be used in design processes an example is presented that focuses on façade materials for an office building (Table 8.3). Is a massive façade covered with natural stone with 50% openings for windows favourable over its live cycle compared with a skeleton façade completely covered with glass?

- Period under consideration:
  If a façade out of natural stone panels is constructed properly it can last about 50 yr without damage. This is the reason to choose 50 yr as period under consideration.
- System boundary:
  A façade completely made out of glass with no exterior sunblind necessitates air conditioning in order to guarantee adequate indoor climate. At times where energy saving is regarded as of high importance one could set air conditioning aside, in the volume of a massive wall with only 50% of openings. Instead of air conditioning, cooling panels will be planned. Thus the system boundary is not limited to the construction of the façade but will include a share of the cooling or air conditioning facilities. Processes at the end-of-life will be omitted due to the fact that the cost of disposal or the value of aluminium in 50 yr is extremely uncertain.
- Time value of money: A NPV is calculated with a discount rate of 5.0%. This includes a general inflation rate of 1.5% and a specific inflation rate for energy of 4.0%.
- Sources of risk and uncertainty: The future development of prices for energy is uncertain. It may increase with a higher or lower rate. Therefore a best-worst case analysis will be carried out for a best case with 1.5% and a worst case with 8.0% inflation rate for energy prices. Also the life time expectancy of the air conditioning facilities may be longer or shorter than the expected 15 yr. A best case with 25 yr and a worst case with 10 yr will be considered as well.

Figure 8.5 shows the cumulated LCC for both alternatives. The investment includes costs for façade construction and for air conditioning or cooling facilities. The costs for cooling are calculated for 2 m² of floor space per square metre

Table 8.3   Performance criteria of alternative façade design.

	Alternatives	
Performance criteria	**Glass only**	**Stone and glass**
Sun protection	Indoor sunblind	Exterior sunblind
Consequences for technical equipment	Air conditioning	Additional cooling
Life time expectancy of solid façade elements (yr)	—	50
Life time expectancy of glass (yr)	25	25
Life expectancy of air conditioning/ additional cooling (yr)	15	25

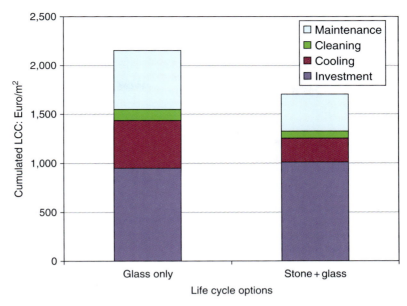

Figure 8.5 LCC for two alternative façade constructions including the consequences of air conditioning/cooling.

Table 8.4 Results of best-worst case analysis, expressed as the difference between the two alternatives ("glass only" minus "stone + glass").

Modification of variable	Best case	*Realistic case*	Worst case
Inflation rate for energy	443	*450*	462
Life time expectancy for air conditioning	280	*450*	648

of façade. Cleaning is more expensive for the "glass only" alternative because auxiliary devices (hoists) are needed. On the other hand, the windows in the stone façade can be opened for cleaning. Another clear difference between the alternatives is to be seen in the cost for maintenance: this is due to the costs for renovation of the air conditioning installation every 15 yr. In this comparison, the decision should be taken in favour of the stone and glass alternative because the WLC is lower.

The consequences of changed inflation rate for energy and of modified life time expectancy for the air conditioning equipment are shown in Table 8.4 as the difference between the "glass only" and the "stone + glass" façade.

The results show some difference with the changes in assumptions but do not lead to a change in the ranking of the alternatives: the "glass only" façade still has higher LCC. So, even though price tendencies and life cycles cannot be forecasted or predicted with precision, in this case it is very probable that the massive façade is favourable when using a 50 yr perspective.

Finally, Figure 8.6 illustrates that the break-even point of the higher initial costs for the "stone + glass" alternative is reached in year 8, under the above assumptions.

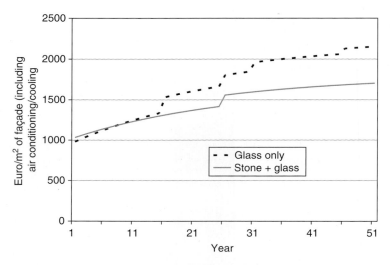

Figure 8.6   Cumulative LCC for alternative façade constructions.

8.6   Inclusion of carbon emissions into WLC modelling

In all conventions for LCC the costs for energy consumption are included in the calculation. However, there are more consequences due to energy consumption than costs for the consumer. The burning of fossil energy sources (coal, petroleum or petroleum gas) results in carbon emissions. Carbon emissions are causing damage to the environment via global warming and other effects. The costs for carbon emissions are at the point of being integrated into economic processes because this is seen as the most effective way of achieving an ideal allocation of productive resources. In Europe, carbon emission trading started in 2005. Voluntary emission trade started in 2003 in USA and Canada. New South Wales in Australia initiated a binding emission trade also in 2003, New Zealand followed in 2008, and then Tokyo city in 2010.

In Europe, industry producing carbon is obliged to buy emission certificates. The price associated with emitting 1 ton of CO_2 is very volatile (Daskalakisa et al., 2009). For example, it changed from a €14 starting point to a highest value of about €30; the lowest value was €1 in March 2012 (boerse.de, 2012). One reason for this volatility seems to be the fact that there are too many certificates on the market (handed out by the governments).

On the other hand, the damage caused by carbon emissions is estimated to be €15–280 per ton of CO_2 (Krewitt and Schlomann, 2007).With the integration of the costs for carbon emissions as external costs LCC becomes WLC. Then it shows not only cash flows that are expected during the life cycle but also consequences for the whole society. The concept of an inclusion could be simple: according to different primary sources for energy the CO_2-intensity (CO_2 per kWh) can be defined. Then a price for CO_2 emissions is to be chosen. It should be not lower than the actual price in carbon emission trade. However, it could also reflect the societal damage costs of CO_2. This price should again be subject to a best-worst case analysis.

8.7   Limitations of WLC

Another value that could be included into WLC is income. No building is constructed to save money, on the contrary, it is built to serve people's needs and thus to produce income. The integration of income is always essential when the alternatives being considered create or have different benefit profiles (Pelzeter, 2007). This can be the case when image or comfort of the designed facility is concerned. On the other hand, differences in income due to a different façade (image) cannot be forecasted or predicted with the same probability as resulting costs. Also, overestimated effects on income will influence the result of WLC comparisons in a very different way. This is the reason why income is not (yet) integrated into the German standards from GEFMA and DGNB.

What WLC, with or without carbon emissions and/or income, cannot reflect is every effect of a design decision (or a final investment) that cannot be monetised at all, for example the consequences of the conversion of farmland into a construction site. Therefore in certifications for sustainability in construction LCC can only be one (if at all!) of a whole bundle of criteria that are quantified with a cost/value benefit analysis.

8.8   Concluding remarks

WLC is defined in ISO 15686-5 as the sum of LCC, non-construction costs, externalities and income. LCCs include the costs for construction, operation, maintenance and end-of-life. LCCs are implemented for decisions in design and/or investment when alternatives not only differ in initial costs but also in consequential or running costs. In most cases, LCC can show which alternative results in lowest costs over the life cycle. However, it is not always possible to cover the whole life cycle of a facility with the forecast of LCC as non-construction costs, externalities and income have to be factored into the equation. To determine LCC, it is necessary to define the period under consideration, system boundaries, time value of money and sources of risks, insecurity or uncertainty have to be stated in a transparent way. For most buildings, costs for construction, maintenance, energy consumption and cleaning are relevant in an optimisation process. Building design can address these costs with different principles:

- Fulfil needs in an efficient way, for example implement combined heating and cooling plant systems.
- Fulfil needs in an effective way, for example an exterior sunblind prevents high indoor temperatures to a certain level and reduces the need for air conditioning.
- Question needs, for example is it necessary to cool indoor air down to 18°C in hot summers, when up to 24°C can still be regarded as comfortable? (This allows for smaller installations with lower consumptions during operation.)

For the inclusion of carbon emissions into WLC a price has to be defined. This can mirror the prices in emission certificate trade or even the estimated costs of damages by CO_2 emissions which are estimated much higher than the prices in the trade or carbon market. Also, income may be important for a successful WLC comparison: if comfort or image or other benefits differ between compared alternatives. However, not every possible benefit can be monetised. This is the clear limitation of WLC. If non-monetisable criteria are important for a decision, the method of WLC may be integrated into a benefit analysis. Yet for optimising construction design in most cases LCC may be very helpful in making key decisions such as creating scenarios that help to decide if a façade lift is saving LCC even though a high additional investment is incurred.

References

boerse.de (2012) CO_2 Emissionsrechte. http://www.boerse.de/rohstoffe/Co2-Emissionsrechte/XC000A0C4KJ2 (accessed 23 March 2012).

Daskalakisa, G., Psychoyiosb, D. and Markellos, R.N. (2009) Modelling CO_2 emission allowance prices and derivatives: evidence from the European trading scheme. *Journal of Banking & Finance*, **33**(7), 1230–1241.

Ellingham, I. and Fawcett, W. (2006) *New Generation Whole-Life Costing – Property and Construction Decision-making under Uncertainty*, Taylor & Francis, New York.

Krewitt, W. and Schlomann, D. (2007) Externe Kosten der Stromerzeugung aus erneuerbaren Energien im Vergleich zur Stromerzeugung aus fossilen Energieträgern. http://www.erneuerbare-energien.de/files/erneuerbare_energien/downloads/application/pdf/ee_kosten_stromerzeugung.pdf (acessed 23 March 2012).

Pelzeter, A. (2007) Building optimisation with life cycle costs – the influence of calculation methods. *Journal of Facilities Management*, **5**(2), 115–128.

Pelzeter, A. and Sigg, R. (2011) Ermittlung von Lebenszykluskosten. *Handbuch Facility Management*, **30**, 1–64.

Chapter 9
Procurement and Contract Strategy: Risks Allocation and Construction Cost

John Adriaanse and Herbert Robinson

9.1   Introduction

Significant investment is required for the development of construction projects. An appropriate procurement and contract strategy is therefore necessary to protect such investment and to achieve value for money for clients and other stakeholders including investors who have a stake over the long-term use of such facilities. However, clients are often faced with the dilemma of how to procure construction projects and what type of contract to use to ensure that projects are delivered in the most cost effective and efficient manner, and most significantly to ensure that the facilities are fit for the purpose intended. In deciding what procurement strategy to adopt, it is important to understand the key drivers of change or factors that will or are likely to affect the particular sector or industry the project is operating in.

The central dilemma for clients is often how to select consultants and contractors (with an information advantage) that can act in their best interests in the project. This information asymmetry gives rise to two fundamental problems. The first is what is usually referred to as 'incomplete contracts', lacking sufficient precision to define the project, cover the entire scope of work including all possible risks. The greater the completeness of contract documentation, the lower the potential for dispute, abuse and opportunism but it is difficult to achieve this in real world contracts. Secondly, it creates the potential for abuse and mistrust in the relationships between various parties.

This chapter examines the link between procurement and contract strategy, the role of contracts in allocating risks, and the relationship between parties in various types of procurement as well as the implications for construction cost. It starts with an overview of the need for an effective procurement and contract strategy. The key risks in relations to the different forms of contract, mechanisms for the allocation of risks from traditional architect-led approaches to integrated fast-track methods of delivering projects are discussed.

Design Economics for the Built Environment: Impact of Sustainability on Project Evaluation, First Edition.
Edited by Herbert Robinson, Barry Symonds, Barry Gilbertson and Benedict Ilozor.
© 2015 John Wiley & Sons, Ltd. Published 2015 by John Wiley & Sons, Ltd.

9.2   Procurement strategy and contract selection

Procurement strategy and the form of contract chosen can have a significant impact on the cost and time overruns in major projects. Procurement involves *actors* (public and private organisations including design and construction firms) and the organisation of *activities* (from planning, design, construction to operation and maintenance) to deliver an asset (Howes and Robinson, 2005). The project stages are generally sequential but there are overlaps between certain activities such as design and construction and sub-activities or packages such as concreting and formwork. The degree of overlap depends on the choice of procurement method. For example, in traditional procurement, there is very little overlap between the design and construction stages compared with 'design and build' where there is a significant overlap.

Within each stage of the procurement process, different actors involved have various expectations which can sometimes conflict, create risks with potential cost consequences. For example, whilst an architect might have a preference for an innovative approach which poses additional risks, the structural engineer's proposed solution could affect the architect's desired shape, form of the design and the appetite for risks. Similarly, the services engineer might have a preference for running service ducts in a particular way or direction but this could affect the work of the fit-out specialists in terms of patterns or space configuration. The nature of these interactions and the risks involved in the design process could also affect the construction cost or budget set by the quantity surveyor or cost engineer.

The nature of interaction of the key actors depends on the procurement option chosen which defines the actors involved, their role, at what stage they are involved and the precise relationship between, for example, the client, architects, structural engineers, quantity surveyors and fit-out specialists. In a traditional procurement system, the builder and facilities management consultant has no input in design which could affect the way the building is constructed and operated. This is in sharp contrast to other approaches such as 'design and build' procurement with a major input on 'buildability' at an early stage or public–private partnership (PPP)/UK private finance initiative (PFI) procurement system, where the facilities management consultant has a significant input in the design as a greater emphasis is placed on the functionality and flexibility of the buildings during use. There is a need to select an appropriate procurement and contract strategy to govern the relationships between key actors, clients and consultants, contractors and specialist sub-contractors. In the process of selection, it is important to ensure that these two elements are carefully considered as they are intrinsically linked. First, the procurement strategy provides the 'direction and speed of travel' and secondly, the selection of the appropriate contract forms reflects the client's appetite for risk and 'level of comfort in the journey' to deliver a construction project. These are major decisions which the client cannot afford to get wrong as it will have significant implications on the costs and the time taken to deliver a project.

The key to a successful procurement strategy and contract selection is to identify the priorities in the client's objectives and to plan a journey or path that will deliver the project on time, within budget and to the required specification. Procurement routes depend on a range of factors such as the type of project, expected project completion date, relationship between key parties and the allocation of risks. For example, it is widely accepted that the traditional procurement method places more risk on the client but the 'design and build' approach allocates more risk to the contractor. The type of contracts also plays a key role in the allocation of risk to different parties. Selecting a construction contract is dependent on a number of criteria which include but are not limited to the type of work, value of work, expertise of client and amount of control required, programme requirements and importance of price certainty. The process of choosing a contract type should therefore satisfy the requirements of the project to ensure that risk allocation is appropriate based on the needs of stakeholders – clients, consultants, contractors and sub-contractors.

9.3   Wembley stadium case study

In October 2000, Wembley Stadium was closed and it was demolished in late 2002 for major redevelopment of the iconic stadium. The aim of the Wembley project was to design and build a state-of-the-art national stadium to be the home of English football and to host large events such as cup finals, music events and athletics. The roof structure covers 11 acres, 4 acres of which are movable. The 90,000 seat capacity makes it the second largest stadium in Europe at the time after the Nou Camp in Barcelona with a capacity of 98,000, but it is one of the largest in the World to have a covering roof. There are 310 wheelchair spaces and increased capacity for other physically impaired spectators. In addition, there are 400 media seats, 2,618 toilets and four banqueting halls, the largest of which can accommodate 2,000.

The procurement system used for the development of the Wembley Stadium project was 'design and build' and the form of contract used was the guaranteed maximum price (GMP). According to Construction Manager (2006, p. 12), "Politicians perceive GMP contracts as a panacea for all cost overrun". The managing director of Wembley National Stadium Limited (WNSL) was certain that the project was well defined and was not going to be subject to many changes (Construction Manager, 2006, p. 13). This type of contract was said to be good for clients who want to make sure they will spend no more money than they have budgeted for. In February 2000, some of the potential bidders such as Sir Robert McAlpine and Bouyges pulled out of the bidding process because of serious concerns about the form of contract proposed.

Key parties

The client was WNSL and the architect employed was Foster and Partners and HOK Sport, one of the world's leading design firms and provider of project delivery services. They employ about 1,600 professionals across America, Europe and Asia. The main contractor was Multiplex Ltd, an Australian based contractor, part of a group employing over 2,000 people with established operations in Australia, New Zealand, the UK and the Middle East. There were also a number of specialist sub-contractors involved. The original steel sub-contractor was Cleveland Bridge but they were eventually replaced by Hollandia (designbuild-network.com, 2007). The mechanical and electrical contractor was Emcor Drake & Scull and the building services engineering was carried out by Mott MacDonald. There were a number of other engineers and consultants involved.

Project implementation

According to sportengland.org (2007), Multiplex Ltd had secured a 'design and build' contract to build the new Wembley stadium. The total construction cost was put at £326.5 million but by the time the bid had been signed, the cost had increased to £445 m. In September 2000, just after WNSL had announced that it had chosen Multiplex Ltd as the contractor, there were delays as politicians argued over what Wembley should be used for. They also had trouble attracting investors, costs began to escalate and the design went back to the drawing board several times (Construction Manager, 2006, p.13). The initial target completion date was May 2003 but work started in September/October 2002. The revised completion date set was March 2006 before the May 2006 FA Cup final, by which time the price of the stadium had risen to £757 million. During the implementation, the relationship between Multiplex Ltd and its steel sub-contractors, Cleveland Bridge, was problematic. As a result, Cleveland Bridge left the site in 2004 because they did not believe they would be paid for materials and there were serious difficulties between the two parties (designbuild-network.com, 2007). According to Construction Manager (2005, p.19) both parties appeared in a court case on 26 April 2006. Cleveland Bridge claims that there were serious problems as a result of late and incomplete design which caused cost increases and a delay of 50.5 weeks. According to designbuild-network (2007), the two companies sued each other for breach of contract. Multiplex Ltd sued for £45 million and Cleveland Bridge sued for £22.5 million. In *Multiplex Construction (UK) Ltd v Cleveland Bridge UK Ltd & Anor* (2008) EWHC 2220 (TCC), Mr Justice Jackson (as he was then) found in favour of Multiplex Ltd.

In November 2005, the Construction Manager, in one of its articles titled "More than a stadium" included an interview with Mike Richardson of WNSL (Construction Manager, 2005; also cited in building.co.uk, 2005). The tone of an unhappy client was cited as Mike Richardson said:

> We cannot tell them how to build a job or who to use. They've got a fixed price for this job…we are not a risk transfer organisation. Every time we change our minds they have the opportunity to say that wasn't in our original price. So the trick is not to change your mind too often.

In 2006 Multiplex Ltd had estimated losses rising to £106 million according to Construction Manager (2006, p. 13). Multiplex Ltd intended to claim £150 million from WNSL for 560 design changes and according to Construction News (2006, p.2) Multiplex Ltd confirmed that its losses on the Wembley contract were at £180 million. The stadium was completed in March 2007 and was ready for the 2007 FA Cup final. The final construction cost was £798 million with an additional time overrun of 12 months.

9.4   Allocation of risks and forms of contract

The construction of the new Wembley stadium is a landmark project which has been controversial and fascinating but raises important issues about how to allocate risks appropriately. The parties were highly experienced and had access to high level legal advice. Despite this it spawned at least 25 cases and numerous adjudications. It raises important questions about the procurement strategy (design and build) and the secondary process of actual construction. It is imperative that the form of contract is used to support the procurement method ensuring that it is sufficiently detailed and clear to accommodate both the client and the contractor's appetite for risks. If appropriate decisions about risks allocation are not taken during the early stage of a project, it can have a significant impact on final cost and duration of a project.

Risks are varied in construction work. Some are known at the time of contracting, for example when the completion date is crucial (e.g. for a supermarket or an Olympic stadium). In others, the financial consequences of late completion are known: in *Masons (A Firm) v WD King Ltd & anor* (2003) EWHC 3124 (TCC), the parties knew that if the completion date was not met losses for the year would amount to £600, 000. In *Copthorne Hotel (Newcastle) Limited v Arup Associates and anor* (1996) 12 Const LJ 402, the risk created by a medieval wall close to the site of piling meant all the piling sub-contrcators inflated their prices to cover the risk.

There are other factors that can also affect the progress of the work. Chief of these are the unforeseen risks such as: unexpected ground conditions; unpredicted weather conditions; shortages of material and skilled labour; accidents, whether by fire, flood or carelessness; and innovative design that does not work or proves impossible to construct. As a result, the allocation of these risks is a very important part of the contract since these factors always result in additional costs being incurred. This inevitably raises the question of who should pay (Adriaanse, 2010).

Risk and uncertainties are part of construction work regardless of the size of the project. The selection of a particular form of contract can affect the balance of risks or can shift particular risks towards one of the stakeholders. There are various forms of contracts which define the relationship of the parties in a construction contract and risks allocation through contract clauses and conditions. For example, there are clauses relating to modification of design or quality where circumstances in which the work is carried out changes and the consequences for such changes, valuation of the cost consequences due to design variations, claims where contractors are entitled to additional payments, dealing with unforeseen events, interim payments for regular work progress, extension of time, and the cost implication of extension of time which are all essential in contract management to ensure the smooth running of construction projects. The precise wording of the clauses is also important as they are potential sources of conflicts between different parties involved and can lead to delays and cost overruns.

An appropriate contract form is therefore required to work in conjunction with the chosen procurement method. The selection of the procurement route will directly affect the contract options available. There are numerous standard forms of contract available. However, the most commonly used contracts are the Joint Contracts Tribunal (JCT) and the New Engineering Contract (NEC). The International Federation of Consulting Engineers (FIDIC) which is similar to the ICE contract is used mostly on international projects. Contracts can also be bespoke or tailored to the specific requirements of the parties but can be a costly and time consuming process compared with utilising the standard forms.

9.5   Risks and construction costs

There are a number of approaches used to deal with the problems of risks; for example, by choosing a particular form of contract, incorporating special provisions in a contract or by creating or increasing the contingency funds available.

Care has to be taken when considering bespoke amendments to standard forms of contract. They are usually drafted by a committee drawn from the industry and as such represent a consensus on the allocation of risk. Amendments need to be made sensitively in order not to disturb that allocation.

At the heart of the standard form is the doctrine of privity of contract and ways of dealing with it. At tender stage provisions need to be inserted to provide protection for the employer and third parties. For the employer, direct warranties are needed to allocate rights against sub-contractors and suppliers (these are additional to that owed by the contractor to the employer). The JCT family of contracts also allows for the provision of Collateral Warranties or the use of the Contract (Rights of Third Parties) Act 1999. These also provide protection against contractor insolvency by providing 'step-in' rights for the developer or client should this occur. In addition it provides for termination of the employment of the contractor for specified defaults and most importantly for the financial consequences that follow. The contractor is also required to take out joint insurance to provide cover against construction activities causing damage to third parties.

There are other types of risk indirectly affecting a project such as disruption to third party business (e.g. traffic, noise, business closure, etc.), industry risk which might affect the entire industry (e.g. national strike involving building workers) and corporate activities of the design and construction firms involved in a project which may have serious business consequences.

Risks should be examined at an overall project level as well as at a detailed level looking at specific clauses of a contract. At a project level, if a client is interested in price certainty, the lump sum (fixed price) contract would be the most appropriate as the final cost is determined prior to the start of the construction. This contract can be agreed either 'with quantities' or 'without quantities'. Regardless of the agreed lump sum cost for the work, any variations to the project will result in additional costs. There is also the option of using 'measurement contracts' where the client retains the risks as the final contract sum is not agreed until the project is completed. This type of contract is often used in projects, particularly refurbishment projects, where the extent of the work cannot be quantified or is not clearly defined at the tender stage. The client takes the risk regarding the final quantity. If this increases substantially then the agreed rate for the work would result in a considerable cost increase to the client. Cost reimbursement (sometimes called cost plus) contracts are also used based on the cost for labour, plant and materials and an agreed overhead and profit percentage is added to this amount. This type of contract places a high level of risk on the client to pay for construction work without sufficient or detailed knowledge of the true cost which can sometimes provide little or limited incentive for the contractor to keep construction costs down. There is an alternative incentive based contract such as guaranteed maximum price (GMP) contracts or pain-gain sharing sometimes called target cost contracts to provide an incentive for contractors to keep costs down by sharing the savings made during construction with the client.

Risks should also be examined at the detailed level looking at specific contract clauses which can affect project performance, design quality and the budget of a project. Risks can sometimes be eliminated by a complete redesign of schemes or elements of the design by the client's architect, minimised by choosing a traditional design rather an innovative design solution, transferred, for example, through 'design and build' contracts or passed on to specialist sub-contractors. The options to manage risks will consider the potential costs for risk transfer, containment and risk reduction policies such as insurance. In practice, the distribution of risk is analysed through a comprehensive risk matrix or register and appropriate project costs are allowed for or added (value of risks) during the estimating process. Examples of risk may include design errors, inaccurate cost estimates, inflation, poor site and ground conditions, variations in weather, labour availability, hidden defects and quality of materials. Risk response strategy and appropriate mitigating instruments will be required (Table 9.1 and Table 9.2).

Risks should be analysed in an organised and systematic way considering the full impact on programme time and cost overruns. Too often risks are dealt with in an arbitrary way or ad hoc fashion by simply adding a percentage for contingency on the estimated project cost. Irrespective of the type of contract chosen (whether fixed price, cost plus or GMP), there will always be a need for the client or contractor to avoid cost and time overruns in construction projects. Adopting a systematic risk management approach to itemise and quantify all the risks involved, whether it is in the form of measured works in the bill of quantities or special conditions of contract, is essential to manage project time and cost effectively.

Table 9.1   Examples of risk response strategy.

Risk strategy	Examples
Eliminate/avoid	Eliminate potential threats to project, for example review design processes, change design, adopt alternative construction technique or use new materials
Distribute/share	Introduce a pain/gain sharing mechanism, for example guaranteed maximum price contract
Transfer	Does not reduce criticality; move to another party best able to manage it, for example contractual provisions or insurance Use lump sum (transfer to contractor) or cost plus contracts (transfer to client)
Accept/retain/reduce	Other options are undesirable, uneconomical or impossible, for example can reduce risk by in-depth site investigation

Table 9.2   Examples of risks mitigating instruments.

Risks	Mitigating instruments
Design	Professional indemnity insurance
Cost overruns	Standby credit
Inflation	Contingency funds
	Contract clauses
Delays	Liquidated damages
Hidden defects	Contractual arrangements
Force majeure	Insurance

9.6   Procurement systems and contract issues

Architect-led or traditional system

In the architect-led or traditional system, contractors are invited to tender on the basis of complete drawings, specification and other contract documents defining the scope of the work (Figure 9.1). Competitive tendering is based on complete design such as key design drawings, specification, bills of quantities. Co-ordination by the architect or lead designer is absolutely critical in relation to specialist contractors responsible for the building structure (core and shell), services, and the fit-out work or configuration of spaces and building elements. In the figures, co-ordination only links at the design level are shown as dotted lines and the other lines reflect the contractual and co-ordination links.

The contractor is selected based on price, and other technical factors such as time for completion, sustainability, health and safety strategy and project programme. The lowest tender or best value bid is traditionally accepted, provided all other technical requirements are met. This process of not involving the contractor until a later stage removes the benefits of the builder's input to the design. The design teams may not be able to effectively maximise cost and programme savings because the majority of the design is complete.

The JCT Standard Form contract (JCT 11) now provides for the provision of a 'Contractor designed portion'. This allows the employer to place design liability on the contractor for specified work. In allowing for this, part of the criticism made above is addressed.

Changes in the work scope (variations) and employer delay are the substantial drivers in unexpected costs. The JCT 11 deals with delay caused by the employer and its consequences, as well as delay due to variations, in the same way. Delay due by the contractor is dealt with by levying Liquidated Damages. The scheme to protect the right to such sums is provided by the provision of 'relevant' events (clause 2.29), which enable the contractor to make a claim for an extension of time. These are events which should they occur is at the expense of the employer. What this should mean in practice is the likelihood of the risks occurring is borne by the employer and should attract a lower price from the contractor who has not

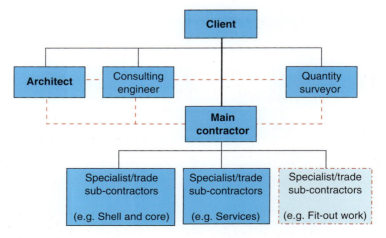

Figure 9.1   Traditional system.

had to price for it. In addition, the contractor is entitled to claim for costs arising which are not covered by any other provision in the contract. Clause 4.24 lists the relevant matters that entitle the contractor to these costs. Keating states that it is difficult to conceive of circumstances where the employer under a JCT contract could lose its right to Liquidated damages. As the Privy Council stated in *Phillips Hong Kong v the Attorney General of Hong Kong* (1993) 61 BLR 41 'in building contracts…parties should know with reasonable certainty what their liability [for damages] *is* under the contract'. The scheme for dealing with damages in construction contracts has been comprehensively analysed by Adriaanse (2008) in *The rule in Hadley v Baxendale* (1854) *and the standard forms of contract.*

The approach taken by the NEC 3 contract is radically different to that usually found in other construction contracts. For a start, The Engineering and Construction Contract (NEC 3) in its Secondary Option clause: Option X7 provides for the employer to opt to levy Delay (Liquidated) Damages (instead of relying on its common law rights). Delay and its consequences is managed through 'compensation' events. A compensation event is one that arises through no fault of the contractor. It compensates for any *delay not* caused by the contractor (a potentially wide exposure to risk). Clause 60.1 comprises 19 events and, unlike other contracts such as the JCT 11, deals with the consequences of the delay and the costs associated with it at the same time. Quite substantial workload is caused by the operation of the procedure required to operate the system. In addition the compensation event can also involve the requirement to give early warning (clause 16) and risk reduction meetings required by clause 16. Compensation events require:

1. Notification of the event.
2. Submission of alteration to the accepted programme together with quotations for carrying out the work.
3. Preparation and submission of quotations (also time and money under clause 63).
4. Implementation of the compensation events clause 63.

Limiting the design liability in the JCT 11 is done by limiting the contractors' design liability to that of a professional person and the limiting of consequential loss arising from such a failure is limited to a sum fixed in the appendix. The NEC 3 also adopts the same approach.

Design and build

Design and build, sometimes called a 'package deal', is significantly different from the traditional system. The approach involves a significant overlap between design and construction activities (Figure 9.2). The client appoints a 'design and build' or 'develop and construct' contractor so there is a single point of responsibility and more significantly crucial construction input is provided at an early stage, unlike the traditional approach.

If the design is carried out by the contractor, it limits the client's responsibility for design defects by transferring risk from the client to the principal contractor and their consultants. The client must determine whether the additional cost spent on the contractor taking the additional risk can be offset by the potential savings. The variant termed 'develop and construct' is where the client has a concept or scheme design which is completed by the contractor. The use of this variant of the

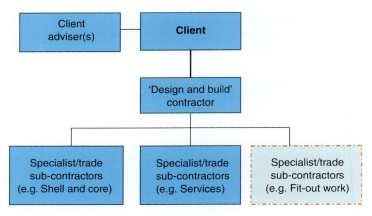

Figure 9.2   'Design and bBuild' approach.

'design and build' procurement method whereby clients use consultants to design a project up to a particular stage and then have contractors price the project is sometimes criticised as it can create complex issues.

Some clients may not wish to become too involved in the design process. The key in the early stages of a 'design and build' project is to appoint an adviser to develop appropriate briefing and design concepts to ensure that flexibility in design is achieved through better coordination of structure, services and equipment, fit-out and space planning elements. If an adequate brief and a statement of client's requirements are prepared, design and build could offer an integrated package to ensure good co-ordination between the design and construction teams. The 'design and build' contractor is usually appointed on the basis of pre-qualification and negotiation leading to the agreement of a GMP for a finalised design and specification. Alternatively, the appointment can be made on the basis of competitive bids usually consisting of design proposals and lump sum prices. This procurement method does not offer so much flexibility compared with that of the traditional approach and changes after the design has been agreed are usually more expensive to incorporate. Often the client is exposed to the risk of receiving a finished product that satisfies the Contractor's Proposals in the contract but does not give the client the desired end product originally anticipated.

'Design and build' contracts are available for the 'design and build' procurement route. The JCT Design and Build Form contract (D&B11) also provides for 'Changes' in section 5. These are variations in the traditional contract. Changes in the work due to employer delay are, as pointed out earlier, the main drivers in unexpected costs. Delay due by the contractor is dealt with by levying liquidated damages. The scheme to protect the right to such sums is provided by the provision of 'relevant' events (clause 2.26). These are events that should they occur are at the expense of the employer. What this means is that the *risks* of these is borne by the employer and should attract a lower price from the contractor who does not have to price for them. In addition the contractor is entitled to claim for costs arising which are not covered by any other provision in the contract. Clause 4.20 lists the relevant matters that entitle the contractor to these costs.

A key issue here is design liability. In the traditional contract, the interface between design and workmanship is called the 'fuzzy edge'. Placing design liability on the contractor is meant to resolve this issue. In practice, the contractor whether employed under a traditional contract or under a 'design and build' contract neither designs nor builds but instead manages the process (Adriaanse, 2007). This view is confirmed by H.H.J. Lloyd who said in *Birse Construction Ltd v Eastern Telegraph Company Ltd* (2004) EWHC 2512 (TCC) that:

> On virtually all building contracts of any magnitude, the role of the contractor is to use his management know-how not only to procure the requisite skills but also to know whether and to what extent they are being provided adequately to meet the requirements of the contract.

The Standard forms of contract (JCT/NEC) limit the liability of the contractor in design to that of the exercise of reasonable care and skill. By contrast the design liability of a contractor at common law is that of fitness for purpose. In practice the contractor will sub-contract the design and construction to either professional consultants or specialist design sub-contractors (who will in turn employ professional consultants). The result is that the fuzzy edge is pushed down the chain. The consequence is best summed up by *Cliffe (Holdings) Ltd v Parkman Buck Ltd* (1997) 14-CLD-07-04, where H.H.J. Wilcox QC, OR observed that:

> The contractor may supplant the architects in some of their traditional roles as exemplified under a full (JCT) agreement. In particular, involvement in the choice and the co-ordination of the specialist systems and sub-contractors, the integration of those systems, the approval of all drawings and site supervision. The result may achieve economies and render the tendering contractor more economic. It does however put a premium upon the strength of the organisation and experience of the contractor in the enhanced role it plays in such a contract. Merely by employing an architect in a restricted role, such in this case, can it expect the architect to compensate for the shortcomings of the other sub-contractors.

The employer in fact has to balance what it is receiving in its procurement system from what it *thinks* it is receiving. If parties are aware of the risk allocation, they will price their risk and exposure better. This may well mean that the client could abandon the project. In the well known case of *Pacific Associates Inc and anor v Baxter and ors* (1988) 44 BLR 33, hard materials in the underlying soils were not found during the site investigation. Although the contractor won damages in the resulting arbitration its losses far exceeded its winnings. Had the client known the true nature of the site, it might well have decided to abandon the project at the start.

Management contracting

Management contracting is also a slight variation from the traditional system. In this method, the design process is separated from the construction work but there are overlaps to speed up the procurement process. The client appoints a design team and also appoints a 'manager' often called a management contractor as a professional consultant at an early stage of the project usually a contractor who is paid a fee for co-ordinating the work of sub-contractors split into packages (Figure 9.3).

The management contractor supports the client's design team with construction expertise to improve the design solutions and the 'buildability' of the scheme. The

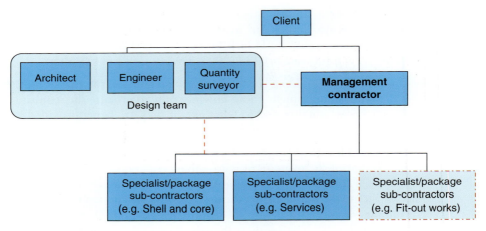

Figure 9.3   Management contracting approach.

management contractor is expected to appoint works contractors to undertake discrete packages of work. The contractual arrangement will be between the management contractor and the works contractor, however it is important to realise that the client is duty bound to reimburse the management contractor for all costs incurred. Under this arrangement, the client mainly carries the risk. A major criticism of this approach has been the risk aversion by some clients attempting to unload significant risks to the management contractor. One way of doing this is by insisting that the reimbursement of package sub-contractors will be subject to a maximum tender figure, hence any cost overrun, other than agreed variations, will be the responsibility of the management contractor. This has led to adversarial relationships, which has affected the popularity of this route in recent years. This procurement route has therefore largely fallen out of favour with some clients in the global construction market but where conditions are right it may provide an appropriate solution.

Construction management

This is a form of procurement which is similar to management contracting. The main difference is that the client places *direct contracts* with each of the specialist contractors and suppliers (Figure 9.4).

The expert construction management specialist is appointed as a consultant whose task is to effectively manage and control the construction process, as well as integrating the design team. In this manner the client has more direct control over works contractors and this helps to mitigate the risk associated with management contracting. This implies that the client is familiar with construction and has a close relationship with the professional team. The expert construction manager is therefore appointed as a consultant whose task is to facilitate the integration of the design team and effectively manage and control the construction process. Overriding influences in the selection of this route concern the complexity of the project and the strength of the risk involved.

Both the management contracting and construction management methods of procurement are in use but their use has declined due to the risks not being clearly

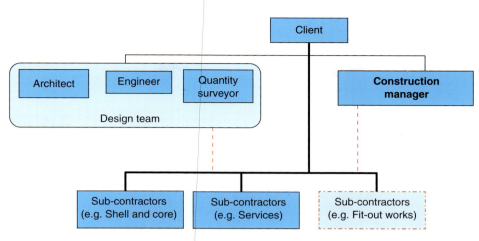

Figure 9.4   Construction management approach.

allocated. In management contracting, the employer engages the management contractor to partake in the project at an early stage. Normally an experienced builder, the contractor is employed not to undertake the work but to manage the process. All the work is sub-contracted to works contractors who carry it out. Construction management differs from management contracting in that the employer enters into a direct contract with each specialist. The employer engages the construction manager to act as a 'consultant' to coordinate these contractors. It is the legal uncertainty and the exposure to unexpected costs that has made these unattractive to employers.

9.7   Alternative forms of procurement

The procurement approaches discussed in the previous sections have limitations in terms of the durability of the relationships between clients, consultants and contractors. For example, in a traditional procurement, the liability of the con-tractor for rectification of defects after practical completion is normally restricted to a shorter period, usually 12 months. There has been a trend towards other types of procurement and forms of contract that fosters long-term collaborative relationships and transferring operating risk for a much longer period during the operational stages of the facilities or assets (Robinson and Scott, 2009). In long-term procurement approaches such as PPPs or the UK PFI model, the contractor is liable for the delivery of the assets and a wide range of hard and soft facilities management (FM) services as well as the associated operating risks during the performance period spanning 25–35 years (Figure 9.5).

The public sector client and private sector consortium (SPV) need to have a full range of skills to complete a PFI/PPP contract. Legal, technical and financial advisors are appointed by the public sector to help define business requirements, develop the business case, deal with risk transfer, cost and affordability, payment stream, managing the procurement process, and to negotiate the best contract for the client.

Gruneberg et al. (2007) argued that 'if a supplier has a responsibility for how something performs, then his or her contractual liability must extend into the

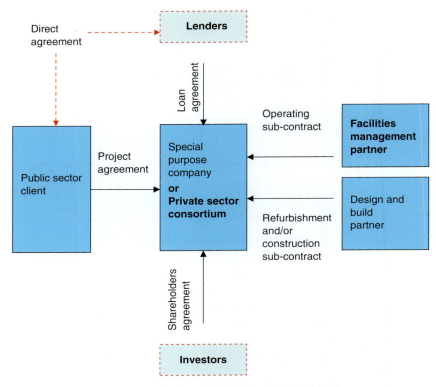

Figure 9.5   Contractual relationship and agreement between parties in PFI/PPP.

performance period'. This approach has clearly shifted and increased the risks on the PFI contractor as 'liability is inevitably extended under performance-based contracts' (Gruneberg *et al.*, 2007). The PFI process involves exploring risk allocation and the value of risk transfer compared with the traditional route. There are various risks associated with planning and design development, construction, and operation of PFI/PPP projects. For example, design and construction risks are retained by the SPV/contractor but such risks are transferred to the 'design and build' sub-contractor where it is a separate firm. Operational risks relating to life-cycle costs, innovation and technological changes are transferred to FM companies/sub-contractors. Table 9.3 shows a simple example of how risk of cost overrun is valued based on the probability and the cost of the event. The project estimated construction cost is £200 million and the likelihood of cost overrun reflecting various risks is shown to amount to £17 million.

Risk is valued based on (A) the probability of the event occurring and (B) the costs should the event occur. For each risk event, the process is repeated to arrive at an estimate of the cost or financial consequences. In this type of project, the distribution of risk is carefully analysed and appropriate project costs are allowed for in the bid.

Typically, PFI projects seemed to value risk transfer at around 30–35% of construction costs (ACCA, 2002). The UK PFI model is seen as a durable procurement approach as the responsibility for long-term use and flexibility of the building is transferred to the private sector based on the output requirements of the client. The implications are significant for the way design, construction and operating

Table 9.3   Value of risk in PFI/PPP projects.

Scenario	Probability of event (A)	Cost of event (B) (£ million)	Value of risk (C) = (A) x (B) (£ million)
Projected completed below budget by £10 million	0.10	−10	−1.0
Project completed on budget	0.20	0	0.0
Project overrun by £20 million	0.40	+20	+8.0
Project overrun by £30 million	0.20	+30	+6.0
Project overrun by £40 million	0.10	+40	+4.0
Risk adjustment to project cost			+17.0

Source: Robinson *et al.* (2010).

risks are managed which has created challenges for some and opportunities for others such as facilities management firms.

9.8   Concluding remarks

The building of Wembley Stadium used a 'design and build' approach for design and construction. It is therefore a typical example of a 'modern' contract: one that shows how outmoded the traditional contract is when it comes to modern methods of procurement. The disputes that occurred were lower in the contractual chain rather than between the employer and the contractor. It was still a very costly exercise for all parties. The employer also lost revenue. Perhaps all it demonstrates is that a fully prepared scope of work will always be superior to making it up as you go along. No matter the method of procurement, having a design prepared will always be better than finding out your costs as you go along. A good example is *Plymouth & South West Co-Operative Society Ltd v Architecture, Structure & Management Ltd* (2006) EWHC 5 (TCC). The parties operated with no clear means of monitoring their costs. This resulted in an overspend of at least £2 million in excess of the Society's estimate of about £6.3 million. Knowing who takes the risk matters. The Wembley project was mired in dispute about delay and the resulting costs. Ensuring the parties who carry the risk have priced for it is a key to a successful project. As Professor Uff (2002) said in the 90th Thomas Hawksley Memorial Lecture citing Godfrey: 'a full identification of construction risks and their management…is part of the management process'.

References

ACCA (Association of Chartered Certified Accountants) (2002) PFI: Practical Perspectives, April, Certified Accountants Educational Trust, London.

Adriaanse, J. (2007) *The Contractors Liability for Workmanship and Design – a matter of Status or Competence?* Proceedings of the CIB World Building Conference, 14–18 May 2007, Cape Town.

Adriaanse, J. (2008) *The Rule in Hadley v Baxendale (1854) and the Standard Forms of Contract.* Built Environment Education & Research Conference – BEAR 2008, CIB W089, 10–15 February 2008, Kandalama, Sri Lanka.

Adriaanse, J. (2010) *Construction Contract Law – the Essentials,* 3rd edn, Palgrave, London.
building.co.uk (2005) They think it's all over, 8 November 2005. http://www.building.co.uk/they-think-it%E2%80%99s-all-over/3058730.article (accessed 31 July 2014).
Construction Manager (2005) They think it's all over. December, pp. 12–19.
Construction Manager (2006) No room for manoeuvre. May, pp.12–13.
Construction News (2006) Multiplex ditches White City but will stay in UK. 24 August, p. 2.
Designbuild-network.com (2007) Wembley Stadium. http://www.designbuild-network.com/projects/wembley/ (accessed 27 March 2007).
Gruneberg, S., Hughes, W. and Ancell, D. (2007) Risk under performance-based contracting in the UK construction sector. *Construction Management and Economics,* **25**, 691–699.
Howes, R. and Robinson, H. (2005) *Infrastructure for the Built Environment: Global Procurement Strategies,* Elsevier Butterworth-Heinemann, Oxford.
Robinson, H., Carrillo, P., Anumba, C. and Patel, M (2010) *Governance and Knowledge Management for Public Private Partnerships,* Wiley-Blackwell, Oxford.
Robinson, H.S. and Scott, J. (2009), Service delivery and performance monitoring in PFI/PPP projects. *Construction Management and Economics,* **27**(2), 181–197.
Sportengland (2007) Wembley Stadium. http://www.sportengland.org/new_wembley.pdf (accessed 27 March 2007).
Uff, J. (2002) Risk in Engineering: is it Such a Bad Thing?, 90th Thomas Hawksley Memorial Lecture. Institution of Mechanical Engineers.

Chapter 10
Sustainable Design, Investment and Value

Thomas Lützkendorf and David Lorenz

10.1 Introduction

Whenever design and construction works are agreed upon it is usually assumed that an essential starting point is the description and specification of the functional and technical requirements. These requirements may either result from the investor's/client's (or future users') ideas and preferences or may also be based on the current legal and normative framework. These are the requirements, implied characteristics and attributes which a building or construction work should possess during its subsequent life cycle. The traditional view, particularly popular among designers and planners, does not fully represent the complexity of investment decisions and their underlying rationale within the construction sector. Indeed, investors and project developers also specify requirements regarding a building's functional and technical quality. However, depending on their role and perspective they also integrate these requirements in early phases of a project with their ideas on a budget, financing possibilities, financial and other risks, as well as the expectations concerning future cash flows, returns, value stability and development. In addition, investors and project developers take into account the respective location and market situation and its specific dynamics. Increasingly, they also consider the impact of their investment decisions on society and the environment and try to contribute to sustainable development. The reasons for this include the decision to take responsibility, marketing gains and the reduction of reputational risks. Within this context, investors increasingly recognise and acknowledge the direct relationship between a building's quality and sustainability performance together with its marketability and lettability which in return directly impacts on its current market value, value stability and likely value development.

This chapter examines how design and sustainable investment in property influences the value of buildings using property performance assessment techniques from a German perspective. The key question for many stakeholders is how

Design Economics for the Built Environment: Impact of Sustainability on Project Evaluation, First Edition.
Edited by Herbert Robinson, Barry Symonds, Barry Gilbertson and Benedict Ilozor.

investment in sustainable design affects (property) value. The chapter starts with a discussion of the economic goals of projects and how they can be formulated during the design and project development phases. In addition, the complex relationships between key characteristics and attributes of buildings and aspects of economic performance are explored, analysed and assessed to show how they influence the early phases of a project. This is crucial to identify which building characteristics and attributes impact on a building's sustainability performance or upon sustainability assessment results. The chapter also discusses the performance based approach which enables investment decision processes to be analysed by taking into account the perspective and interests of future users and buyers such as corporate tenants, private tenants, institutional and private investors. The chapter also discusses property rating systems using examples such as the system for property analyses developed by The Association of German Public Banks (Bundesverband Öffentlicher Banken Deutschlands, VÖB) and the German certification system for sustainable buildings used by the German Sustainable Building Council (DGNB). Finally, examples of the economic benefits of buildings or impact on selling price/property value due to sustainable credentials are provided from Germany as well as other countries with similar sustainability assessment systems such as Switzerland, Netherlands, and USA.

10.2   Formulation of project goals

At project inception it is important to consider, analyse and target the interplay between capital (including economic goals), location (including lot, market and environmental situation) and project idea (building type and usage including functional and technical requirements). It is self-evident that besides functional and technical requirements, economic goals are also pursued by investors from the beginning of a project. These economic goals, however, depend on the investor's role and goals (owner-occupation, trade, etc.), the investor's perspective (short, medium and long term) as well as attitude towards risk (risk–return profile, etc.). Investors increasingly attempt to link the realisation of their individual and institutional goals with their (assumed) responsibility towards society and the environment. As a consequence, project development, design and realisation increasingly follow the principles of sustainable design, construction and management of buildings. These, in turn, are based on a specific translation of sustainable development principles to the construction sector.

Also in the context of sustainability assessments of buildings, the description, assessment and targeting of economic criteria is therefore intensively discussed. For example, the European standard EN 15643-1: 2010 (CEN, 2010) explicitly states that besides technical and functional requirements of a construction project, requirements concerning its environmental, social and economic performance also need to be formulated and realised. The economic performance requirements are often deduced from the overarching goals of a sustainable development. Amongst others, these include the minimisation of life-cycle costs (LCCs) (see, e.g. Pelzeter, 2006) and the protection of capital in the sense of maintaining the market players' (economic) capacity to act. Also from a societal perspective, economic performance categories include the reduction and avoidance of subsidies, the minimisation of negative external effects and, specifically within the housing sector, the safeguarding of affordable housing.

Within the construction and real estate industry various groups of actors exist with often diverging roles, duties and goals. Even within the group of institutional and private investors and construction clients different goals and objectives prevail which both impact on the definition and conception of projects and become the measure for evaluating design solutions and variations. It is traditionally assumed that functional and technical requirements of legislators, investors and/or clients as well as of future users form the starting point for initial design considerations and project development. However, this is not always the case, particularly in relation to the conception of investment (income-generating) properties, where economic goals and requirements provide the criteria for development. In this context, the available budget as well as expectations regarding marketing times and achievable returns make up the key framework conditions for projects. The following economic aspects might play a role within the context of investment decision making:

- Overall budget: maximum investment sum, maximum construction cost
- Financing possibilities and costs (particularly project financing)
- Lettability, marketability, marketing risks
- Overall project risks
- Achievable rents/required minimum rent
- Investment return and yield expectations
- Risk–return profile
- Stability, security of the cash flow, risk of losing the tenant(s)
- Level of operating costs attributable to tenants/level of operating costs non-attributable to tenants
- Total costs of ownership (in the case of owner-occupiers)
- Value, value stability, value development potential.

The specific individual and institutional interests and goals differ between groups of actors. Figure 10.1 indicates which economic aspects are of particular relevance and interest for different groups of property market players.

The considerations regarding the aforementioned economic aspects are usually coupled with more general goals such as image and reputational gains, a positive market appearance or the reduction of reputational risk.

Within the design domain various approaches including traditionally used methods are adopted such as:

- Design to cost
- Design to LCC
- Design to value.

These sensibly complement the other planning and design strategies like performance based building, design for the environment, eco-design, design for deconstruction, and so on.

European standardisation activities at CEN (the European Committee for Standardisation) also follow and support an approach of formulating economic goals in the early stages of a project – this particularly applies within the context of sustainability assessments of buildings. For example, the European standard EN 15643-1: 2010 (CEN, 2010) allows – in addition to technical and functional requirements – for specifying goals relating to a building's environmental, social and economic performance as a basis for design, decision making and assessment (Figure 10.2).

Aspects of interest

Actors and their roles

	Risk - return ratio	Investment performance / total return	Construction cost / additional construction cost	Life cycle cost / total cost of ownership / full cost	Level of operating costs attributable to tenants	Level of operating costs non-attributable to tenants	Rent level	Value / stability & development of value	Risk (asset specific)
Individual & institutional investors with medium-to long-term interests	✔	✔			✔			✔	✔
Individual & institutional investors with short-term interests	✔	✔						✔	✔
Project developers			✔				✔	✔	✔
Landlords / awarding authorities and buyers of rental assets	✔	✔	✔			✔		✔	✔
Awarding authorities and buyers/ owners of self-occupied assets			✔	✔	✔			✔	✔
Tenants					✔		✔		
Financers	✔		✔						✔
Fund managers	✔	✔	✔		✔	✔	✔	✔	✔

Figure 10.1 Key economic performance aspects and their relevance to different players. Source: Lützkendorf and Lorenz (2009, p. 26).

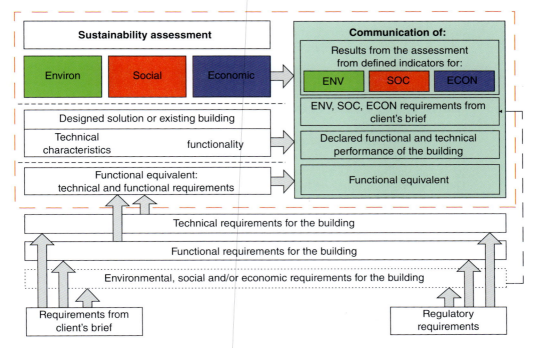

Figure 10.2   The overall concept of sustainability assessment of buildings (CEN, 2010).

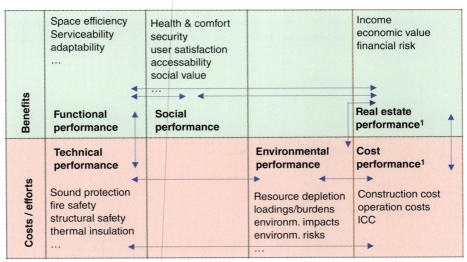

[1]as part of economic perfomance (and whole life cost WLC)

Figure 10.3   Relationships between partial aspects of building performance.

Pursuing and achieving economic goals requires both detailed knowledge about the respective location and market situation as well as a profound analysis and influencing of the relationships between a building's economic performance aspects and its other characteristics and attributes (Figure 10.3). Within this context, it is worthwhile noting that the relationships between a building's sustainability performance and its current market value, value stability and likely value development are currently intensively debated at the national and international levels.

10.3   Identifying value-related characteristics

From the inception of a project and the agreement upon design objectives it is important to bear in mind how the relationship between economic performance goals and the building's key characteristics and attributes can be taken into account. These interrelationships are highly complex (Figure 10.4). Therefore, the procedure for taking this into account during the planning and design phase is somewhat different than that of a sustainability assessment. In planning and design the interdependencies between partial aspects of building performance have to be considered at the commencement of the project. In order to deal with this complexity, it is better to treat the complexities by addressing partial questions (Figure 10.4) and run through several iterative processes. The success of such endeavours can then be assessed, described and portrayed in a structured manner through conducting an overall sustainability assessment of the building under investigation.

In this respect, knowledge and awareness of the methods and procedures for property valuation and risk analysis are also necessary. A closer inspection of these methods and processes reveals which characteristics and attributes of buildings directly or indirectly impact upon a building's market value, value stability and risk profile. These characteristics and attributes therefore deserve special attention during the concept and design phase.

As a result of the increasing importance of sustainability considerations within the property and construction industry it is an additional requirement to analyse

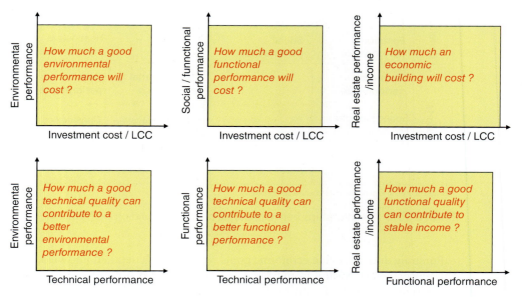

Figure 10.4   Handling complexity by addressing partial questions.

which building characteristics and attributes impact on a building's sustainability performance or upon sustainability assessment results, respectively. In this respect, knowledge of the various sustainability assessment systems and schemes for buildings is necessary. This may show that many of those characteristic and attributes which are traditionally taken into account within the scope of property valuations are also relevant for the sustainability assessment of buildings. In addition, investment decisions are based on an evaluation and assessment of market and locational risks. Since buildings have comparatively long lifespans, an analysis of the market and property location requires an assessment of future developments. For example, an assessment of both demographic developments as well as future impacts of climate change increasingly becomes part of standard locational analyses. On the other hand, market analyses increasingly include assessments of trends in workplace design as well as dynamics in living styles and preferences. Consequently, and with a focus on value stability and value development, those characteristics and attributes which contribute to a long or prolonged technical lifespan, as well as to a long or prolonged useful economic life, deserve attention.

The technical lifespan of a building is, amongst other issues, determined and influenced by the type and quality of the technical solution as well as by the building's ability to respond to current and future locational risks. Key characteristics in this regard are: durability, resilience, ease of conducting maintenance and inspection works, and ease of retrofitting or upgrading the building with innovative systems (retrofitability). The question whether and in how far additional capacities of a building (e.g. load bearing capacities) should be technically realised, for example in order to contribute to easing retrofitting or upgrading, depends on an assessment of the type and direction of market risks as well as on the likeliness of changes in the type of usage. Such decisions are always associated with a certain degree of uncertainty.

The useful economic life of a building is, amongst other issues, determined and influenced by its functional quality and the associated ability of the building to guarantee long-term lettability and/or marketability as well as by the building's flexibility and adjustability to respond to changes in market participants preferences in particular and to changes in the overall market conditions in general (e.g. legislative changes). Key characteristics in this regard are: ability to fulfil current and future user requirements, adjustability, expandability, and flexibility regarding the type of usage. In this context, the technical and functional qualities of a building are closely interrelated since the type of the technical solution is a key impact factor regarding the building's flexibility and retrofitability. The interrelationships between the characteristics and attributes of a building on the one hand and the locational and market risk on the other hand as well as the linkages to the overall investment and financial risk are conceptualised as shown in Figure 10.5.

In order to proceed further, it is now required to examine how key characteristics and attributes of buildings can be described, assessed, portrayed and communicated.

10.4   The performance approach

One important condition for the achievement of economic goals lies in safeguarding long-term lettability and marketability. This leads to the question of how the advantages and qualities of buildings can best be signalled and communicated to

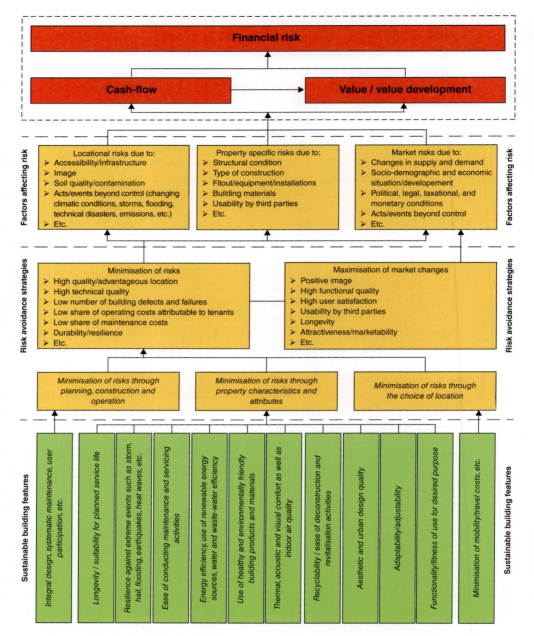

Figure 10.5   Sustainable building features linking to assessments of risk. Source: Lorenz and Lützkendorf (2011).

potential tenants and buyers and which characteristics and attributes of buildings significantly impact on tenants' and buyers' decision making processes. Even though the location has a key impact, the following discussion focuses on characteristics and attributes of buildings.

When analysing investment decision processes it is important to take into account the perspective and interests of future users and buyers. Future users and buyers can include corporate tenants, private tenants, institutional and private investors, and so on. Their interests, motivations and decisions are not homogenous. For example, in comparison with tenants, buyers usually have a higher willingness to pay for those characteristics and attributes of buildings which impact on the durability and value stability of a building (Knapp, 2011). These issues are less important from the perspective of tenants (Kippes and Symonds, 2012).

The goals and interests of tenants and buyers are less focused on the description of the technical solution; instead, their interests are focused on the achievable functional performance (user/tenant perspective) as well as on the economic performance (investor perspective). Investors, however, need to also have the user/tenant perspective in mind in order to safeguard lettability and marketability.

In a dialogue between project developers, users and investors the so-called performance approach has been developed (see, e.g. http://www.pebbu.nl). Within this approach the starting points are the requirements regarding the usage purpose and usability of a building. In this context, questions are addressed concerning the quality of living and working as well as the spatial realisation of functional processes (space allocation plan). In addition, economic requirements are also formulated to reflect the target group's financial expenses (e.g. budget for construction and acquisitions, willingness to pay for rent and operating costs, etc.). Within current applications of the performance approach, there are also the requirements concerning environmental and health compatibility or sustainability which are to be agreed upon. The current development in this regard is to move from requirements like usage of renewables, availability of green roofs, rainwater usage, and similar, towards performance-oriented requirements such as indoor air quality or resource use and environmental impact assessed on the basis of a life-cycle assessment (LCA).

Within a second stage of the performance approach these usually diffuse and vague goals and framework conditions need to be translated into detailed and precise requirements, for example regarding the thermal, visual and acoustic comfort, the required indoor air quality, the space allocation plan and its flexibility, requirements concerning safety and accessibility, and so on. Subsequently, these requirements need to be transformed into design variables, technical solutions and construction works. However, users or buyers usually require proof that the proposed or realised technical building solution meets their primary interests and requirements. This is typically done by using selected performance criteria which also directly feed into a potential buyer's assessment of project specific risks and financial implications.

In order to signal qualities within the property market, particularly in relation to sustainability, quality and assessment certificates have been proven to be of value. These certificates condense several characteristics and attributes of a building into a highly aggregated overall result concerning the quality of the property under investigation.

A good example of these are sustainability assessment certificates. However, it has also become apparent that an unquestioned or unquestionable uptake of highly aggregated assessment results can also lead to new problems and risks. This is because market players usually do not have enough knowledge and expertise regarding details of these assessment systems, their assessment logic, assumptions and limitations. For example, two buildings with very different characteristics and attributes can achieve identical overall assessment results. As a consequence, working with aggregated assessment results usually hinders a critical evaluation of the assessment results. For this reason it is not only recommended (for designers and other professionals) to learn more about these assessment systems but also to additionally use disaggregated assessment results as an additional source of information.

Concerning performance measurement it is important to distinguish whether the basis for the measurement have been planning and design information or if actual information obtained during the operational or usage phase have been used. In any case, a life cycle accompanying performance measurement is highly recommended (e.g. in the form of post occupancy evaluations).

Besides the quality of the building and of its location, an additionally important factor that impacts on the realisation of economic goals is the quality of processes: quality of the planning and design process, quality of the construction process, quality of the management process. Therefore, not only safeguarding quality during the planning and construction phase by making use of quality management systems is an important issue but also creating a process of continuous improvement through, for example, controlling of operating costs, energy consumption monitoring as well as post occupancy evaluations.

In particular, the quality of the planning and construction phase can be signalled to the target group (buyers or investors) and then fed into financial risk assessments. For further reference see, for example http://www.pebbu.nl.

Originally, the performance approach was focused on aspects of functional quality. However, during recent years more and more aspects such as economic requirements, questions relating to environmental protection or design and urban aesthetic quality have been integrated into the performance approach. As a consequence, the performance approach and sustainability assessments are moving closer towards one another and are increasingly linked within a so-called second generation of methods and systems for sustainability assessments of buildings (see, e.g. Lützkendorf et al., 2005; Lützkendorf and Lorenz, 2006).

10.5   Use of sustainability assessment systems

In the real estate and real estate related finance sectors in Germany, systems for risk and real estate analysis are used. Examples are the so-called property rating systems; for example the system for property analyses developed by The Association of German Public Banks (Bundesverband Öffentlicher Banken Deutschlands, VÖB) as well as for sustainability assessment [e.g. the building sustainability assessment system (BNB) used by the Federal Ministry of Transport, Building and Urban Development (BMVBS), and the German certification system for sustainable buildings used by the German Sustainable Building Council (DGNB)]. This situation is applicable to other countries as well such as BREEAM used in the UK and LEED in the USA discussed elsewhere (see Chapter 12). Often these systems are used only upon completion of a building or for already existing buildings. These systems usually establish a link between characteristics and attributes of the building and real estate risks as well as the sustainability of a building. The assessment principles, criteria and standards used for risk and sustainability assessment are suitable, in principle, for the formulation of building requirements to minimise financial risks, ensure sustainability and support value stability and economic performance. In this sense, these systems also constitute a checklist for the early stages of planning and are suitable for supporting goal setting and agreement between principals (e.g. client, investor) and agents (planner or general contractor). Investors are therefore advised to obtain and systematically evaluate the relevant information from existing systems for risk and real estate analysis and sustainability assessment.

Furthermore, it is recommended for designer and planners not to formulate an overall risk or sustainability level as a design goal but to set specific requirements on individual features and characteristics that are relevant to sustainability, economic value and financial risk. Germany has developed highly acceptable methods of pre-assessing the sustainability in the early stages of planning, both in the field of private economic activities as well as in public projects. Based on initial planning decisions and statements of intent it is analysed, whether and to what extent a building that exhibits all the envisaged and planned characteristics and features would comply with the requirements on sustainability and other goals (e.g. the technical and functional quality). Therefore, it is possible to analyse in the early stages of planning the consequences of planning decisions on technical, functional, ecological, social and economic quality of buildings. Where appropriate, a target adjustment is possible. While for the completed and existing buildings the only priority is the assessment, the identification of the different interactions is achieved in early stages of the planning process. What is examined is how the improvement of the technical quality of a building leading to increased users' satisfaction, less environmental impacts, as well as reduced operational and LCCs affects its overall value, value stability and value development. In a sustainability assessment system developed in collaboration with the housing industry, the criterion 'long-term value of the investment' was introduced. The prospective investment costs are weighted against the prospective capitalised revenues at completion and/or the accounted investment costs against the capitalised revenue at the time of handover and commissioning.

Clearly there is a need for better exchange of information between investors and developers, banks, analysts, designers and sustainability. The relevant assessment systems should be known to key actors to facilitate understanding and to be passed on to the planners in case of need as a requirement. The planners themselves should introduce these tools in an accompanying planning phase for monitoring the success of individual planning steps.

This also provides new information flows and opportunities for planners and designers. For example, within the scope of applying for a loan, a sustainability-engaged project developer can present detailed documentation regarding the project's anticipated or actual sustainability performance. Alternatively, an engaged bank may request additional documentation from the developer to check whether or not the project meets energy efficiency and/or sustainability requirements. Within property and construction markets such a practice would greatly increase the demand for verifiable information (energy performance certificates, building files or detailed sustainability assessments). In addition to this, banks can advise and influence the developer regarding the project's sustainability features. This requires that banks develop competencies in assessing and advising on construction projects (Lützkendorf et al., 2011). The actor constellations and flows of information between them as described above are graphically displayed in Figure 10.6.

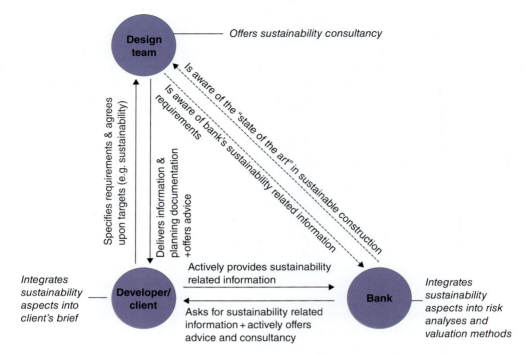

Figure 10.6   Actor constellations and flows of information. Source: Lützkendorf et al. (2011).

All this has profound consequences for designers: design teams must be more aware of the information needs and requirements of financial stakeholders such as banks and insurance companies and others such as valuation professionals. Designers also need to be better positioned to actively create and offer developers corresponding information and documentation. This can significantly improve the competitive position of designers (Lützkendorf and Lorenz, 2011).

An important condition for further motivating and engaging banks and investors in strengthening the demand for sustainable buildings and design solutions lies in empirically demonstrating their economic advantages. This is briefly discussed in the following section.

10.6   Relationship between sustainable credentials and value

Currently in the real estate industry it is intensively discussed which features and characteristics should be included in a sustainability assessment, and how in turn these results have an influence on the value or valuation. Two trends are discernible:

1. Large parts of the previously developed and used sustainability assessment systems focus on aspects such as energy efficiency, environmental and health compatibility. In particular, the interrelation between the value and energy quality and energy performance of buildings has been analysed in several empirical studies demonstrating such influences (Table 10.1).

Table 10.1 Empirical evidence on the economic benefits of buildings with sustainable credentials (Lützkendorf and Lorenz, 2011).

Study/Authors	Country	Property Type	Sustainable Credentials	Observed impact on	+/-	Magnitude
Brounen and Kok, 2010	The Netherlands	Residential Homes	Energy Performance Certificate (Class A, B, or C)	Selling Price	+	2.8%
City of Darmstadt, Rental Index, 2010	Germany (Darmstadt)	Residential multi-family houses	Primary energy value below 250 kWh/m²a	Rental Price	+	0,38 €/m²
			Primary energy value below 175 kWh/m²a			0,50 €/m²
Eichholtz, Kok and Quigley, 2010	USA	Office Buildings	LEED	Selling Price	+	11.1%
				Rental Price	+	5.9%
			Energy Star	Selling Price	+	13%
				Rental Price	+	6.6%
Fuerst and McAllister, 2010	USA	Office Buildings	LEED	Occupancy Rates	+	8%
			Energy Star		+	3%
Fuerst and McAllister, 2008	USA	Office Buildings	LEED, Energy Star	Selling Price	+	31%–35%
				Rental Price	+	6%
Griffin et al, 2009	USA (Portland/ Seattle)	Residential Homes	Built Green, Earth Advantage, Energy Star, or LEED	Selling Price	+	3%–9.6%
				Selling/Marketing Time	–	18 days
Pivo and Fischer, 2010	USA	Office Buildings	Energy Star, close distance to transit, location in redevelopment areas	Net Operating Income	+	2.7%–8.2%
				Rental Price	+	4.8%–5.2%
				Occupancy Rates	+	0.2%–1.3%
				Market Value	+	6.7%–10.6%
				Income Returns (Cap Rates)	–	0.4%–1.5%
Salvi et al, 2008	Switzerland	Residential Homes	MINERGIE Label	Selling Price	+	7%
		Residential Flats		Selling Price	+	3.5%
Salvi et al, 2010	Switzerland	Residential Flats	MINERGIE Label	Rental Price	+	6%
Wameling and Ruzyzka-Schwob, 2010	Germany (Nienburg)	Residential Homes	Primary energy demand per m² and year (kWh/m²a)	Selling Price	+	1,26 €/m² per reduced kWh/m²a
Wiley, Benefield and Johnson, 2008	USA	Office Buildings	LEED, Energy Star	Rental Price	+	7%–17%
				Occupancy Rates	+	10%–18%

However, it should be noted that the results of international studies cannot easily be transferred to other regions. Crucial for a (positive) effect on the value is the relative "distance" of the energy-efficient variant from the market average. In countries with an already high level of legal and normative requirements on the energy quality the concentrative effect on value is increased as a general tendency. This becomes a new trend. Energy efficient buildings will be able to gain additional value, whereas buildings with below-average energy quality will suffer a value deduction. A systematic maintenance and adaptation of the energy quality of the building in line with the technical progress through regular modernisation are thus part of a strategy to maintain the value of buildings or property.

2. Newer systems for sustainability assessment (e.g. in Germany DGNB and BNB) include in addition to aspects of environmental and health compatibility also the functional, social, technical and economic quality as assessment criteria. They are moving closer to the content of the valuation and risk analysis systems and these tend to blend into each other. So far, there have been some analysis of willingness to pay carried out for sustainability-certified buildings. However, information from actual transactions is very limited.

In order to improve the availability of data and empirical evidence on the economic advantages of sustainable buildings, efforts are required on the side of designers and planners as well as on the side of banks, investors and valuation professionals. Designers and planners can contribute to this – besides improving planning itself through, for example, the application of methods of integral planning – by improving/extending planning documentations and, in doing so, realising a more profound and detailed description of buildings in the market place. Through an increased usage and analysis of improved documentations of planning results, it is then possible to facilitate information gathering processes for property valuation and sustainability assessments. On the other hand, banks, investors and valuation professionals should include additional information on sustainability-related property characteristics and attributes within their data collection, transaction analyses, valuation and risk assessment procedures. Similar suggestions are now also put forward by the large professional organisations for property professionals, notably by the Royal Institution of Chartered Surveyors (RICS) (RICS, 2009, 2011).

10.7   Concluding remarks

The common element between the economically dominant considerations of investors and an effort to contribute to sustainable development is the long-term perspective. Through adaptable buildings with low resource use and environmental impact meeting the needs of present and future users, better rental values and marketability can be achieved. This in turn leads to improved value stability and positive value development potential. However, buildings need not only be planned, constructed and operated in this manner but also their quality must be appropriately described, evaluated and reported. It is likely that a lack of relevant information (which, in part, could be easily obtained during the planning phase) will become a risk factor in itself and may lead to value reductions. Previously, the building design, the sustainability assessment and the property valuation were considered as entirely separate activities. In the future, these processes should be considered together in an integrated approach. The fulfilment of user requirements in the strict sense, a positive contribution to the development of society and economy as well as the preservation and enhancement of capital become components of an overall approach.

It becomes also apparent that an appropriate/improved formulation of design/planning objectives in combination with responsible and longer term focused investments in the property sector has not only the potential to safeguard the value and value stability of buildings but also to contribute positively to a more sustainable development.

References

CEN (2010) EN 15643-1:2010 – Sustainability of construction works – Sustainability assessment of buildings – Part 1: General framework. CEN.

Kippes, S. and Symonds, B. (2012) The attitudes of tenants, home owners, vendors and buyers concerning environmental questions – an empirical survey on the importance of sustainable housing in Germany. ARES Conference paper, 17–21 April 2012, St Petersburg, FL.

Knapp, O. (2011) Design-to-Value – Increasing product profitability, Stuttgarter Gespräche. http://www.rolandberger.com/media/pdf/Roland_Berger_Design_to_Value_Stuttgarter_Gespraeche_20110504.pdf (accessed 30 June 2014).

Lorenz, D. and Lützkendorf, T. (2011) Sustainability and Property Valuation – Systematisation of existing approaches and recommendations for future action. *Journal of Property Investment & Finance*, 29(6), 644–676.

Lützkendorf, T., Fan, W. and Lorenz, D. (2011) Engaging financial stakeholders: opportunities for a sustainable built environment. *Building Research & Information*, 39(5), 483–503.

Lützkendorf, T. and Lorenz, D. (2006) Using an integrated performance approach in building assessment tools. *Building Research & Information*, 34(4), 334–356.

Lützkendorf, T. and Lorenz, D. (2009) Key Financial Indicators for Sustainable Buildings. UNEP-FI/SBCI'S Financial & Sustainability Metrics Report, pp. 16–55.

Lützkendorf, T. and Lorenz, D. (2011) Capturing sustainability-related information for property valuation. *Building Research & Information*, 39(3), 256–273.

Lützkendorf, T., Speer, T., Sziheti, F. *et al.* (2005) A comparison of international classifications for performance requirements and building performance categories used in evaluation methods. CIB Joint Symposium, 13–16 June 2005, Helsinki.

Pelzeter, A. (2006) *Lebenszykluskosten von Immobilien – Einfluss von Lage, Gestaltung und Umwelt*, Rudolf Müller Verlag, Cologne.

RICS (Royal Institution of Chartered Surveyors) (2009) *Valuation Information Paper 13, Sustainability and Commercial Property Valuation*, RICS, London.

RICS (2011) *Sustainability and Residential Property Valuation*, IP 22/2011, RICS, London.

Chapter 11
Carbon Reduction and Fiscal Incentives for Sustainable Design

Paul Farey

11.1 Introduction

The subject of tax relief could easily warrant a book of its own: the UK is blessed with "one of the most complex and opaque tax codes in the world" (BBC, 2010) with all aspects of personal life and business taxed, either directly or indirectly and sometimes both. In March 2011, The Office of Tax Simplification (OTS) published its final report highlighting 1,042 reliefs, allowances and exemptions within the system (OTS, 2011). Only a handful of these have subsequently been abolished.

The challenge of working within the tax rules becomes yet more demanding when the behemoth that is the UK tax system is applied to the brave new world of sustainability and the drive towards low energy and carbon reduction. Carbon itself is a relatively new commodity for the system to tackle and new measures have been created to tax consumption, as well as fiscal encouragement to reduce its use. The Government's ability to balance the carrot of incentive against the stick of levy has been undoubtedly hamstrung by the scale of UK debt, as evidenced through the accelerated reduction of feed-in tariffs and the effective taxation of carbon through the Carbon Reduction Commitment (CRC) Efficiency Scheme.

This chapter examines the key drivers for building owners and occupiers, particularly around the triple bottom line of social, economic and environment to achieve the goals of sustainability. The UK Government's current aim to achieve a significant reduction in carbon emissions by 2020 through reducing demand for energy in buildings, decarbonising heating and cooling supply. The fiscal drivers through Government's policy, statutory and regulatory framework in the form of taxation and levies, reliefs and allowances such as plant and machinery allowances, enhanced capital allowances (ECAs), subsidies and other incentives are then explored.

Design Economics for the Built Environment: Impact of Sustainability on Project Evaluation, First Edition.
Edited by Herbert Robinson, Barry Symonds, Barry Gilbertson and Benedict Ilozor.
© 2015 John Wiley & Sons, Ltd. Published 2015 by John Wiley & Sons, Ltd.

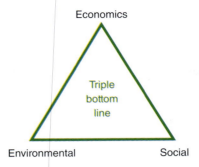

Figure 11.1   The triple bottom line.

11.2   Key drivers of owners and occupiers

Many drivers exist for building owners and occupiers, particularly around the triple bottom line of social, economic and environment; and it is now necessary to design, engineer, construct and operate buildings in a much smarter way (Figure 11.1). Traditional above the line measures such as capital cost are giving way to more sophisticated appraisal techniques that reflect life cycle, cash flow and indirect costs, such as taxation. Occupier demand for high quality residential or commercial accommodation with significant services content must be balanced against building regulations, planning rules, energy sources and the liquidity or marketability of the resulting space.

Traditional commercial drivers are also being diluted through the added burden of regulations, current and future Government policy, governance, corporate social responsibility and shareholder action. Put simply, not only is there a requirement to add value to real estate assets, but also in *being seen to add value*, from the perspective of many project stakeholders during the life cycle. Within this chapter, some of the key drivers, legislation, taxes and incentives associated with carbon reduction are examined to highlight how efficient design can work within the current framework to reduce the capital and operating costs of buildings. All data and rates quoted within this chapter are accurate as at July 2013.

11.3   Reducing demand for energy in buildings

The Government's current stated aim is to achieve a 34% reduction in carbon emissions by 2020, in accordance with the 2009 Low Carbon Transition Plan. Further, the Government has stated that by 2030, up to half of the heat used in buildings may come from low carbon technologies (HM Government, 2011). This policy will be achieved through a series of measures, backed up with statute and instruments.

Improving the heat efficiency of buildings

The majority of energy used in residential property in the UK (HM Government, 2011) is for space and hot water heating. Improving heat efficiency is therefore a major component in reducing energy and carbon. Some of the measures put forward on this include:

- The Green Deal – introduced in 2012.
- Energy Act 2011 – including the requirement for a minimum standard for leasing private residential and commercial property from April 2018.

- Energy Company Obligation (ECO) – additional funding to support solid wall insulation and assisting low-income households with basic heating.
- Building Regulations – ongoing improvements to the statutory new-build standards. Future improvements to Part L will also focus on energy efficiency measures in existing buildings.

Improving the electrical efficiency of lighting and appliances

White goods, lighting and electrical powered equipment (such as televisions) account for some 14% of the UK's carbon consumption (HM Government, 2011). The market has quickly established a regime that removes the most inefficient products, whilst promoting greener technologies through visual labelling.

This process continues to gather pace with minimum EU performance standards and labelling conventions to be evaluated for most domestic and commercial appliances.

Changing behaviour to reduce demand

Clearly technology and built solutions are important aspects of the drive to reduce carbon; however such measures are ineffective without a corresponding shift in the behaviours of building owners and occupiers. Some of the key drivers here include:

- Smart Meters – a plan for every home to have these by 2019.
- Energy Performance Certificates (EPCs) – these are required on the sale, rent or construction of a building, providing a rating between A (very energy efficient) through to G (not energy efficient).
- Display Energy Certificates (DECs) – for public buildings which are larger than 1,000 m².
- CRC Energy Efficiency Scheme – a mandatory scheme aimed at improving energy efficiency and cutting emissions in large public and private sector organisations.

Heating and cooling supply

After the relatively light-touch approaches outlined above, the next task is to tackle the embedded technologies within buildings to ensure that energy requirements are as low as possible. In engineering terms, this extends to using heating systems with renewable energy sources, such as biomass, solar thermal or ambient heat from the ground or air. Some of the key Government policies include:

- Payments to subsidise the relatively higher investment cost in these technologies through Feed-in Tariffs (FITs) and the Renewable Heat Incentive (RHI).
- Renewable Heat Premium Payment (RHPP), a single payment of up to £2,300 for residential owners and occupiers installing approved low carbon heat technologies. The second phase of the scheme was extended after which applications are made under the Domestic Renewable Heat Incentive (Domestic RHI).

- Capital allowances provide tax relief for non-residential property owners and occupiers, with a 100% first year allowance (ECAs) for eligible technologies.

Network-level technologies

The final part of the jigsaw relates to the central generation and distribution of heat and power. The technologies include combined heat and power (CHP), biomethane gas injection and district heating systems. These infrastructure networks are highly capital-intensive and face a number of challenges through land ownership and leasing, wayleave agreements, regulation, maintenance and charging structures. Given that many of the practicalities for these barriers exist at local Government level, no single incentive or measure has yet been implemented. The Government has published summary evidence [DECC (Department of Energy and Climate Change), 2013) confirming that there are 1,765 CHP networks installed across the UK as of March 2013.

The Low Carbon Buildings Sectoral Plan states:

> In 2009, 37% of UK emissions were produced from heating and powering homes and buildings. By 2050, all buildings will need to have an emissions footprint close to zero. Buildings will need to become better insulated, use more energy-efficient products and obtain their heating from low carbon sources (HM Government, 2011).

11.4   Fiscal drivers

Whilst the Government's policy filters down through the statutory and regulatory framework, there are a number of other factors that impact on building owners and occupiers. As a result, these naturally affect the demands on the design and construction process. The direct economics associated with the construction and management of property are likely to be more practical behavioural influencers to clients, designers, contractors and occupiers. The traditional property tenet of reducing capital cost and maximising sale price or rental value still applies, albeit that more sophisticated models are now used, often reflecting whole life cost. European methods also reflect a 'below the line' approach which also takes into account aspects of direct and indirect taxation.

Life cycle costing is not a new science and many surveyors and construction professionals will be familiar with college textbook examples, such as the comparison of uPVC windows against lower cost but higher maintenance timber alternatives. Yet despite this, the method is still rarely applied in modern commercial appraisals; maybe unsurprising given current procurement methods and the relative short-term interest of developers and contractors. Arguably, this mindset extends to designers, advisors and even for owner-occupiers. Given modern appraisal methods and risk management techniques, this situation will need to change in the future. In an economic environment where funding is not readily available, a cultural shift has started with more rigour already demanded by lenders within appraisals, particularly in terms of assumptions of income and expenditure, input variables and timing.

Both sides of the equation can be enhanced to reflect additional income sources, such as FITs or the RHI; with taxation, net of allowances naturally, impacting on the longer-term cost. Timing of payments or relief can be factored in using discounted cash flow, using much more explicit assumptions than traditional yield capitalisation methods. Capital cost can even be reduced when payable credits are surrendered for tax losses created through Land Remediation Relief and ECAs. Conversely, where input tax for VAT purposes is irrecoverable, then this becomes a real cost burden for which unplanned funding is necessary. Applying risk factors to particular inputs is therefore an essential requirement to sensitivity analysis that may form part of the decision-making process for developer, investor or lender alike.

Taxation versus incentive

The tax system has been consistently used by the Government as an enforcer of policy since William Pitt the Younger's income tax of 1798: collecting cash is a near guaranteed method of focusing behaviour and getting a message across. The stick of legislation, supported by taxation, has occasionally been tempered with the carrot of incentives, with some of the Government measures outlined above supplemented with tax reliefs, allowances and credits.

However, despite best intentions, the system is still multifarious, with the framework and prevailing rates regularly reviewed and occasionally manipulated for socio-political purposes through annual budgets. With interest rates at a structural long-term low and conservative loan-to-value ratios imposed by lenders, borrowers realistically find themselves in a tax paying position more readily: the ability to hedge taxable income through interest charges for long-term periods is less prevalent.

In order to make property developments work and improve upon marginal returns, building owners and occupiers need to be more sophisticated in their approach to tax. This requires an awareness of the current rules as well as the ability to capitalise upon available reliefs as a means of minimising tax liability. However, HMRC rigorously enforce the tax laws to the extent that frivolous claims for relief resulting in underpayments of tax are strictly penalised. The taxpayer therefore needs to tread a fine line between compliance and commerciality; ideally taking professional advice.

Taxes and levies

It is all the more remarkable perhaps, that given the costs and inevitability of tax within the development process, that there is not more consideration given to capital and operating cost on a post-tax basis, that is factoring the quantum and timing of tax, as well as planning opportunities for relief.

The UK tax system is one of the most established and the most complex in the world having evolved since the late seventeenth century. The current regime is split between direct and indirect taxes, as shown in Table 11.1, along with the significant rates on profits and gains as at the date of publication.

In addition, taxes are also raised by local government through measures such as business rates and council tax. The most significant tax implications for property owners and occupiers are listed in the following, as well as the key allowances and reliefs.

Table 11.1   Taxes levied in the UK, with 2013/14 rates (HMRC, 2013).

Direct taxes	Indirect taxes
Income Tax	Value Added Tax
Basic rate (to £32,010): 20%	Standard rate: 20%
Higher rate (to £150,000): 40%	Reduced rate: 5%
Top rate (over £150,000): 45%	Zero rate: 0%
Dividends: up to 37.5%	
Corporation Tax	Customs Duty, Tobacco, Petrol and Other
Small profits rate (to £300,000): 20%	Expenditure Taxes
Higher rate (to £1.5 million): 23.75%	
Top rate (profits over £1.5 million): 23%	
National Insurance	Climate Change Levy
Capital Gains Tax (max. 28%), Inheritance	Landfill Tax and Aggregates Levy
Tax and Stamp Duty Land Tax (top rates)	
Residential: 7%	
Commercial: 4%	
Shares: 0.5%	

Income tax

Income tax is levied on individuals, partnerships and non-resident landlords (Figure 11.2). The rates are varied from time to time for basic and higher rates, with other bands occasionally introduced and adjusted to suit Government fiscal policy. Taxpayers are subject to self-assessment based upon their taxable income for relevant chargeable periods. For owners and occupiers of property, rental income or trade profits are taxed at the appropriate rate, less expenditure on any interest, fees and charges incurred wholly and exclusively in connection with the trade. Such expenditure is known as revenue expenditure and is fully deductible against tax in the year it is incurred as a trade expense. Developers who hold property as trading stock are deemed to incur revenue expenditure. They are therefore able to fully relieve their costs against proceeds upon disposal.

For property investors, occupiers and owner-occupiers, costs incurred in constructing, acquiring or upgrading a property (as a fixed asset) is considered to be capital expenditure and is not tax deductible. Some relief may be given through capital allowances.

Corporation tax

UK companies are liable to corporation tax for adjusted accounting profits, including gains (Figure 11.3). As with income tax, the rates and profit bands are subject to fiscal policy and are regularly reviewed within Budgets. Accounting profits are adjusted to add back certain expenditure that is not tax deductible, such as depreciation and refurbishment works that improve or upgrade fixed assets.

Capital gains tax

Capital Gains Tax (CGT) is levied upon UK individuals disposing of fixed assets. Non-resident landlords and offshore investors are not liable to CGT. The rates charged are banded and reviewed annually, but may be relieved or deferred through

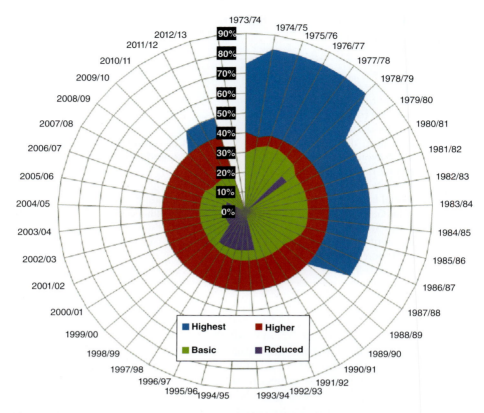

Figure 11.2   UK income tax rates since 1973 (HMRC, 2012).

an annual exemption, entrepreneur's relief or rollover relief. UK companies are not taxed separately on gains for disposed assets; these are included within the overall profits chargeable to corporation tax (PCTCT).

Value added tax

Value Added Tax (VAT) is a European tax on the consumption of certain goods and services introduced into the UK in 1973. It is an indirect tax administered by traders with the final cost borne by the end consumer. Sometimes dubbed a 'stealth tax', the legislation relating to VAT on construction projects and property transactions has compounded since introduction, with voluminous VAT Tribunal and European Court of Justice decisions, to evolve into the intricate regime currently operated. Accordingly, clients, advisors and occupiers need a sound knowledge of the rules and compliance issues if they are to avoid the risk of costly errors. The law has been subject to many changes by HMRC to counter avoidance and antiforestalling through rate changes. The current rates are detailed in Table 11.2, showing the impact to businesses of VAT paid on goods and services (input tax) relative to the nature of supplies made and VAT charged (output tax).

Relevant residential (RRP) and some relevant charitable (RCP) use of buildings benefit from zero-rating and a reduced rate can apply to certain renovation and conversion projects. The supply of other construction services will generally be standard rated. Commercial property can be exempt from VAT or standard rated. For exempt

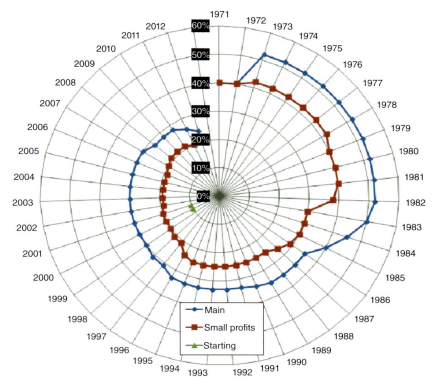

Figure 11.3   UK corporation tax rates since 1971 (HMRC, 2012).

Table 11.2   Effective VAT rates (HMRC, 2012).

Rate	Supply Output Tax	Input Tax
Standard	20%	Recoverable
Reduced	5%	Recoverable
Zero	0%	Recoverable
Exempt	Outside of VAT regime	Not recoverable

transactions, such as the transfer of a going concern (TOGC), the decision can be made to 'opt to tax', bringing the transaction within the charge to tax at the standard rate. Whilst VAT is charged on the sale prices, this decision also enables a vendor or landlord to recover VAT costs incurred (input tax) that would have otherwise been irrecoverable. It also means that VAT must be levied on any rent received; the tenants might themselves be able to recover this VAT, unless they are outside the scope (e.g. banking). In other circumstances, it is the use or intended use of the property that will determine the extent to which input VAT may be reclaimed. The reduced rate of VAT is applicable to certain energy-saving materials for residential properties.

Table 11.3 shows the impact of VAT to various parties as a result of a building owner installing a new boiler in a property. Each party pays input tax on procured materials and goods; and charges output goods on the subsequent sales cost. The difference is paid to HMRC as part of a quarterly VAT return.

Table 11.3 VAT implications of construction works on supply chain.

Party/Function	Net Cost	VAT	Input Tax	Sales	VAT	Output Tax	Pays HMRC
Manufacturer							
Manufactures boiler							
Materials[(i)]	£ 1,500	20%	£ 300				See[(i)]
Direct Labour	£ 2,500		£ -				
	£ 4,000		£ 300	£ 5,000	20%	£ 1,000	£ 700
Supplier							
Supplies boiler							
Purchase of boiler	£ 5,000	20%	£ 1,000				
Transport to Sub-Contractor[(ii)]	£ 250	20%	50				See[(ii)]
	£ 5,250		£ 1,050	£ 6,500	20%	£ 1,300	£ 200
Sub-Contractor							
Installs boiler							
Parchase of boiler	£ 6,500	20%	£ 1,300				
Sub-contracted labour[(iii)]	£ 2,100	20%	£ 420				See[(iii)]
	£ 8,600		£ 1,720	£ 10,000	20%	£ 2,000	£ 280
Main Contractor							
Overall building works							
Payment to sub-contractor	£ 10,000	20%	£ 2,000				
Other sub-con attendances[(iv)]	£ 1,000	20%	£ 200				See[(iv)]
	£ 11,000		£ 2,200	£ 12,000	20%	£ 2,400	£ 200
							£ 1,430
HMRC Receives:							Met by
From traders' transactions above, plus							£ 1,430
(i) Manufacturer's material supplier							£ 300
(ii) Supplier's transport							£ 50
(iii) Sub-contracted labour							£ 450
(iv) Other sub-contractors							£ 200
							£ 2,400

Figure 11.4 illustrates how the ultimate cost of £2,400 is borne by the client and paid by the traders as part of the supply chain process.

CRC energy efficiency scheme

The CRC Energy Efficiency Scheme (originally introduced as the Carbon Reduction Commitment within the Climate Change Act 2008) is a mandatory scheme focused on improving energy efficiency and cutting emissions in large public and private sector organisations. The scheme features a range of reputational, behavioural and financial drivers, which aim to encourage organisations to develop energy management strategies that promote a better understanding of energy usage. Originally intended to be a revenue recycling 'cap and trade' scheme, it effectively became an indirect tax on carbon as part of the Government's spending review in October 2010: The 2012 Budget announced additional consultation with a view to simplifying the system and reducing the administrative burden. In the event that efficiencies cannot be delivered, the Government intends to rationalise the revenues in favour of a direct environmental tax. Whilst there are no specific allowances or reliefs against this tax, the only real way of mitigating the impact is to reduce energy and carbon consumption.

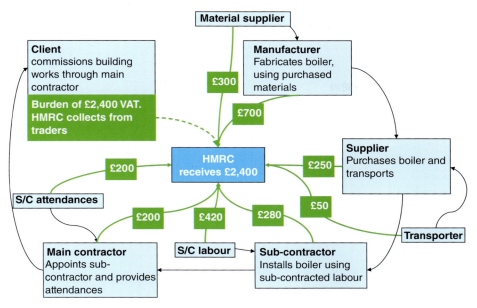

Figure 11.4 How HMRC collects VAT from traders.

Table 11.4 CCL rates from 1 April 2014 (HMRC, 2013).

Supply	Rate
Electricity	0.541 p per kWh
Natural gas supplied by a gas utility	0.188 p per kWh
Petroleum and hydrocarbon gas in a liquid state	1.210 p per kg
Solid fuels, that is coal, lignite, coke and semi-coke	1.476 p per kg

Climate change levy

The climate change levy (CCL) was introduced in 2001 as a tax on businesses based upon their carbon emissions. It applies to non-domestic users and is designed to encourage the use of energy from renewable sources; electricity provided this way is not taxed. The levy applies as shown in Table 11.4.

The tax on these energy sources is included in energy bills paid by the users at the time of supply, with the suppliers effectively collecting the tax (like VAT). Non-business use by charities as well as domestic usage is exempt, although a *de-minimis* limit applies for small quantities of fuel and power.

11.5 Reliefs and allowances

Capital allowances

Capital allowances provide tax relief for qualifying capital expenditure incurred by taxpaying property owners and occupiers. Depreciation of fixed assets is a non-allowable expense and is unable to be deducted from income when calculating

Lanesra Estates Ltd	**Year Ended 31 December 2012**				
Schedule A	£1,53,56,109	(W1)			
Schedule D Case III (Mortgage interest)	(1,23,06,575)				
Chargeable Gain	1,24,561				
Profits Chargeable to Corporation Tax	£31,74,095				
Corporation Tax on Profits	£7,61,783	<< tax saved = £3,07,766 on CAs			

Workings

(W1) Net income - per accounts		£1,57,91,795
Add: Depreciation	7,62,096	
s33B "Repairs"	84,576	8,46,672
Less: Capital Allowances (W2)	(12,82,358)	(12,82,358)
Schedule A		£1,53,56,109

(W2) **Capital Allowances Computation**	AIA/ ECAs	Main Pool 18%	SR Pool 8%	Allowances Given	Notes
Pool brought forward - 1.1.12		£30,35,678	63,54,161		
Additions					
- Alpha House		51,343	2,05,371		
- Beta Court		97,524	3,90,098		
- Gamma Square - refurbishment					Refurbishment
: General pool		48,942			
: ECAs	1,45,000				
: Integral features			3,59,872		
Disposals					
- Delta Street		(1)	(1)		s198 Election
- Epsilon Way		(1,08,860)	(4,35,440)		No s198 - full DV
	£1,45,000	£31,24,626	£68,74,061		
Annual Investment Allowance	25,000				
	£1,70,000	£31,24,626	£68,74,061		
AIA/ECAs - 100%	(1,70,000)			1,70,000	
Main Pool - 18% (Hybrid rate)		(5,62,433)		5,62,433	
Integral features - 8%			(5,49,925)	5,49,925	
Total Allowances Given				£12,82,358	
Pools carried forward - 1.1.13	Nil	£56,86,820	£1,31,98,197		

Figure 11.5 Sample capital allowances computation.

taxable profits. Instead of depreciation, capital allowances are given for expenditure on certain capital assets as a tax allowable expense. As they reduce taxable profits they represent a real and often overlooked cash saving to taxpayers. For capital expenditure incurred on property the primary forms of capital allowances include:

- Plant and machinery assets, including both fixtures and chattels.
- Renovation of a business premises in certain disadvantaged areas.
- Capital expenditure on research and development.
- Enterprise Zone Allowances, in designated zones where there is a focus on high value manufacturing.

Historically, allowances were also available for industrial buildings, agricultural buildings and hotels, but this regime was phased out from 2007, with the scheme finally abolished in 2011. The 100% relief for Flat conversion allowances was withdrawn in 2012.

Figure 11.5 shows how capital allowances are utilised within a corporation tax computation for a company, Lanesra Estates Ltd. At a corporation tax rate of 24%, the allowances generate a cash saving of over £300,000.

Plant and machinery allowances

In practical terms, plant and machinery allowances are by far the most common form of relief for commercial property. They are not available for dwelling houses, ruling out all but common parts of residential blocks. The allowances are available on capital expenditure incurred on the plant and machinery within a building, including those assets fixed to the building, such as heating systems, lifts and fire alarms. Because there is no statutory definition for fixed plant in buildings, it is therefore down to the taxpayer to identify what is 'plant' as this varies by property and the deemed 'business' use. Plant can best be described as equipment which facilitates the trade being conducted and applies to fixed plant, within the building contract. In addition, the method of affixation to the building can affect eligibility. The Capital Allowances Act 2001 (CAA2001) does not define plant and machinery in absolute terms, so reference must be made to current HMRC practice and interpretation of precedent case law. In addition, recent changes to legislation have created additional tests for background plant, together with a new pool for 'integral features'.

Special rate pool

The Finance Act 2008 introduced the concept of dual pooling for plant and machinery fixtures. The special rate pool provides relief for capital expenditure incurred on the provision of:

- Long life assets, that is those with a useful economic life of greater than 25 years. Offices and retail shops are specifically excluded by s93(1) CAA2001.
- Thermal insulation installed to existing buildings (s28 CAA2001).
- Integral features, as defined by s33A CAA2001.

Capital expenditure incurred on assets within these categories is pooled separately and subject to an annual writing down allowance of 8% (since April 2012) per annum on a reducing balance basis. Assets which are considered to be 'integral features' include:

- Electrical systems (including lighting)
- Cold water systems
- Lifts, escalators and moving walkways
- External solar shading
- Space or water heating systems, powered systems of ventilation, air cooling or air purification, and any floor or ceiling comprised in such systems.

Main pool

The main pool captures all other eligible plant and machinery fixtures and assets which do not fall into the special rate pool. Capital expenditure incurred on assets within this pool will be subject to an annual writing down allowance of 18% (since April 2012) on a reducing-balance basis. Assets within this pool would typically include furniture, fittings and equipment, sanitary fittings, ambient and decorative features, computer equipment, carpets and other trade-related assets.

Enhanced capital allowances

The tax system recognises and incentivises investment in certain energy- and water-saving technologies through ECAs. ECAs were introduced in 2001 as part of the UK Government's commitment under the Kyoto Agreement to reduce UK

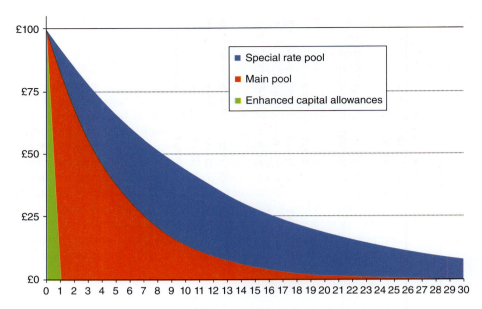

Figure 11.6   Effect of £100 of qualifying expenditure claimed over 30 years.

carbon emissions, initially for energy-saving equipment, with water technologies added in 2003. The regime is regularly reviewed with major changes announced in the Budget and smaller changes amending eligibility criteria issued in the summer. Under the scheme, businesses are able to claim 100% first year allowances on investment in energy and water-saving plant and machinery. Key features include:

- All businesses can claim ECAs, regardless of size, industrial or commercial sector or location.
- Only investments in new and unused plant and machinery can qualify. This also includes unused assets acquired from a developer.
- ECAs permit the full cost of the investment to be relieved for tax purposes against taxable income in the relevant chargeable period.
- Qualifying technologies have to meet defined energy- or water-saving criteria published in a list administered by the DECC (2012), the Carbon Trust (2012) (Table 11.5) and Defra (2012) (Table 11.6).
- The list of technologies and criteria are regularly reviewed and updated, with the lowest-rated products removed from the list.
- The list is predominantly model-specific and is time sensitive too. Some technologies' eligibility is determined by performance specification.
- There are no territorial restrictions on manufacturers wishing to place their products on the list, or the source of products.
- There are presently 16 energy- and 14 water-saving technology categories on the approved lists.

The effective rates of recovery for the main pool, special rate pool and ECAs are shown in Figure 11.6.

The technology category for compact heat exchangers was removed from the Energy Technology List on 8 October 2010.

Table 11.5 Examples of energy-saving technologies (Carbon Trust, 2012).

Technology	Sub-technology	Overview
Air to air energy recovery		This process involves recovering energy from air expelled into the atmosphere, and using it as supply air. This means not as much energy is needed to heat the supply air, so less is used and emissions are reduced Various devices can be used, including plate heat exchangers, thermal wheels, run-around coils, heat-pipe generators and regenerators
AMT equipment	• Component based products • Portable products	AMT equipment can be portable or fixed (component based). Both types monitor how much energy a site is using, identifying any areas where it's being wasted. This helps companies introduce energy-saving measures in the areas where they will provide the most benefit
Boiler equipment	• Automatic boiler blowdown control equipment • Biomass boilers and room heaters • Burners with controls (273) • Combustion trim controls • Condensate pumping equipment • Condensing economisers • Flue gas economisers • Gas-fired condensing water heaters • Heat recovery from condensate and boiler blowdown • Hot water boilers • Localised rapid steam generators • Optimising controllers for wet heating systems • Retrofit burner control systems • Sequence controls • Steam boilers	For many sites, the boiler plant uses more energy than any other equipment. Fortunately, this is one of the areas where it is easiest to make substantial energy savings. The technologies in this category all reduce the amount of energy a boiler uses, and its carbon emissions

(continued)

165

Table 11.5 (continued)

Technology	Sub-technology	Overview
CHP	Performance	CHP is the simultaneous generation of heat and power (usually electricity) in a single process. CHP Schemes are by their nature bespoke and approval of a given CHP manufacturer or product would not provide sufficient assurance of environmental benefit. With CHP, case by case Certification is needed to ensure support is provided for 'good quality' CHP. Certification is achieved using the CHPQA programme. Further information about CHP eligibility criteria and the CHPQA programme can be found at www.chpqa.com
Compact heat exchangers		This technology category was removed from the Energy Technology List on 8 October 2010
Compressed air equipment	• Electronic drain traps • Energy saving controls for desiccant air dryers • Flow controllers • Master controllers • Refrigerated air dryer with energy-saving controls	Compressed air systems are found on many industrial premises and can often be used as an alternative to a direct electricity supply Compressed air is often referred to as the fourth utility after electricity, gas and water. Unlike the other three, it is generated onsite. So users have much more control over costs and the amount they use. Compressed air represents one of the best opportunities for immediate energy savings on any site, and often requires only a modest level of investment
Heat pumps	• Air source: gas engine driven split and multi-split (including variable refrigerant flow) • Air source: packaged • Air source: single-duct and packaged "double-duct" • Air source: split and multi-split (including variable refrigerant flow) • Air to water hHeat Pumps • Ground source: brine to air • Ground source: brine to water • Heat pump dehumidifiers • Water source: packaged • Water source: split and multi-split (including variable refrigerant flow)	Heat pumps transfer heat from external sources (air, water or ground temperature) to spaces that require heating. They also provide cooling by reversing the process In many applications, heat pumps offer a more efficient way to provide heat. They are particularly useful in areas with no natural gas supply or where only low levels of heat are needed, such as underfloor systems

HVAC equipment	• Close control air conditioning • HVAC zone controls	HVAC systems control the temperature, humidity and quality of air in buildings to a set of chosen conditions. To achieve this, the systems need to transfer heat and moisture into and out of the air as well as control the level of air pollutants either by directly removing them or by diluting them to acceptable levels By considering HVAC systems as individual elements rather than as an interacting system, it would be easy to overlook a major area of energy wastage that one component might impact on another. For example, it would be wasteful to increase heating inside a building whilst the cooling system is fighting to reduce temperatures. It is therefore useful to look at how each element of an HVAC system complements the other and fine tune each part to save energy and money
High speed hand air dryers		Hand air dryers are widely used in washrooms to dry hands after washing, as an alternative to paper or linen hand towels. They use an electric blower to produce one or more jets of air that are used to dry hands placed under, or into, the hand air dryer unit. Some models heat the air jets prior to use with electrical heating elements or by passing it over the electric motor that drives the blower
Lighting	• High efficiency lighting unit • Lighting controls • White light emitting diode lighting units	For many buildings, lighting uses up more energy than anything else. Substantial savings can be made in this area by designing and installing the right units and controls Lighting is a major energy consumer in buildings. It is estimated energy consumption for commercial lighting accounted for 47.9 TWh of electricity in 2005. Substantial energy savings are available through use of the most efficient equipment. Substitution of standard units by high efficiency lighting units can typically generate savings of between 20% and 70%.

(continued)

Table 11.5 (continued)

Technology	Sub-technology	Overview
Motors and drives	• Integrated motor drive units • Multiple speed motors • Permanent magnet synchronous motors • Single speed induction motors • Switched reluctance drives • Variable speed drives	Motors and drives often account for a large percentage of a site's energy consumption, particularly on industrial premises. As well as reducing this, energy-saving motors and drives will often lower the ongoing maintenance costs, as the equipment itself usually lasts longer
Pipework insulation		Distribution losses from a heating or cooling system can account for as much as 20% of the total energy used. Insulating the pipework effectively can reduce these losses It is also important to look out for leaks in valves and test points. These are often forgotten when insulating pipework but can account for 5% of the energy used if not properly sealed
Refrigeration equipment	• Absorption cooling and other heat driven cooling and heating equipment • Air-cooled condensing units • Automated permanent refrigerant leak detection systems • Automatic air purgers • Cellar cooling equipment • Commercial service cabinets • Curtains, blinds, doors and covers for refrigerated display cabinets • Evaporative condenser • Forced air pre-coolers • Liquid pressure amplification • Packaged chillers • Refrigerated display cabinets • Refrigeration compressors • Refrigeration system controls	For many sites, particularly retailers and businesses that generate a lot of heat, refrigeration is the highest energy cost. However, it is also one of the areas where it is easiest to make substantial savings Up to 10% of the energy used in a refrigeration plant can be due to leaks in the pipework, which can be resolved by insulating it effectively

Technology	Description
Solar thermal systems	These systems capture the sun's energy and use it to heat water. They differ from photovoltaic systems, which capture solar energy and convert it to electricity The thermal systems include a collector, storage mechanism, pipework (plus valves, insulation, etc.) and controls For larger, industrial applications, the system may not always generate enough to get the water to the required temperature. If this is the case, the water will need some top-up heat to get it to the right temperature
Uninterruptible power supplies	Uninterruptible power supplies are products that are specifically designed to maintain the continuity and quality of a power supply to electrical appliances or electrically driven equipment. When the mains electricity supply is operating, they charge up an energy storage device, which can be used to provide electrical power for a defined period when the mains electricity supply is interrupted. They are widely used throughout industry and commerce to maintain the safety critical and business critical systems located in process control stations, computer rooms, data centres and server areas
Warm air and radiant heaters	• Biomass fired warm air heaters • Packaged warm air heaters • Radiant heating equipment These heaters are more efficient than older boiler based systems, particularly in larger building spaces. If used in the right way, they can completely eliminate traditional hot water systems that use far more energy

AMT, automatic monitoring and targeting; CHP, combined heat and power; CHPQA, combined heat and power quality assurance; HVAC, heating, ventilation and air conditioning.

Table 11.6 Technologies covered by the ECA Water Scheme (Defra, 2012).

Technology	Sub-technology
Cleaning-in-place equipment	Monitoring and control equipment Spray devices
Efficient showers	Aerated showerheads Auto shut off showers Flow regulators Low flow showerheads Thermostatic controlled showers
Efficient taps	Automatic shut off taps Electronic taps Low flow screw-down/lever taps Spray taps
Efficient toilets	Low flush toilets Retrofit WC flushing devices Urinal controls
Efficient washing machines	Efficient commercial washing machines Efficient industrial washing machines
Flow controllers	Control devices Flow limiting devices
Leakage detection equipment	Data loggers Pressure reducing valve controllers Remote meter reading and leak warning devices
Meters and monitoring equipment	Flow meters Water management software
Rainwater harvesting equipment	Monitoring and control equipment Rainwater filtration equipment Rainwater storage vessels Rainwater treatment equipment – removed August 2008; historical purposes only
Small-scale slurry and sludge dewatering equipment	Belt press equipment Centrifuge equipment Filter press equipment
Vehicle-wash water reclaim units	Partial or full reclaim system
Water efficient industrial cleaning equipment	Scrubber/driers (walk-behind machines) Scrubber/driers (ride-on machines) Steam cleaners
Water management equipment for mechanical seals	Seal water recycling units Internal flow regulators Monitoring and control units
Water reuse systems	Efficient membrane filtration equipment Wastewater recovery and reuse Due to their bespoke nature, these technologies are both eligible for enhanced capital allowances via a certification scheme for each individually installed system, rather than the standard product list used for other technologies on the list Water reuse involves reusing suitably treated wastewater from one process for a different purpose. Water reuse technology reduces the demand on drinkable sources of freshwater and reduces the volume of wastewater discharged to sewer. Water reuse can be an economical way to reduce your costs

Enterprise zones

The constant change of the tax system can be illustrated by the reintroduction of first year allowances for Enterprise Zones under Finance Act 2011. A 100% allowance was available for all commercial buildings within the Enterprise Zones created by the Conservative Government in the early 1980s. The regime was included within the Industrial Buildings Allowances rules that were finally abolished in 2011. The reintroduction of Enterprise Zones supports the Government's policy of encouraging investment in certain designated disadvantaged areas. Unlike the previous system, the first year allowance is now only available within certain zones where there is a focus on high-value manufacturing for capital expenditure incurred by trading companies on qualifying plant or machinery. Subject to EU state aid approval, the allowance has an initial 5-year life (from 1 April 2012 to 31 March 2017). The relief is not available for leased assets and the plant or machinery cannot be used outside of a designated zone for a period of 5 years.

Land remediation relief

A key part of the Government's urban regeneration strategy is the redevelopment of brownfield sites. In 2001 a 150% tax relief was introduced for companies incurring expenditure on land remediation. Subsequent Finance Acts have updated the legislation including the 2009 introduction of additional relief for long-term derelict sites. As with other areas of the tax legislation, the relief is highly conditional with many criteria to satisfy and technicalities to overcome before the benefit can be realised. The relief is only available to UK companies in remediating contaminated land. The primary purpose is to facilitate the supply of developable land within the context of the brownfield agenda whilst meeting the UK's housing needs and is one of a number of measures introduced to encourage urban regeneration and remove 'blights' from the built environment. For developers and investors, Land Remediation Relief (LRR) provides major savings on construction expenditure but should not be confused with landfill tax exemption. The cash benefits are arguably significant enough to make this a genuine incentive to invest: at a 23% corporation tax rate, developers, incurring revenue expenditure, will recover 11.5% of their qualifying land remediation expenditure. Capital expenditure incurred by investors and occupiers may be worth up to 34.5% of cost. Loss-making companies can surrender the relief for a payable credit but at a rate of 16%. Because the relief can only be utilised against corporation tax, it is not available for remediation expenditure incurred by individuals, partnerships and non-resident landlords. Anti-avoidance also applies in that the claimant cannot have polluted the site or contributed to contamination during the period of ownership (Table 11.7).

The relief was originally highlighted for abolition by the Office of Tax Simplification in December 2010, however it has subsequently been retained.

Payable credits

In addition to the levels of tax relief identified, the legislation also recognises that companies may not be able to immediately utilise the enhanced cash flow benefit of both ECAs and LRR. Therefore UK corporate taxpayers who have tax losses

Table 11.7   Definitions of land remediation, contaminated and derelict land.

	Purpose or definition
Relevant land remediation	Preventing or minimising, or remedying or mitigating, the effects of any relevant harm, or any pollution of controlled waters, by reason of which the land is in a contaminated state
Contaminated land	Substances on or under the land are causing relevant harm, or there is the significant possibility of relevant harm being caused to: • The health of living organisms • Ecological systems • Quality of controlled waters • Property • Contamination due to former 'industrial activity', since April 2009
Derelict land	Unlike the interpretative definition for contaminated land, qualifying works for derelict land is highly prescriptive and does not work by analogy. Relevant derelict land remediation is specified as the removal of: • Post-tensioned concrete heavyweight construction • Building foundations and machinery bases • Reinforced pile caps • Reinforced concrete basements • Redundant services below ground

can surrender unclaimed relief in exchange for payable credits – effectively a cheque back from HMRC. ECAs are subject to a 19% payable credit, subject to a cap of either £250,000 or the company's total PAYE and National Insurance obligations. For LRR, where the relief creates a loss, then the company may either carry this forward, or claim a tax credit of 16% of the loss surrendered. The decision as to whether to take the tax credit, or carry forward losses for future chargeable periods will depend on the company's particular cash flow requirements and the prospects of moving into a tax paying position in the future, particularly where the corporation tax rate fell to 21% in 2014 and possibly lower in the future. Other options available include group or consortium relief.

Reduced rate of VAT

In many cases, the question of VAT tends only to give rise to cash flow issues; where a company or individual, within the charge to tax, will account for VAT on a quarterly basis; paying (or claiming) the difference between output and input tax. However, where exempt supplies are made, or the trader is below the current VAT registration threshold of £79,000, then the input tax paid to suppliers becomes irrecoverable and a real cost. Expenditure on land and buildings is traditionally problematic with many instances where supplies are deemed to be exempt. Therefore any opportunity to mitigate the impact of VAT paid, by use of reduced or zero rating can deliver significant savings (Table 11.8).

Table 11.8   Examples of reduced or zero rated VAT supplies for property (HMRC, 2013).

Reduced rate	Zero rate
• Altering the number of dwelling houses within a residential property • Residential alterations where the property has been vacant for a period of more than 2 years • Certain energy-saving materials within a residential property • Installation of mobility aids in residential property for the elderly	• The first sale or grant of a major interest in a newly constructed residential or charitable building • Converting a commercial property to relevant residential or charitable purpose • 'Approved alterations' to listed buildings (until 30 September 2015) • Certain goods and services supplied to disabled persons

11.6   Subsidies and incentives

Feed-in tariffs (FIT)

There has been much controversy surrounding the UK FIT regime. The scheme was originally introduced on 1 April 2010, following on from the Energy Act 2008. The tariff provides cash payments for the self-generation of electricity plus bonus payments for any surplus electricity exported back into the national grid for a 20-year fixed period. The tariff is intended to encourage the use of low-carbon electrical generation. Recognising that significant capital expenditure is required in innovative technologies, the FIT provides a guaranteed cash subsidy as an attempt to make the net cost of power at a similar rate to traditional fossil fuel techniques. It is primarily aimed at 'small-scale' installations, that is less than 5 MW with technologies covered including:

- Anaerobic digestion
- Hydro
- Micro CHP
- Solar Photovoltaic (PV)
- Wind.

The Government accelerated plans to reduce FIT rates through a 'fast-track' review in 2011, which has been challenged through the courts (The Guardian, 2012). Nonetheless, the rates will reduce as the market cost aligns with traditional power generation (Figure 11.7).

Renewable heat incentives (RHI)

The RHI is unique in that it is the first scheme that pays for the transmission and secondary use of heat generated from non-fossil fuels. The initiative was first announced in March 2011 and only operates in England, Scotland and Wales. There are two initial phases. Phase 1 introduced Renewable Heat Premium Payments for residential properties plus the RHI for non-domestic installations in the industrial, business and public sectors. The scheme works alongside the Green Deal. Phase 2 extends the RHI to the domestic sector for eligible installations

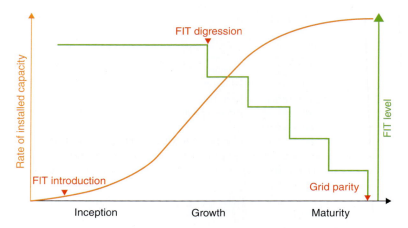

Figure 11.7   Feed-in tariff and market maturity.

installed after 15 July 2009, as well as those with additional technologies and fuels for non-domestic properties. Payments are made quarterly over a 20-year period by Ofgem and the scheme is subject to regular reviews and digression as equipment costs fall and particular pre-set points are triggered (DECC, 2012).

Renewables obligation certificates

The Renewables Obligation (RO) is the Government's primary mechanism for supporting large-scale generation of renewable electricity. Although it was first introduced in 2002, the regime was updated in April 2009 to introduce banding reflecting a more thorough appraisal of the various technologies and scales available within the market. A 4-year review took place with the revised bands running for a further 4 years from April 2013. The RO is expected to be closed to new projects from April 2017. The current supported technologies include:

- Advanced Conversion Technology
- Bioliquids
- Biomass, including conversion
- Enhanced co-firing
- Energy from waste
- Geothermal

- Hydro
- Onshore and offshore wind
- Photovoltaics
- Sewage gas
- Tidal
- Wave.

The RO places obligations on electricity suppliers to source a proportion of electricity from renewable sources. Failure to meet the specified target, which increases annually, triggers a fine. The obligation in England and Wales for 2013/14 is 0.206 ROCs/MWh. Like the RHI, it is administered by Ofgem, who issue Renewables Obligation Certificates (ROCs) to generators of renewable electricity. The number of ROCs granted varies depending upon the nature, scale and location of the energy generation – follwoing the 4-year review, tidal and wave technologies receive between 2 and 5 ROCs/MWh, with hydro landfill gas only receiving 0.1 ROCs/MWh (DECC, 2014).

The green deal

The Green Deal is at the forefront of the Government's plans to tackle energy consumption (HM Government, 2011). Cleverly presented as a means of reducing energy bills, the measure seeks to achieve its end by highlighting the economic benefit of using less carbon-intensive energy. As part of the sectoral plan for energy efficiency, the Green Deal mitigates the impact of initial capital expenditure through subsidy, recovered through the savings obtained on consumers' future energy bills. Launched in October 2012, it is initially targeted at the non-residential sector; the Green Deal is a cash-flow friendly method of encouraging investment in improving energy performance. It removes the upfront costs to the consumer, as the investment, paid for by the energy suppliers, is recovered through energy bills. However, the Green Deal has received a negative press and has not been well-received by the UK property industry. At the time of writing, some of the detail is still to be clarified including how the interaction with Capital Allowances works for the user and energy supplier. Further, small and medium-sized enterprises may have cash flow issues, particularly where they are committed to short-term leases.

11.7   Conclusion

Despite best intentions, the UK tax system is more complicated than at any time in its history. Tax payers and advisers have to contend with statute for direct and indirect taxation that runs to tens of thousands of pages: the legislation, rates and reliefs are reviewed and adjusted annually, resulting in additional risk when trying to model financial appraisals some years into the future. At present the regime taxes only monetary income and gains, or the financial equivalent of gifts and benefits; however the time is approaching where carbon consumption will be taxed in its own right.

This weight of legislation, coupled with strict penalties imposed by HMRC for getting it wrong, creates a difficult environment in which to do business. However, this does not need to be the case. Property owners and occupiers are under increasing statutory pressure to reduce energy consumption within the built environment, yet a number of fiscal incentives exist to make the changes less painful. There does seem to be a real intention from the Government to use the tax and tariff system to reward tax payers who embrace the required changes, resulting in reduced costs, improved cash flow and even income generation. Debate will always exist as to the optimum and relative sizes of stick and carrot, but both are clearly key components of the UK's drive towards low energy.

REFERENCES

BBC (2010) Tax systemto be simplified to encourage investment'. http://www.bbc.co.uk/news/uk-politics-10691779 (accessed April 2012).

Carbon Trust (2012) Energy Technology List. http://etl.decc.gov.uk/etl (accessed April 2012).

DECC (Department of Energy and Climate Change) (2012) http://www.decc.gov.uk (accessed June 2012).

DECC (2013) Summary of evidence on District Heating Networks in the UK, July 2013 (URN 13D/183).

DECC (2014) A table summarising the banding levels for the banding review period (2013–17) in England and Wales. https://www.gov.uk/government/uploads/system/uploads/attachment_data/file/211292/ro_banding_levels_2013_17.pdf.

Defra (2012) Water efficient technologies. http://www.businesslink.gov.uk (accessed April 2012).

HM Government (2011) The Carbon Plan: Delivering our Low Carbon Future. HM Government, December 2011.

HMRC (2012) http://www.hmrc.gov.uk (accessed April 2012).

HMRC (2013) http://www.hmrc.gov.uk (accessed July 2013).

Office of Tax Simplification (2011) Review of Tax Reliefs. Final report, March 2011.

The Guardian (2012) UK government loses solar feed-in tariff bid. http://www.guardian.co.uk/environment/2012/mar/23/uk-government-solar-feed-in-tariff?newsfeed=true (accessed April 2012).

Chapter 12
Environmental Assessment Tools: An Overview of the UK's BREEAM and the US's LEED

Ina Colombo, Benedict Ilozor and Herbert Robinson

12.1 Introduction

The Building Research Establishment Environmental Assessment Method (BREEAM) is to the UK what the United States Green Building Council's (USGBC's) Leadership in Energy and Environmental Design (LEED) is to the US. Design economics is intrinsically linked to the environmental aspects of buildings and often involves a complex relationship establishing a trade-off between the initial capital cost, life cycle cost and environmental cost. BREEAM and LEED provide the means to better appreciate these relationships. Case studies show that to achieve a BREEAM environmental rating of 'very good', that is to reduce environmental cost or increase environmental scores, an uplift of 0.2% on capital cost is required. To achieve 'excellent' rating, 0.7% increase in capital cost is necessary. Although no such relationships have been specifically drawn between capital costs and the LEED rating, there is likely to be a close comparison in capital cost uplifts for LEED ratings of Certified, Silver, Gold and Platinum. The initial capital cost comprises the cost of design, construction, supply and installation of building components. The life cycle costs relate to the daily, weekly and annual maintenance costs for cleaning, repair, redecoration, replacement and energy consumption. The environmental costs include the costs associated with resource utilisation and environmental pollution, extraction, manufacturing, transportation of materials, use and disposal of buildings, as well as carbon emission associated with the use of materials, ground water, indoor environmental quality and energy. Sustainable buildings can provide additional benefits for building owners and users in terms of savings relating to the use of energy and water, wastewater, health and productivity, operations and maintenance of the facilities, as well as the reduction of carbon costs associated with greenhouse gases.

This chapter focuses on the key features and operationalisation of two of the most widely used assessment tools which are the UK's BREEAM and the US's

Design Economics for the Built Environment: Impact of Sustainability on Project Evaluation, First Edition.
Edited by Herbert Robinson, Barry Symonds, Barry Gilbertson and Benedict Ilozor.
© 2015 John Wiley & Sons, Ltd. Published 2015 by John Wiley & Sons, Ltd.

LEED. Examples of other tools abound such as life cycle assessment (LCA), Green Globe, Environmentally Sustainable Design (ESD), material flow accounting (MFA), material input/output service unit (MIPS), cumulative energy requirements analysis (CERA), Building for Environmental and Economic Sustainability (BEES) and Ecohomes (now replaced by the Code for Sustainable Homes) and environmental input/output analysis (IOA), environmental risk (ERA), and check lists for eco-design. Little effort has been made before now to view these two most widely used tools side by side. It is envisaged that doing so as this current work would assist potential users in both hemispheres, and others in other countries without a similar assessment system but desirous to assess the environmental performance of their buildings. In addition, it can facilitate interchangeable use of the tools whereby some in the US may prefer and go for BREEAM assessment and certification, and those in the UK preferring LEED certification, can opt for it. The chapter starts with a brief context, the rationale for design to reduce carbon emission, followed by an outline and a discussion of the key features of environmental assessment tools; the main structure of BREEAM and LEED tools are explained; the key assessment stages, as well as the mitigation principles, scoring and rating systems are discussed.

12.2   Context and the need to design to reduce carbon emission

It is increasingly recognised that good design should focus on reducing carbon emission. In response to the challenge of reducing emissions of carbon and greenhouse gases to the atmosphere, developed countries signed the Kyoto Protocol in December 1997, which proposed to reduce six greenhouse gases (UNEP, 1997; UNFCCC, 1997). The European Union member states adopted a collective target to reduce EU emissions, and each member state has a legally binding target, with the UK undertaking to reduce its emissions significantly by some 60% by 2050 (Defra, 2006).

Today, a significant proportion of carbon emission (almost half of the UK's carbon emissions) come from the use of buildings whether domestic or commercial buildings. The Worldwatch Institute (an independent American research organisation that focuses on innovative solutions to global environmental, resource, and economic issues) estimates that buildings consume at least 40% of the energy utilised in the world each year, generate one-third of the CO_2, and two-fifths of acid-rain-causing compounds. In the US, buildings annually consume more than 30% of the total energy and more than 60% of the electricity used (USGBC, 2010). Commercial buildings accounted for more than one billion metric tons of carbon dioxide produced in 2006, and this represents an increase of over 30% over the 1990 levels (Energy Information Administration, 2006). Five billion gallons of potable water are used daily to flush toilets (USGBC, 2010). Americans generate 1.6 million tons of household hazardous waste per year, and an average home can accumulate as much as 100 lb of household hazardous waste, which when improperly disposed of, creates a potential risk to people and the environment (Bonneville County, 2011). A major part of the energy used in developed countries is for the purposes of heating or cooling (DTI, 2007). In the UK both the retail and housing sectors are currently under significant pressure to reduce their energy requirements and environmental impacts. Indeed, recent building regulations in the UK stipulate that both will need to be 'zero carbon' by 2019 (DCLG, 2007). In recent decades, the development of environmental or green (sustainable) buildings rating systems has become a significant initiative to address global warming and its associated negative environmental impact due to the carbon emission. Various environmental assessment tools have been developed with differences in their structure, technical details and evaluation methodology but they all have common objectives of targeting the environmental aspects of design to determine a building's carbon footprint. They mitigate the effects of buildings on the environment and provide a measure indicating the extent to which environmental issues are addressed.

12.3   Key features of environmental assessment tools

Environmental assessment tools should have a number of characteristics. First, it should have a scoring system, designed to evaluate the performance of buildings, based on selected criteria reflecting environmental issues relating to the design, construction and operation of a building. The rating system should use measurable indicators to demonstrate the extent of sustainable design incorporated into the building. For example, in the BREEAM tool, buildings are evaluated with performance criteria and scoring system set by the BRE with 'credits' awarded based on the level of performance and an overall environmental performance rating which is pass, good, very good, excellent or outstanding. LEED follows the route with prerequisites and credits, and performance rating levels categorised as Certified, Silver, Gold and Platinum. Secondly, there is the issue of applicability: Can the rating system be applied to all type of buildings (e.g. commercial, residential, offices, hospitals, etc.)? Thirdly, there is the issue of availability: Is the rating system easily adaptable to other markets? For instance, can a UK residential house be certified by means of a US LEED rating system or vice versa? BREEAM can be used to assess buildings anywhere in the world. For example, the new building of the European Investment Bank in Luxembourg and the Van der Kamp bakery building in Los Angeles City College, CA, USA have BREEAM ratings. Clients in India, Dubai, Qatar, Spain, and many other countries have also expressed interest in using the scheme. The LEED format has been adapted in other countries such as Green Building Council Italia (GBC Italia), and Canada Green Building Council (CaGBC), and LEED certification has been used for projects in Mexico, China, India, Brazil, Arab Emirates, and so on. Making LEED adaptable internationally is one of the core objectives of LEED Version 4.0 released in late 2013. Fourthly, there is the issue of the robustness and maturity of the methodology. Is the methodology based on sound concepts, technical standards and legislation, and does it reflect changes in national and international agreements on environmental or sustainability standards? The concepts underpinning the methodology should be based on an acceptable and tested idea of how to achieve sustainability such as life-cycle thinking, John Elkington's triple bottom line (profit, people and planet), design for better environment, and cleaner technology. It could also be based on the concept of green architecture, embodied energy, embodied water, embodied carbon, or ecological design principles which provide specific ways of minimising energy and materials use, reducing pollution, preserving habitat and fostering community. The fifth aspect is its usability, which is a key to success: Is it practical and easy for the user to apply and understand? Dilemmas often occur in putting any new idea into practice, so it must be simplified to increase its usability. The sixth aspect is the communicability relating to the reporting style of certification at the end of the evaluation process and its recognition and acceptance. Some tools are usually supported by technical elements, such as models, software and simulation tools to facilitate the decision process, and communication based on the implementation data. The final aspect is the cost. This criterion is very important to the user (building developer, owner and occupant) and consists of all the costs arising during the building's certification process.

12.4   The BREEAM tool

BREEAM is UK's leading environmental building assessment tool. It sets the standards for best practices in sustainable building design, construction and operation, and has become one of the most comprehensive and widely recognised measures of a building's environmental performance. A BREEAM assessment uses recognised measures of performance, which are set against established benchmarks, to evaluate a building's specification, design, construction and use. The measures used represent a broad range of categories and criteria from energy to ecology. They include aspects related to energy and water use, health and well-being, pollution, transport, materials, waste, ecology and management processes. BREEAM addresses wide-ranging environmental and sustainability issues and help developers, designers and building managers to demonstrate the environmental worthiness and attributes of their buildings to clients, planners, funders and other initial parties. According to BRE (BRE, 2011), BREEAM:

- uses a straightforward scoring system that is transparent, flexible, easy to understand and supported by evidence-based science and research;
- has a positive influence on the design, construction and management of buildings;
- defines and maintains a robust technical standard with rigorous quality assurance and certification.

Just as LEED in the US, BREEAM is currently not mandatory in the UK. It is a voluntary commitment from clients and developers to do right by the environment. However, many local planning authorities now require a BREEAM assessment and certification.

Evolution of BREEAM

Since its creation in 1990, several versions of BREEAM have been introduced and updated regularly to align it with the UK building regulations. In August 2008, a significant review of BREEAM was carried out to cover different types of building such as retail offices, education, prisons, courts, healthcare, industrial, including the creation of the BREEAM Bespoke method and multi-residential for specialised buildings. The main differences between previous versions and BREEAM 2008 is the introduction of mandatory credits (similar to LEED's prerequisites), a new rating level of BREEAM – Outstanding, and a new two-stage assessment processes of the design stage (DS) and the post-construction stage (PCS). BREEAM 2008 was updated to BREAM 2010 with minor changes but the 2011 version had a major adjustment particularly relating to the carbon emissions of buildings. The BREEAM 2011 New Construction (NC) scheme was updated to simplify the assessment process which is now just one scheme that covers 49 common sustainability issues across nine categories through the project's design and construction phases. BREEAM NC covers the following type of non-domestic buildings: commercial buildings (offices, industrial and retail), public buildings (education, healthcare, prisons and law courts), multi-residential accommodation buildings and others buildings (residential institutions, non-residential institutions, assembly and leisure).

Assessment stages and mitigation principles of BREEAM NC

The BREEAM NC aims to mitigate the life cycle impacts of new buildings on the environment in a vigorous and cost effective manner. A certificated BREEAM assessor trained under the UK Accreditation Service accredited competent person scheme is responsible to evaluate the performance of buildings against the best practice of environmental issues and to provide a strategy to achieve the assessment measures. The BREEAM assessor is generally engaged at the pre-assessment stage of the design and procurement stages in line with the clients, design team, principal contractor and other stakeholders to ensure that realistic targets are assigned and can be met without adverse effect on the design and budgets. The assessor is in charge of evaluating the information provided by the stakeholders against the BREEAM guidelines and advises if further actions should be undertaken on the environmental issues relating to the design and the award of credits. After completion of the evaluation, the assessment is forwarded to the National Scheme Operator (NSO) to provide formal verification. If the amount of credits provided is not sufficient, the BREEAM assessor has to respond to the NSO enquiries within a timescale to justify the credit assessments. In some cases, credits may not be awarded if the evidence provided does not comply with the BREEAM guidelines and criteria. After the final approval from the NSO, a certified BREEAM rating, that is the label, is issued that provides quality assurance of the building performance based on the BREEAM standard.

The engagement process for the BREEAM pre-assessment stage should not be later than stage B of the RIBA Outline Plan of Works and ideally sooner as shown in Figure 12.1.

As shown in Figure 12.1, there are two types of BREEAM certifications: the interim design stage (DS) and the post-construction stage (PCS). These assessments can be done individually or together depending on the client's requirement. The PCS confirms the performance of building as built and can be reviewed at the interim DS assessment. In cases where the DS assessment has not been done previously, a full construction stage assessment will be required. In both cases, the PCS will be certified. For each environment issues, there is a different assessment criterion for the DS and PCS with different range of ratings.

BREEAM scoring and rating procedures

To evaluate the sustainability performance of the assessed buildings, a BREEAM benchmark rating is applied in terms of a percentage score. Figure 12.2 summaries the principles of the rating and scoring of BREEAM and shows the 9 environmental

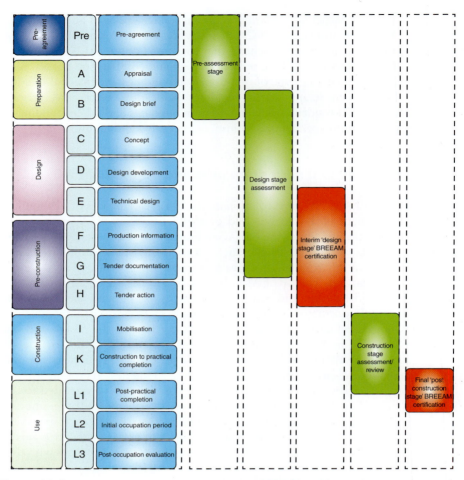

Figure 12.1   RIBA Outline Plan of Works vs the BREEAM assessment stages (BRE, 2011).

categories with their related 49 environmental issues that need to be addressed. The column "Credit Available" indicates the achievable maximum credit of each environmental issue. The BREEAM NC guideline explains in detail the requirement to achieve the credits. The "Credit Category" varies slightly according to the type of building assessed. Based on the total credits achieved, an overall percentage rating is obtained which corresponds to a categorisation that varies from 'Unclassified' to 'Outstanding' ratings. Below 30% rating, the building is considered 'Unclassified' meaning that the assessed building is not compliant with BREEAM by failing to meet minimum standards of performance set by BREEAM. In order to ensure that the scoring is consistent among categories, there is a mandatory minimum standard level of 14 environmental issues and associated credits to be achieved based on the ratings targeted as shown in Figure 12.2. The letters O, E, VG, G and P indicated in the BREEAM ratings table show the credits to be awarded to achieve the mandatory minimum standard level. For instance, if an assessed building has a target rating of "Excellent", it will be required to achieve

BREEAM rating

	BREEAM rating	%
O	Outstanding	85
E	Excellent	70
VG	Very good	55
G	Good	45
P	Pass	30
U	Unclassified	<30

Environmental issues

Categories	Code	Issue	Credits available	Credits category	Credits %	Minimum standard level	Rating
Management	Man 01	Sustainable procurement	8	22	12%	1 or 2* credit(s)	O*, E, VG, G, P
	Man 02	Responsible construction practices	2			1 or 2* credit(s)	O, E
	Man 03	Construction site impacts	5				O, E
	Man 04	Stakeholder participation	4			1 credit (building user info)	O, E, VG, G, P
	Man 05	Service life planning and costing	3				
Health and wellbeing	Hea 01	Visual comfort	2	10	15%	Criterion 1 only	O, E, VG, G, P
	Hea 02	Indoor air quality	2				
	Hea 03	Thermal comfort	2			Criterion 1 only	
	Hea 04	Water quality	1				O, E
	Hea 05	Acoustic performance	1				O, E
	Hea 06	Safety and security	2				O, E
Energy	Ene 01	Reduction of CO_2 emissions	15	30	19%	10 or 6 credits	O*, E, VG, G
	Ene 02	Energy monitoring	1			1 credit (first sub-metering)	O, E, VG, G
	Ene 03	Energy efficient external lighting	1				
	Ene 04	Low or zero carbon technologies	5			1 credit	O, E
	Ene 05	Energy efficient cold storage	2				
	Ene 06	Energy efficient transportation systems	2				
	Ene 07	Energy efficient laboratory systems	1				
	Ene 08	Energy efficient equipment (process)	2				
	Ene 09	Drying space	1				
Transport	Tra 01	Public transport accessibility	5	9	8%		
	Tra 02	Proximity to amenities	1				
	Tra 03	Cyclist amenities	2				
	Tra 04	Maximum car parking capacity	2				
	Tra 05	Travel plan	1				
Water	Wat 01	Water consumption	5	9	6%	1 or 2* credit(s)	O*, E, VG, G
	Wat 02	Water monitoring	1			Criterion 1 only	O, E, VG, G
	Wat 03	Water leak detection and prevention	2				
	Wat 04	Water efficient equipment (process)	2				
Materials	Mat 01	Life cycle impacts	5	12	13%		
	Mat 02	Hard landscaping and boundary protection	1				
	Mat 03	Responsible sourcing of materials	3			Criterion 3 only	O, E, VG, G, P
	Mat 04	Insulation	2				
	Mat 05	Designing for robustness	1				
Waste	Wst 01	Construction waste management	4	7	8%	1 credit	O
	Wst 02	Recycled aggregate	1				
	Wst 03	Operational waste	1			1 credit	O, E
	Wst 04	Drying space	1				
	Wst 05	Speculative floor and ceiling finishes	1				
Land use and ecology	LE 01	Site selection	2	10	10%		
	LE 02	Ecological value of site/protection of ecological features	1			1 credit	
	LE 03	Mitigating ecological impact	2				
	LE 04	Enhancing site ecology	3				
	LE 05	Long term impact on biodiversity	3				
Pollution	Pol 01	Impact of refrigerants	3	13	10%		
	Pol 02	NOx emissions from heating/cooling source	3				
	Pol 03	Surface water run-off	5				O, E, VG
	Pol 04	Reduction of night time light pollution	1				
	Pol 05	Noise attenuation	1				
Innovation	Inn 01	Innovation (additional)	10	2	100%		

Figure 12.2 BREEAM rating road map.

all 14 minimum standard credits except the "construction waste management" credit. The credits with asterisk (*) indicate that these credits are essential for achieving the 'Excellent' rating. Also, there is a minimum standard credits for the "reduction of CO_2 emission" issues which is 6 credits for "Excellent" rating, and 10 credits for "Outstanding" rating. There are 10 credits for the Innovation additional category. These credits can be awarded for new technologies, designs or construction projects that meet the BREEAM eligibility criteria for Innovation credits.

12.5   The LEED tool

LEED is the US premier environmental building assessment tool. LEED recognises the negative impacts of buildings on the environment, and posits that responsible green building practices can eliminate or reduce these impacts through various high-performance design, construction and operation approaches. The added benefits of green practices to all stakeholders are reduced operating costs, elevated building marketability, increased occupant productivity and reduced potential for liability associated with sick-building syndrome and various building-related illnesses.

Based on extensive studies of existing green metrics and rating systems such as that by the American Society of Heating, Refrigeration, and Air-Conditioning Engineers (ASHRAE), LEED developed a rating system for buildings. As LEED matured, it undertook further initiatives expanding into rating systems that go beyond LEED for New Construction (Building Design + Construction) (LEED NC BD + C) to include LEED for Existing Buildings: Operations & Maintenance (LEED EB O + M), LEED for Commercial Interiors (Interior Design + Construction) (LEED CI ID + C), LEED for Core & Shell (LEED CS), LEED for Schools, LEED for Retail, LEED for Healthcare, LEED for Homes, LEED for Neighborhood Development (LEED ND), and SITES – LEED for Landscape proposed by the American Society of Landscape Architects (ASLA).
LEED-certified buildings are designed to:

- Lower operating costs and increase asset value
- Reduce waste sent to landfills
- Conserve energy and water
- Be healthier and safer for occupants compared with non-LEED-certified counterparts
- Reduce their harmful greenhouse gas emissions
- Qualify for owners tax rebates and incentives, zoning allowances in some municipalities
- Demonstrate owner's commitment to environmental stewardship and social responsibility.

LEED assesses five environmental categories and two bonus categories. The environmental categories are:

1. Sustainable sites
2. Water efficiency
3. Energy and atmosphere
4. Materials and resources
5. Indoor environmental quality.

The two bonus assessment categories are:

1. Innovation in design
2. Regional priority.

The sustainable sites category encourages and awards credits to building project stakeholders seeking LEED certification for:

- Selecting and developing their sites wisely
- Reducing emissions associated with transportation through their development
- Planting sustainable landscapes
- Protecting surrounding habitats
- Managing storm-water runoff
- Reducing the heat island effect
- Eliminating light pollution.

The water efficiency category encourages and awards credits to building project stakeholders seeking LEED certification for:

- Monitoring their water consumption performance
- Reducing indoor potable water consumption
- Reducing water consumption in a way that it saves energy and improves environmental well-being
- Practicing water-efficient landscaping.

The energy and atmosphere category encourages and awards credits to building project stakeholders seeking LEED certification for:

- Tracking building energy performance – designing, commissioning and monitoring their developments
- Managing refrigerants used in their developments to eliminate chlorofluoro-carbons (CFCs)
- Using renewable energy.

The materials and resources category encourages and awards credits to building project stakeholders seeking LEED certification for:

- Selecting sustainable materials
- Practicing waste reduction
- Reducing waste at its source
- Reusing and recycling
- Using local materials
- Using rapidly renewable materials
- Using third-party certified wood.

The indoor environmental quality category encourages and awards credits to building project stakeholders seeking LEED certification for:

- Improving indoor ventilation
- Managing air contaminants
- Specifying less harmful materials
- Allowing occupants to control desired settings
- Providing daylight and views.

The innovation in design (exemplary performance) category encourages and awards credits to building project stakeholders seeking LEED certification for building performance that greatly exceeds what is required in an existing LEED credit(s).

The regional priority category encourages and awards credits or gives exceptions to building project stakeholders seeking LEED certification depending on a consideration of regionally important issues, such as water conservation in the arid and desert regions. The category acknowledges the importance of local conditions in determining best environmental design, construction, and operation practices. LEED is not mandatory in the US. However, some owners in the public and private sectors request LEED certifications of various levels for building projects financed by them.

Evolution of LEED

Created in 1993, the USGBC soon found the need to develop metrics to measure performance of green buildings far more extensive than what were available then. To focus squarely on attaining this objective, it came with LEED to take charge of formulating the assessments. The initial efforts led to the launching of LEED Version 1.0 in August 1998. Following various updates, revisions and errata, LEED Version 2.0 was released in March 2000. Version 2.1 followed in 2002 and Version 2.2 came in 2005. Version 3.0 was launched in April 2009. There have been successive errata and addenda, with major revisions appearing in Version 4.0 released in November 2013.

The initial four levels of LEED certification were: Certified (26–32 credit points), Silver (33–38 credit points), Gold (39–51 credit points) and Platinum (52–69 credit points). However, with the revisions came increased credit points requirement but the four levels of certification remain. Currently they are: Certified 40–49 credit points, Silver 50–59 credit points, Gold 60–79 credit points and Platinum 80 or more credit points. Certified is the lowest level of certification, while Platinum is the highest.

The reason for the updates and revisions is an acknowledgement that LEED is continually evolving, and will continue to improve. While the new version is not going for net zero, it is seeks to be "net positive" to move in the right direction in terms of entire market transformation. It opens up LEED to a wider range of buildings, including manufacturing industries, to deliver green building practices throughout the building supply chain.

Version 4.0 changes include extension to new market sectors such as hospitality facilities like hotels, existing schools and existing retail, warehouses and distribution facilities, data centres, and LEED for mid-rise homes; increased technical rigor involving revisions to credit weights, new credit categories focusing on integrated design, life cycle analysis of materials, and an increased emphasis on measurement and performance; and streamlined services including an improved LEED user experience that makes the LEED Online platform more intuitive and introduces tools to make the LEED documentation process more efficient (USGBC, 2013).

The credit categories will expand from 7 to 9, and there are more prerequisites – increasing from 9 to 15. The 9 credit categories are: Integrated Process, Location and Transportation, Sustainable Sites, Water Efficiency, Energy and Atmosphere, Materials and Resources, Indoor Environmental Quality, Performance Innovation, and Regional Priority. Besides the new categories, the intent is to also better align the credits among the existing categories, while borrowing existing credits for the new categories. Some of the existing credits will also be combined.

LEED assessment

The Green Building Certification Institute (GBCI) is a separately incorporated entity established in 2008 to administer credentialing and certification programmes related to green building practice. Assessments operate through rating systems supported by reference guides. Each rating system is organised into the five environmental categories, plus the two bonus categories previous outlined.

All prerequisites must be met before any project can seek credits. The reference guide outlines for all prerequisites and credits the intent, requirements, benefits and issues to consider, related credits, the referenced standards, implementation, timeline and team, calculations if any, the documentation guidance, examples, exemplary performance, regional variations, and resources.

For project certification, those who consider LEED right for their developments must prepare an application and register their project with GBCI online otherwise called LEED Online. A registration fee of US$450 applies for member organisations of USGBC, while US$600 has to be paid by non-members. If in doubt whether a credit can be obtained or not, they can consult Credit Interpretation Rulings (CIRs) or go for Credit Interpretation Requests. Credit Interpretation Request applies if a new issue arises, while Credit Interpretation Ruling applies to previous decisions and rulings. LEED for Core & Shell requires precertification application, and all (LEED for New Construction, Core & Shell, Schools, etc.) must complete certification applications. The certification applications can be reviewed in one or two stages; namely, Design Phase Review and Construction Phase Review. Certain features are amenable for Design Phase Review or Construction Phase Review or both. One can seek review of its design-related prerequisites and credits before completion and then apply for construction-related credits after the project is finished. Alternatively, one can wait until the project is complete to submit documentation for all the credits being pursued. The credits are warded after this review.

The certification fee charged is based on the project size, and the rating system under which the project will be considered for certification. The fee is paid when the documentation is submitted for review through LEED Online. Table 12.1 and Table 12.2 show the current certification review rates for all projects applying for certification after 1 January 2010. The fees apply to single-building projects only; special rates may apply to multiple-building projects, and the determination rests with GBCI who will provide a quote prior to submitting an application for certification.

The score for each LEED credit is estimated based on the carbon footprint for a typical LEED building. A building's carbon footprint is the total greenhouse gas emissions associated with its construction and operation, with its construction and operation (USGBC, 2010):

- energy used by building systems;
- building-related transportation;
- embodied emissions of water (electricity used to extract, convey, treat, and deliver water);

Table 12.1   Project certification rates (in US$) effective from 1 January 2010.

	Less than 50,000 square feet	50,000– 500,000 square feet	More than 500,000 square feet	Appeals (if applicable)
LEED 2009; New Construction, Commercial Interiors, Schools, Core & Shell full certification	Fixed rate	Based on square footage	Fixed rate	Per credit
Design Review				
USGBC members	2,000	0.04/ft²	20,000	500
Non-members	2,250	0.045/ft²	22,500	500
Expedited fee	5,000 regardless of square footage			500
Construction Review				
USGBC members	500	0.010/ft²	5,000	500
Non-members	750	0.015/ft²	7,500	500
Expedited fee	5,000 regardless of square footage			500
Combined Design and Construction Review				
USGBC members	2,250	0.045/ft²	22,500	500
Non-members	2,750	0.055/ft²	27,500	500
Expedited fee	10,000 regardless of square footage			500

Table 12.2   Project certification rates (in US$) effective from 1 January 2010.

Recertification Review				
USGBC members	750	0.015/ft²	7,500	500
Non-members	1,000	0.02/ft²	10,000	500
Expedited fee	10,000 regardless of square footage			500
LEED for Core & Shell: Precertification	Fixed rate			Per credit
USGBC members	3,250			500
Non-members	4,250			500
Expedited fee	5,000			500
CIRs (for all Rating Systems)				220

- embodied emissions of solid waste (life-cycle emissions associated with solid waste);
- embodied emissions of materials (emissions associated with the manufacture and transport of materials).

LEED 2009 for new construction and major renovations
Project checklist

Sustainable sites	Possible points:	26
Y ? N		
Prereq 1 Construction activity pollution prevention		
Credit 1 Site selection		1
Credit 2 Development density and community connectivity		5
Credit 3 Brownfield redevelopment		1
Credit 4.1 Alternative transportation—public transportation access		6
Credit 4.2 Alternative transportation—bicycle storage and changing rooms		1
Credit 4.3 Alternative transportation—low-emitting and fuel-efficient vehicles		3
Credit 4.4 Alternative transportation—parking capacity		2
Credit 5.1 Site development—protect or restore habitat		1
Credit 5.2 Site development—maximize open space		1
Credit 6.1 Stormwater design—quantity control		1
Credit 6.2 Stormwater design—quality control		1
Credit 7.1 Heat island effect—non-roof		1
Credit 7.2 Heat island effect—roof		1
Credit 8 Light pollution reduction		1

Water efficiency	Possible points:	10
Prereq 1 Water use reduction—20% reduction		
Credit 1 Water efficient landscaping		2 to 4
Credit 2 Innovative wastewater technologies		2
Credit 3 Water use reduction		2 to 4

Energy and atmosphere	Possible Points:	35
Prereq 1 Fundamental commissioning of building energy systems		
Prereq 2 Minimum energy performance		
Prereq 3 Fundamental refrigerant management		
Credit 1 Optimize energy performance		1 to 19
Credit 2 On-site renewable energy		1 to 7
Credit 3 Enhanced commissioning		2
Credit 4 Enhanced refrigerant management		2
Credit 5 Measurement and verification		3
Credit 6 Green power		2

Materials and resources	Possible points:	14
Prereq 1 Storage and collection of recyclables		
Credit 1.1 Building reuse—maintain existing walls, floors, and roof		1 to 3
Credit 1.2 Building reuse—maintain 50% of interior non-structural elements		1
Credit 2 Construction waste management		1 to 2
Credit 3 Materials reuse		1 to 2

Materials and resources, continued	Possible points:	15
Y ? N		
Credit 4 Recycled content		1 to 2
Credit 5 Regional materials		1 to 2
Credit 6 Rapidly renewable materials		1
Credit 7 Certified wood		1

Indoor environmental quality	Possible points:	15
Prereq 1 Minimum indoor air quality performance		
Prereq 2 Environmental tobacco smoke (ETS) control		
Credit 1 Outdoor air delivery monitoring		1
Credit 2 Increased ventilation		1
Credit 3.1 Construction iaq management plan—during construction		1
Credit 3.2 Construction iaq management plan—before occupancy		1
Credit 4.1 Low-emitting materials—adhesives and sealants		1
Credit 4.2 Low-emitting materials—paints and coatings		1
Credit 4.3 Low-emitting materials—flooring systems		1
Credit 4.4 Low-emitting materials—composite wood and agrifiber products		1
Credit 5 Indoor chemical and pollutant source control		1
Credit 6.1 Controllability of systems—lighting		1
Credit 6.2 Controllability of systems—thermal comfort		1
Credit 7.1 Thermal comfort—design		1
Credit 7.2 Thermal comfort—verification		1
Credit 8.1 Daylight and views—daylight		1
Credit 8.2 Daylight and views—views		1

Innovation and design process	Possible points:	6
Credit 1.1 Innovation in design: specific title		1
Credit 1.2 Innovation in design: specific title		1
Credit 1.3 Innovation in design: specific title		1
Credit 1.4 Innovation in design: specific title		1
Credit 1.5 Innovation in design: specific title		1
Credit 2 LEED accredited professional		1

Regional priority credits	Possible points:	4
Credit 1.1 Regional priority: specific credit		1
Credit 1.2 Regional priority: specific credit		1
Credit 1.3 Regional priority: specific credit		1
Credit 1.4 Regional priority: specific credit		1

Total	Possible points:	110
Certified 40 to 49 points	Silver 50 to 59 points	Gold 60 to 79 points	Platinum 80 to 110

Figure 12.3 LEED new construction and major renovation scorecard (USGBC, 2013).

LEED project scoring and rating procedures

LEED project scoring and rating procedures are provided here. This LEED rating is scored on 100 points, plus 10 points of bonus for in the Innovation in Design and Regional Priority categories. The points distribute across the categories thus (Figure 12.3):

1. Sustainable Sites – 26 points
2. Water Efficiency – 10 points
3. Energy and Atmosphere – 35 points
4. Materials and Resources – 14 points
5. Indoor Environmental Quality – 15 points
6. Innovation in Design – 6 points maximum
7. Regional Priority – 4 points maximum.

12.6   Concluding remarks

Considering that LEED and BREEAM compare closely with each other as can be seen from this chapter, it is possible to use them interchangeably. Some in the US and elsewhere complain of how onerous and costly the LEED certification process is, while those in the UK criticise BREEAM for being too broad; these complainants may find use in substituting one for the other, depending on their preferences. Both are good tools but work remains to fine-tune them to better serve people and projects, while pursuing the highest responsible environmental attainments. Perhaps, attempts may be made in the future to marry both environmental assessment systems, rather than have two closely competing systems coexist for the same customers.

References

Bonneville County (2011) Household Hazardous Waste Program. http://www.co.bonneville.id.us/index.php/public-works/solid-waste (accessed 17 January 2013).

BRE (Building Research Establishment) (2011) *BREEAM New Construction, Technical Manual SD5073-V2.00*, BRE Global Ltd, London.

DCLG (Department for Communities and Local Government) (2007) *Building A Greener Future: Towards Zero Carbon Development*, London.

Defra (Department Environment Food and Rural Affairs) (2006) UK Emissions Inventory and Projections. Climate Change, the UK Program 2006. The Stationery Office, Norwich.

DTI (Department of Trade and Industry) (2007) Energy White Paper 2007: Meeting the Energy Challenge. The Stationery Office, London.

UNEP (United Nations Environment Programme) (1997) Industrialized countries to cut greenhouse gas emissions by 5.2%. http://unfccc.int/cop3/fccc/info/indust.htm (accessed 8 June 2007).

UNFCCC (United Nations Framework Convention on Climate Change) (1997) *Kyoto Protocol*. Kyoto, Japan.

USGBC (United States Building Council) (2010) LEED Reference Guide for Green Building Design and Construction. The United States Green Building Council, Washington, DC.

USGBC (2013) The next version of LEED. http://new.usgbc.org/leed/developing-leed/future-versions (accessed 25 January 2013).

Chapter 13
Space Planning and Organisational Performance
Benedict Ilozor

13.1 Introduction

There is a renewed debate on whether work environments engender perceptible organisational change that translates to improved performance. This contemplation has arisen as a result of the substantially changing work settings, in which the dominant trend is towards a more open-plan and democratic style (Duffy and Tannis, 1993; Knight Frank Hooker, 1995; Duffy, 1997; IFMA, 1997; Gillen, 2006; Elsbach and Pratt, 2007; Davis *et al.*, 2011). Although the current discourse seeks to uncover how the built environment promotes or retards organisational change, whether or not significant change arises at all is yet to be definitively established. Hence, a contribution to the school of thought in this direction is considered invaluable.

The role of the physical properties of work setting and environment, as well as management processes over time, in bringing about improved organisational performance in terms of management effectiveness and increased productivity, has been noted by several authors (Williams *et al.*, 1985; Leaman and Bordass, 1993; Uzee, 1999; Steiner, 2005). Office space is a tool that can be leveraged to improve business results and help achieve corporations' objectives, as Mohr (1996) reported. Nevertheless, "organisational ecology" (Steele, 1986), or pattern of relationships between work/workers and the characteristics of work settings, is not well understood, as Brown (1996) claimed.

A study in Sydney investigated organisational performance relative to the innovativeness of over 100 hundred open-plan work settings as change drivers (Ilozor, 1998, 1999a, 1999b, 2000). The aim of the investigation was to determine whether organisational performance, and hence change, are indeed brought about by innovative work settings. It is envisaged that the results of this study would contribute to mapping the direction and the focus of the on-going search for answers

Design Economics for the Built Environment: Impact of Sustainability on Project Evaluation, First Edition.
Edited by Herbert Robinson, Barry Symonds, Barry Gilbertson and Benedict Ilozor.
© 2015 John Wiley & Sons, Ltd. Published 2015 by John Wiley & Sons, Ltd.

to the question, "How does the built environment impact on organisational change?" It may be that work settings induce this change mainly because of their innovative attributes.

This chapter explores how the built environment and its physical properties of work setting and environment, as well as management processes promotes or retards organisational change in terms of management effectiveness and increased productivity. The chapter reviews the literature briefly on space planning, organisational performance and innovative work settings; it presents a number of hypotheses and test results, which is followed by a discussion and conclusions.

13.2   Organisational performance and innovative work settings

The concept of space planning considers optimal use of space relative to workers' task space needs (Marmot and Eley, 2000). The knowledge base of this concept for commercial office buildings appears relatively less developed than that for residential and other buildings of much smaller scales. The treatment of top management as the sole clients in commercial building design and space planning decisions is potentially a misconception. This approach has become obsolete with the maturation of strategic facilities space management and planning, and greater acceptance of broad-based industrial democracy. Despite this development, there is largely a dearth of objective information on the key issues underlying the effectiveness of facilities space planning and management of commercial office buildings. Little effort has been invested on systematic development of comprehensive coherent criteria, to advance superior knowledge and direct decisions in policy and practice.

A discovery of this shortcoming among others motivated the author to research further into some influential factors in facilities space planning and management of commercial office buildings, with the Sydney Central Business District (CBD) as the primary study setting. As the dominant trend in the changing layout of offices is towards a more open-plan space (Knight Frank Hooker, 1995), investigation of open-plan measures in the determination of facilities space planning and management was considered crucial.

Organisational performance was viewed in terms of a focus on effective space management, rate of space accounting, staff understanding of work setting, staff satisfaction with work environment, staff collaboration (desk and job sharing) and productivity – measuring of staff productivity and the level of productivity. The ordinal and interval/ratio scales developed for measuring these variables have been previously presented (Ilozor and Oluwoye, 1999). Innovative work settings were considered in terms of facilitating environment (work settings perceived to be ingenious in terms of their internal physical conditioning), task performance (work settings perceived to be ingenious in facilitating the conduct of work activities), and staff interaction (work settings perceived to be ingenious in fostering mutual staff collaboration). It is recognised that deep misconceptions about workspace settings such as open-plans still abide to this day. With little anecdotal and/or empirical evidence, some still contend that open-plan offices make employees less productive, less happy, and more likely to get sick (Codrea Rado, 2013).

Considering the need to examine these misconceptions to have a more informed perception of workspace planning and settings, there is a need to test certain salient hypotheses.

13.3   Hypotheses and test results

The following null hypotheses were tested, and they derive from perceptions of impacts on organisational performance:

- There is no difference in the mean ranking of rated organisational performance with work settings perceived to be non-innovative, moderately innovative, innovative and very innovative in environmental aspects.
- There is no difference in the mean ranking of rated organisational performance with work settings perceived to be non-innovative, moderately innovative, innovative and very innovative in facilitating task performance.

Table 13.1   Mean rank scores by perceived environment innovative work settings.

Response	Cases	Mean Rank	D.F.	Chi-square χ^2	Significance p
A focus on effective space management ($n=101$)					
1 Non-innovative	17	37.47			
2 Moderate	35	39.64			
3 Innovative	30	62.68			
4 Very innovative	19	65.58			
			3	20.32	0.0001
Staff understanding of work setting ($n=101$)					
1 Non-innovative	17	39.62			
2 Moderate	35	41.74			
3 Innovative	30	58.25			
4 Very innovative	19	66.79			
			3	15.26	0.0016
Staff satisfaction with work environment ($n=100$)					
1 Non-innovative	16	36.56			
2 Moderate	35	47.24			
3 Innovative	30	54.55			
4 Very innovative	19	61.84			
			3	9.34	0.0251
Measuring of staff productivity ($n=101$)					
1 Non-innovative	17	44.26			
2 Moderate	35	42.14			
3 Innovative	30	60.90			
4 Very innovative	19	57.71			
			3	11.87	0.0078
Level of productivity ($n=99$)					
1 Non-innovative	16	40.88			
2 Moderate	35	43.86			
3 Innovative	30	54.10			
4 Very innovative	18	63.22			
			3	9.39	0.0245
Desk sharing practices ($n=101$)					
1 Non-innovative	17	36.47			
2 Moderate	35	42.63			
3 Innovative	30	59.08			
4 Very innovative	19	66.66			
			3	15.22	0.0016

- There is no difference in the mean ranking of rated organisational performance with work settings perceived to be non-innovative, moderately innovative, innovative and very innovative in fostering staff interaction.

The Kruskal–Wallis H test was used to validate these hypotheses, testing whether there are differences in the mean ranking of rated organisational performance with environment, task performance and interaction innovative work settings. However, in order to capture the correlation sense of the results, the explanation of the Chi-squared results is supported with a discussion of the correlation results by Ilozor (1999b). Based on the Chi-squared values and the significance levels with respect to perceived environmental innovative work settings, the null hypotheses were rejected (i.e. significant differences were observed) in the aspects shown in Table 13.1.

Based on the Chi-squared values and the significance levels with respect to perceived task performance innovative work settings, the null hypotheses were rejected in the aspects shown in Table 13.2.

Table 13.2 Mean rank scores by perceived task performance innovative work settings.

Response	Cases	Mean Rank	D.F.	Chi-square χ^2	Significance p
A focus on effective space management ($n=101$)					
1 Non-innovative	14	39.25			
2 Moderate	45	44.82			
3 Innovative	30	59.32			
4 Very innovative	12	67.08			
			3	11.38	0.0098
Staff understanding of work setting ($n=101$)					
1 Non-innovative	14	43.29			
2 Moderate	45	45.10			
3 Innovative	30	53.95			
4 Very innovative	12	74.75			
			3	12.50	0.0059
Staff satisfaction with work environment ($n=100$)					
1 Non-innovative	14	33.16			
2 Moderate	44	46.00			
3 Innovative	30	55.08			
4 Very innovative	12	75.25			
			3	18.73	0.0003
Level of productivity ($n=99$)					
1 Non-innovative	14	35.43			
2 Moderate	44	45.65			
3 Innovative	29	56.33			
4 Very innovative	12	67.67			
			3	12.98	0.0047
Job sharing ($n=101$)					
1 Non-innovative	14	36.00			
2 Moderate	45	49.20			
3 Innovative	30	59.27			
4 Very innovative	12	54.58			
			3	9.84	0.0200

Table 13.3   Mean rank scores by perceived interaction innovative work settings

Response	Cases	Mean Rank	D.F.	Chi-square χ^2	Significance p
A focus on effective space management (n=101)					
1 Non-innovative	17	31.06			
2 Moderate	42	46.27			
3 Innovative	24	63.17			
4 Very innovative	18	64.64			
			3	18.82	0.0003
Staff understanding of work setting (n=101)					
1 Non-innovative	17	37.06			
2 Moderate	42	44.25			
3 Innovative	24	53.69			
4 Very innovative	18	76.33			
			3	22.46	0.0001
Staff satisfaction with work environment (n=100)					
1 Non-innovative	17	32.15			
2 Moderate	41	48.50			
3 Innovative	24	56.98			
4 Very innovative	18	63.75			
			3	14.64	0.0022
Measuring of staff productivity (n=101)					
1 Non-innovative	17	41.29			
2 Moderate	42	46.95			
3 Innovative	24	60.48			
4 Very innovative	18	56.97			
			3	8.26	0.0410
Desk sharing practices (n=101)					
1 Non-innovative	16	33.94			
2 Moderate	42	47.26			
3 Innovative	25	55.70			
4 Very innovative	18	68.36			
			3	13.50	0.0037
Job sharing (n=101)					
1 Non-innovative	17	36.00			
2 Moderate	42	49.81			
3 Innovative	25	58.76			
4 Very innovative	17	57.53			
			3	10.94	0.0120

Based on the Chi-squared values and the significance levels with respect to per-ceived interaction innovative work settings, the null hypotheses were rejected in the aspects shown in Table 13.3.

13.4   Discussion

Perceived environment innovative work setting

The test results as in Table 13.1 showed that, with respect to environment innova-tive work settings, the null hypotheses were rejected in the aspects of a focus on effective space management ($\chi^2 = 20.32$, $p = 0.0001$), staff understanding of work setting ($\chi^2 = 15.26$, $p = 0.0016$), staff satisfaction with work environment ($\chi^2 = 9.34$, $p = 0.0251$), measuring of staff productivity ($\chi^2 = 11.87$, $p = 0.0078$), and the level of productivity ($\chi^2 = 9.39$, $p = 0.0245$), as well as staff desk sharing practices ($\chi^2 = 15.22$, $p = 0.0016$). However, the null hypotheses were accepted in the aspects of job sharing and the rate of space accounting, as ascertained from the statistically insignificant Chi-squared values (values not reported for reasons of brevity). In other words, organisational performance is influenced only in certain aspects.

In order to perform various tasks, staff need environments which support not only physical activities but also thought processes and emotional well-being (Goodrich, 1982). An innovative environment does not suggest comfort but rather, a work setting/environment that has been designed to optimise its potential.

In a study by Ilozor (1999b), perceived environment innovative work setting was positively associated with desk and job sharing, space use intensity, a focus on effective space management, staff understanding of work setting and satisfaction with their work environment, measuring of staff productivity and the level of productivity. It was negatively associated with the rate of space accounting.

However, when management control (the control variable) was partialled out in a partial correlation exercise, Ilozor (1999b) specifically found that, irrespective of the influence of management control, the more a work setting is perceived to be environmentally innovative, the greater the desk sharing practices among staff. If a work setting cuts down office operations and frees staff to spend more time on site with customers (Flemington, 1988), desk sharing may increase with practices such as hot-desking, telecommuting, hotelling and drop-in.

Furthermore, despite the interaction effects of management control in certain categories, the more a work setting is perceived to be environmentally innovative, the higher the level of productivity. As innovative environment may be likened to the optimisation of a work setting, this result can be expected.

Even without management intervention, the more a work setting is perceived to be environmentally innovative, the greater the focus on effective space management and measuring of staff productivity. Partly, management control is needed so that, the more a work setting is perceived to be environmentally innovative, the higher the staff understanding of work setting and satisfaction with their work environment. The role played by management is significant to this result. Acceptance of and satisfaction with open-plan work setting, for instance, improve when management allows staff to participate in work settings' design decisions (Pile, 1978; Hedge, 1982; Forester, 1989; Carnevale, 1992; Uzee, 1999; Kobach, 2000; Segal, 2000). Annunziato (2000) reported that innovative workspace design enhances general staff satisfaction.

These results therefore serve to revalidate the earlier findings. Except in the aspects where the null hypotheses were accepted, the implication of these findings is for the focus to be on innovative work environments that optimise their settings' potential, while aiming at positive organisational change and/or performance. Today's real estate facilities professionals should embrace smart workplace design to improve employee satisfaction and productivity (Uzee, 1999; Annunziato, 2000).

Perceived task performance innovative work setting

The test results as in Table 13.2 showed that, with respect to task performance innovative work settings, the null hypotheses were rejected in the aspects of a focus on effective space management ($\chi^2 = 11.38$, $p = 0.0098$), staff understanding of work setting ($\chi^2 = 12.50$, $p = 0.0059$) and satisfaction with their work environment ($\chi^2 = 18.73$, $p = 0.0003$), the level of productivity ($\chi^2 = 12.98$, $p = 0.0047$), and staff job sharing practices ($\chi^2 = 9.84$, $p = 0.0200$). However, the null hypotheses were accepted in the aspects of desk sharing, measuring of staff productivity and the rate of space accounting (values not reported for reasons of brevity).

In Ilozor's (1999b) study, task performance innovative work setting was positively associated with desk and job sharing practices among staff, a focus on effective space management, staff understanding of work setting and satisfaction with their work environment, measuring of staff productivity and the level of productivity. It was negatively associated with the rate of space accounting. The zero-order correlation was not significant in the case of job sharing.

However, when management control was partialled out, Ilozor (1999b) found that, irrespective of the influence of management control variables, the more a work setting is perceived to be innovative with respect to task performance, the higher the staff satisfaction with their work environment and the level of productivity. Management intervention is needed so that the more a work setting is perceived to be innovative to task performance, the greater the staff understanding of work setting and measuring of staff productivity. If there is a better fit between work setting's design and user tasks, the office can more effectively support work performance and improve productivity (Vischer, 1995).

Partly, management intervention is needed so that, the more a work setting is perceived to be innovative to task performance, the higher the desk sharing practices among staff. Since this aspect of work settings can be so consequential, the implication of this finding is for more effort to be focused on alternative ways of achieving a better fit between work setting's design and user tasks (Vischer, 1995).

There was found no 'real' relationship between perceived task performance innovative work setting, and the rate of space accounting and a focus on effective space management. Any observed association is probably due to the influence of management control. Perhaps, optimal task performance has little to do with the character of a work setting as Vischer (1995) and Carnevale and Rios (1995) suggested.

Perceived interaction innovative work setting

The test results as in Table 13.3 showed that, with regard to interaction innovative work settings, the null hypotheses were rejected in the aspects of a focus on effective space management ($\chi^2 = 18.82$, $p = 0.0003$), staff understanding of work setting ($\chi^2 = 22.46$, $p = 0.0001$) and satisfaction with their work environment ($\chi^2 = 14.64$, $p = 0.0022$), measuring of staff productivity ($\chi^2 = 8.26$, $p = 0.0410$), staff desk ($\chi^2 = 13.50$, $p = 0.0037$) and job sharing ($\chi^2 = 10.94$, $p = 0.0120$) practices, as well as the rate of space accounting ($\chi^2 = 8.87$, $p = 0.0310$). The null hypothesis was accepted only in the aspect of the level of productivity; however, the result of the partial correlation differed considerably in this aspect, probably due to the moderating effects of management control.

In Ilozor's (1999b) study, perceived interaction innovative work setting was positively associated with desk and job sharing practices among staff, a focus on effective space management, staff understanding of work setting and satisfaction with their work environment, measuring of staff productivity and the level of productivity. It was negatively associated with the rate of space accounting.

However, when management control was partialled out, Ilozor (1999b) found that, irrespective of management influence, the more a work setting is perceived to be innovative in terms of fostering staff interaction, the higher the desk sharing practices. This result can be expected, since improved interaction would likely bring about flexible use of available desk resources.

Except in the case of job sharing, where management intervention is needed, the more a work setting is perceived to be innovative with respect to fostering staff interaction, the greater the focus on effective space management, staff understanding of work setting and satisfaction with their work environment. It has been noted that work settings can bring people together, or keep them apart, facilitating or frustrating individual and organisational purposes (Osmond, 1957). When spatial layout facilitates staff interaction, it influences task performance and satisfaction of organisational needs (Carnevale, 1992; Carnevale and Rios, 1995), such as a focus on effective space management, staff understanding and acceptance of work setting as well as general satisfaction with their work environment. The influence of social density, generated in work environment, with similar results on the degree of interaction, has been investigated (Szilagyi and Holland, 1980; Hedge, 1982).

Partly, management intervention is required so that the more a work setting is perceived to be innovative with respect to fostering staff interaction, the less the rate of space accounting. Apart from the moderating effects of management in certain categories, the more a work setting is perceived to be innovative in terms of fostering staff interaction, the greater the measuring of staff productivity and the level of productivity. This result supports Vischer's (1995) finding that a better fit between staff and work settings leads to greater productivity. The implication of these findings is that work settings which frustrate staff interaction have an adverse impact on productivity.

13.5   Conclusions

Although subtle shifts were observed in the aspects of organisational performance that seem predicated on innovative work settings, to some extent the proposition that the physical properties and design of the workplace can influence organisational performance, positively or negatively, was validated. The specific conclusions in this study regarding work settings in built facilities are as follows:

- A greater environment innovative work setting is associated with increased possibilities for desk sharing practices among staff, a focus on effective space management, measuring of staff productivity and a higher level of productivity.
- A greater task performance innovative work setting is associated with higher staff satisfaction with their work environment and the level of productivity.
- A greater interaction innovative work setting is associated with increased possibilities for desk sharing practices among staff, a focus on effective space management, staff understanding of work setting and satisfaction with their work environment, measuring of staff productivity as well as a higher level of productivity with moderating effects of management control.

These conclusions have implications for the direction and focus of the on-going search for answers to the question, "How does the built environment impact on organisational change?" It may be that work settings induce this change mainly because of their innovative attribute. Hence, a work setting should not be adopted simply for style's sake but should be designed and managed with user tasks in mind. This objective can be achieved by deliberately making work settings innovative in terms of their environment, enhancement of task performance and fostering of staff interaction. More innovative work settings tend to do better in these respects as significant differences exist between the various levels of innovative work settings, with regard to many rated organisational performance parameters.

References

Annunziato, L. (2000) Attracting new recruits. *Facilities Design & Management*, **19**(4), 40–43.

Brown, M. (1996) Spaceshuffle. *Management Today*, **66–74**.

Carnevale, D.G. (1992) Physical settings of work: a theory of the effects of environmental form. *Public Productivity & Management Review*, **15**(4), 423–436.

Carnevale, D.G. and Rios, J.M. (1995) How employees assess the quality of physical work settings. *Public Productivity & Management Review*, **18**(3), 221–231.

Codrea Rado, A. (2013), Open-plan offices make employees less productive, less happy, and more likely to get sick. http://qz.com/85400/moving-to-open-plan-offices-makes-employees-less-productive-less-happy-and-more-likely-to-get-sick/ (accessed 21 May 2013).

Davis, M.C., Leach, D.J. and Clegg, C.W. (2011) The physical environment of the office: contemporary and emerging issues, in *International Review of Industrial and Organizational Psychology* (eds G.P. Hodgkinson and J.K. Ford), John Wiley & Sons, Ltd, Chichester, pp. 193–235.

Duffy, F. (1997) *The New Office*, Conran Octopus, London.

Duffy, F. and Tannis, J. (1993) A vision of the new workplace, in *Hot Open Chaotic*, Property Council of Australia Congress, Sydney, pp. 3–15.

Elsbach, K.D. and Pratt, M.G. (2007) The physical environment in organizations. *The Academy of Management Annals*, 1(1), 181–224.

Flemington, R. (1988), Banking: views of the future: National Westminster Bank at heart it's a people business. *Banking World*, **6**(12), 27–28.

Forester, J. (1989) *Planning in the Face of Power*, University of California Press, Berkeley, CA.

Gillen, N.M. (2006) The future workplace, opportunities, realities and myths: a practical approach to creating meaningful environments, in *Reinventing the Workplace*, 2nd edn (ed. J. Worthington), Architectural Press, Oxford, pp. 61–78.

Goodrich, R. (1982), Seven office evaluations: a review. *Environment and Behavior*, **14**(3), 353–378.

Hedge, A. (1982), The open-plan office: a systematic investigation of employee reactions to their work environment. *Environment and Behavior*, **14**(5), 519–543.

IFMA (1997) Open-plan office layouts make up more than half of workplace environments. *Facilities Design & Management*, **16**(3), 12.

Ilozor, B.D. (1998), *Open-plan Measures in the Determination of Facilities Space Management of CBD Commercial Office Buildings*. The 1998 Excellence Award in Facility Management Research, Facility Management Association of Australia.

Ilozor, B.D. (1999a) *Open-plan Measures in the Determination of Facilities Space Management of CBD Commercial Office Buildings*. The 1999 Excellence Award in Facility Management Research, Facility Management Association of Australia.

Ilozor, B.D. (1999b) Open-plan measures in the determination of facilities space management of CBD commercial office buildings. PhD Thesis, The University of Technology, Sydney.

Ilozor, B.D. (2000) *From Open-plan Design Impact on Effectiveness to Policy Issues on Computer-integrated Facilities Management.* The 2000 Achievement in Facility Management Research Award, Facility Management Association of Australia.

Ilozor, B.D. and Oluwoye, J.O. (1999) Open-plan measures in the determination of facilities space management. *Facilities*, **17**(7/8), 237–245.

Knight Frank Hooker (1995) *BOMA (now the Property Council of Australia) Leading Edge Research*, Sydney, Australia.

Kobach, R. (2000) Creating an environment that fosters innovation. *The British Journal of Administrative Management*, **22**, 26–27.

Leaman, A. and Bordass, B. (1993) Building design, complexity and manageability. *Facilities*, **11**(9), 16–27.

Marmot, A. and Eley, J. (2000) *Office Space Planning: Designs for Tomorrow's Workplace*, McGraw Hill Professional, New York.

Mohr, R. (1996), Office space is a revenue enhancer, not an expense. *National Real Estate Investor*, **38**(7), 46–47.

Osmond, H. (1957), Function as the basis of psychiatric ward design. *Mental Hospitals*, **8**, 23–80.

Pile, J. (1978) *Open Office Planning*, The Architectural Press Ltd, London.

Segal, T. (2000) Designed for a comfortable fit. *ABA Journal*, **86**, 76–79.

Steele, F. (1986), *Making and Managing High Quality Workplaces: An Organizational Ecology*, Teachers College Press, New York.

Steiner, J. (2005) The art of space management: Planning flexible workspaces for people. *Journal of Facilities Management*, **4**(1), 6–22.

Szilagyi, A.D. and Holland, W.E. (1980) Changes in social density: relationships with perceptions of job characteristics, role stress, and work satisfaction. *Journal of Applied Psychology*, **65**, 28–83.

Uzee, J. (1999) The inclusive approach: creating a place where people want to work. *Facility Management Journal of the International Facility Management Association*, 26–30.

Vischer, J.C. (1995) Strategic work-space planning. *Sloan Management Review*, **37**(1), 33–42.

Williams, C., Armstrong, D. and Malcom, C. (1985) *The Negotiable Environment: People, White-Collar Work, and the Office*, Consulting Psychologists Press, Ann Arbor, MI.

Chapter 14
Achieving Zero Carbon in Sustainable Communities

Malgorzata Jacewicz and Herbert Robinson

14.1 Introduction

Contemporary urban lifestyle emphasises green living and encourages people to reduce their carbon footprint. Carbon footprint of users and their demand for energy should be measured to explore scenarios for the design of buildings to create sustainable communities. Sustainable communities promote sustainable living, contribute to a high quality of life and recognise the value of living and working to reduce the impact of the environment to the community. The UK Government defines 'sustainable community' as places where people want to work and live, now and in the future. Such communities should be economically, environmentally and socially viable, healthy and resilient but success depends on commitment, good citizenship, leadership in creating a responsible and caring environment (Institute for Sustainable Communities, 2014). A good example of a sustainable community is BedZED, an environmentally friendly housing development in Hackbridge, London using only energy generated from renewable sources on site (Bioregional Solutions for Sustainability, 2014).

This chapter presents an alternative approach for developing sustainable communities using activity-based design principles based on the quantification of the energy demand per activity performed by members of various communities. It starts with an overview of key concepts and principles for developing of sustainable communities to achieve an equilibrium between the supply of local renewable resources (e.g. wind, water, solar) and demand for alternative energy. Making smart use of renewable energy sources (e.g. solar, wind, water and utility grid) is increasingly critical as the cost of energy from conventional sources continues to rise sharply due to the increase in global demand to support socio-economic development. The chapter also discusses how energy consumption analysis

Design Economics for the Built Environment: Impact of Sustainability on Project Evaluation, First Edition.
Edited by Herbert Robinson, Barry Symonds, Barry Gilbertson and Benedict Ilozor.
© 2015 John Wiley & Sons, Ltd. Published 2015 by John Wiley & Sons, Ltd.

provides the foundation for evaluating overall energy need, space and land requirements to create sustainable communities.

14.2   Key concepts and principles

The activity-based approach for creating sustainable communities was developed as part of a project established through collaboration between the Department of Architecture at London South Bank University, UK and Hyperbody Research Group at the Technical University of Delft in the Netherlands. The approach is based on a set of principles to capture the relationship between renewable energy supply within the local environment and energy demand from communities to achieve a balance. First, sustainability is at the heart of the approach as changes in social behaviours through life/work balance, sharing of common public spaces, reducing the reliance and dependency on conventional energy sources to minimise social, economic and environmental costs. Secondly, the approach is based on setting out planning principles for the generation of decentralised systems using local renewable energy sources such as wind, water and solar. Energy availability is captured from meteorological data for water, wind and sun provided by national meteorological institutes. Such data can be used for urban planning to facilitate the distribution of infrastructure across any country. The goal is to influence development of new urban areas and to ensure efficient use of energy capacity through energy trading markets to capture, use and share surplus energy. Thirdly, the approach is underpinned by activity-based design principles capturing user energy needs based on daily activities for optimum energy utilisation to reduce fossil fuel dependency and the capital and life cycle costs associated with it. Energy usage or consumption is calculated based on certain factors. Table 14.1 is an example showing the breakdown of consumption of energy resources per person for selected activities and types of appliances to determine the demand for energy in order to match it with the available supply of renewable resources using water, wind or solar.

Micro power generation tools such as solar panels, micro wind turbines and heat pumps take energy directly from the surroundings and convert it into electricity and heat. Table 14.2 is an example illustrating how different wind speeds are converted to power density and energy. Similarly, conversion for solar and water sources into energy can be determined based on a set of factors.

During the process by which wind or sunlight is converted into electricity no carbon dioxide is emitted. Such energy sources are inexhaustible, truly renewable and are effective solutions to natural resource depletion of fossil fuels such as petroleum and coal to reduce carbon footprint and global warming. Fourthly, the decentralised energy generated locally from different sources is consumed where it is produced and thousands of units provide a small portion of the energy needed, instead of relying on a few power plants providing enormous amounts of energy. The creation of micro power energy clusters feeding into the main utility grid through local energy networks helps to secure clean energy almost entirely harnessed from local renewable sources providing energy security for the future.

Table 14.1   Breakdown of consumption of resources per person.

available renewable energy resources = users demand	rain – water	water – personal hygiene 45 l/day – drinking and cooking 4–5 l/day – toilet flushing 24–30 l/day – dishwashing 30 l/wash – washing clothes 40 l/wash – other uses/cleaning, gardening 10–30 l/use
	wind – electricity	electricity appliances – kilowatt-hours – electronic equipment 4–8 kWh/year – clothes washers – 450 kWh/year – dishwashers 440 kWh/year – fridge/freezer 500 kWh/year – electric boiler 3200 kWh/year – solar boiler 1,900 kWh/year – lightning
	sun – heat – electricity	heat heating – 150 kWh m²/year electricity – as above

Table 14.2   Wind speed classification and energy conversion.

Class	Power	Wind speed (m/s)	Power density (W/m²)
1		0.0–5.6	0–200
2		5.6–6.4	200–300
3		6.4–7.0	300–400
4		7.0–7.5	400–500
5		7.5–8.0	500–600
6		8.0–8.8	600–800
7		> 8.8	> 800

14.3   Key features of decentralised energy networks

In the traditional centralised energy infrastructure, a significant amount of the energy is lost during conversions, transmission, distribution, and transport. With decentralised energy systems, less conversions are required and virtually no transportation is needed which makes the system very efficient and reliable.

In a decentralised energy system, thousands of units are generating electricity so massive power outages can be avoided. When a unit fails, surrounding units will take over and bear the load of the failing unit. Decentralised energy systems also allow for better cost control as most energy is generated locally from renewable sources reducing the risks associated with increasing oil prices. Oil price is based on supply and demand determined in the global markets with geopolitical influences resulting in strong fluctuations and uncertainty leading to higher or an upward trend in oil prices.

In a decentralised system, the dependency diminishes as energy costs become controllable. As a result, the focus shifts to greater efficiency and innovation in extracting energy from the environment through, for example, improved solar panels and wind turbines. Decentralised local networks also allow for scalable infrastructure design compared with conventional power plants that are large with high tension networks and heating systems that do not allow for gradual increments in size. Changes in conventional power plants are often massive, costly and are usually driven by large government schemes or government support requiring significant investment and risks. However, a decentralised energy system is easily scalable as the technical and investment requirements are far less demanding and manageable which provides for small incremental design site by site, building by building, if needed.

Decentralised energy networks embrace the concept of energy capturing from renewable resources. The nature of the design influences the type of shared activities and community services that will be provided. Within the networks, members exchange energy to maximise efficiency as multiple houses or office buildings are connected into a cluster where energy can be exchanged freely.

Decentralised energy grids can be connected to the main utility grid. A local energy network is basically a mini-grid connected to the utility grid through one connection of the cluster. Electricity can be exchanged within the cluster before surplus capacities are sold back to the grid. When the cluster as a whole is not producing enough energy, additional energy will be sourced from the grid. Sharing energy with other members in the community optimises the efficiency of the system as a surplus capacity for one member can address problems of shortages for other members or the main grid. For instance, at noon solar panels and wind turbines are likely to be producing a significant quantity of electricity. However, because cluster users are at work, most appliances would not normally be used except those that are programmed in advance. Under these circumstances the rest of the power that is surplus would be sold back to the grid.

The overall approach is useful in managing surpluses and shortages to ensure that energy is not wasted and most of the capacity is captured and used. The system would therefore allow for efficient management of power demand and supply from various clusters by distributing it over the whole day based on needs. Local networks planned according to sustainable urban principles would create energy trading opportunities and facilitate the development of new markets.

14.4   Activity-based design approach

The activity-based design approach captures the various activities of users and quantifies the energy demand per activity performed on a daily basis. Rather than making assumptions on space requirements, this approach focuses on the specific activity of users and their space requirements. Moving away from traditional design based on function to activity-based design principle allows for space adjustments in terms of energy demand per activity and determining the required space to perform various activities. For example, an activity such as cooking is not related to the function of having a generic size kitchen with generic appliances but to optimise what is the minimum space required for a user to comfortably perform the activity. In design terms, the objective is to assess energy demand per activity driven by user's lifestyle. For example, two single people living on their own may have different needs. The person cooking every day will require higher energy demand than a single person socialising outside of work who requires only basic cooking facilities to warm up food on a daily basis.

Data and information requirements

For decentralised energy systems to be designed effectively, urban planners need to understand environmental constraints of natural gradients and how to capture renewable energy through shape, form, orientation and the application of modern technologies. Available energy will be quantified from information provided by national meteorological institutes. For instance, meteorological data will enable the quantification of the surface area required for any particular settlement with a given energy consumption pattern and to determine the size of the land to meet the demand. For example, wind speed in the Netherlands on average of 15–20 m/s is an equivalent of 600–800 W/m², solar energy is equivalent to approximately 833 kWh/m² in the North and 1199 kWh/m² in the South per year. Figure 14.1 shows different settlement types and their consumption patterns over a 24-h period.

Figure 14.1 shows peak consumption patterns for offices from 09:00 to 12:00 and from 15:00 to around 19:00/20:00. For commercial, the peak consumption is much longer from around 05:00 to 20:00 and for housing there are two small peak periods from around 06:00 to 08:00 and then from 19:00 to 20:00. Just like a good transport management system, the ideal situation would be to spread the load to avoid extremely 'high' load in certain periods to create more energy balance.

Type of settlements and location decisions

The initial analysis of various types of settlements generated a system of rules applied to optimise and develop design principles to explore gradients efficiency and propose a sustainable energy day cycle for decentralised energy networks. The analysis looks at energy consumption and its dependency on lifestyle to

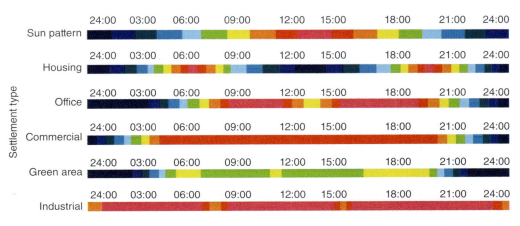

Figure 14.1   Settlement types and their consumption peak times over a 24-h period.

compare it with available energy within the gradient to distribute settlements accordingly. For example, depending on the settlement type it might be an advantage to locate a high energy dependent cluster within the costal gradient to take advantage of wind and sun to generate power to the settlement. For clusters of units with low power consumption but with high water demand it may be suitable to locate them within an appropriate gradient deeper inland. The rationale is to increase and maximise the efficiency of renewable energy for decentralised networks. Further, the gradient should be explored to avoid peaks and troughs within the cluster.

14.5   Key steps in the design process

The design approach for the distributed energy network is to assess the energy demand and locate the cluster within the relevant gradient. The key is to understand the balance between work and lifestyle at the user level. Research was carried out using a group of young people who intended to live and work together in delivering community and art services. The group of eight gathered over a 1-day workshop to design their sustainable living and work environment using a system of rules to reduce their carbon footprint. Every participant had to declare from the start their willingness to share activities and compromise personal needs for a shared benefit to reduce demand for energy and associated costs. The initial research explored assessment of individual demand for energy and space.

Level 1 – Establish activities in the community

The design criteria of Level 1 allowed for every individual to specify the top nine activities they perform on a daily basis. Once this has been achieved, each individual is assigned the space needed for their activity. A choice of three space dimensions was used – small ($6\,\text{m}^2$), medium ($12\,\text{m}^2$) and large ($20\,\text{m}^2$) assigned to each of the top nine activities as shown in Figure 14.2. Level 1 involves establishing a list of all activities to be performed across the settlement in the community.

Level 2 – Defining collectivism and rules to reduce space demand

Level 2 defined collectivism within the group and procedures to follow up a set of rules to reduce space demand. The objective of Level 2 is to reduce personal space if the activity was performed by more than four other users (Figure 14.3). For instance, in the case of cooking which is performed up to various standards, they had to split into two groups and agree a common space dimension for that particular activity. Some were keen on cooking and others more orientated towards take-away food. On this basis, two cooking areas were required, one with full cooking facilities with a dining area where the assigned group would cook together or take turns to cook for each other. Also, a small kitchenette was included for those who did not intend to cook but chose to rely on take-away food. As a result, there were two areas serving the same activity with different energy and space requirements. A similar approach was used for other activities.

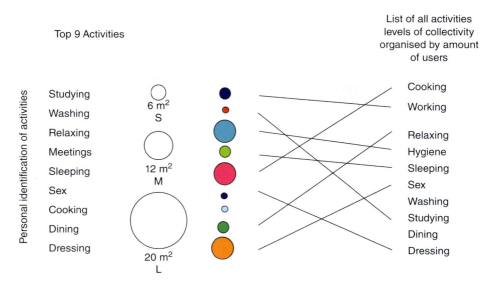

Figure 14.2   Design process – Level 1.

If the activity is shared by 2 to 4 you have to agree on size and shape

Figure 14.3   Design process – Level 2.

Level 3 – Re-assessing energy consumption of key activities

Level 3 involves re-assessing the energy consumption of common activities using manufacturers' guidelines on energy consumption of equipment or appliances to be purchased for shared activities in common areas or single activities in private areas. The power requirements of the items on the required list were quantified per kWh and multiplied by the number of times to be used in a yearly cycle to determine energy demand.

As the project was experimental, relevant location was chosen after the quantification of the users energy demand. This allowed for a proposal of various locations for the group to decide on the land price within their reach. The electricity demand for the group was approximately 48,000 kWh/year. The value is largely influenced by type of appliances and initial capital expenditure for the building. For example, a three-phase electric boiler would consume 3,200 kWh/year. A two-phase solar boiler would reduce this value to approximately 1,800 kWh/year. A similar situation applies to lighting within the house. Depending on the design, if the lights in all areas were to be energy efficient or preferably LEDs, the consumption could be covered by 12 V solar panels and run off batteries (UPS system). This would allow for direct electric current transmission straight from the renewable source. The design process was further explored to assign to each activity an energy value depending on its consumption and percentage split out of the total 100%.

Level 4 – Identify areas with high energy consumption

Level 4 was designed to identify areas with high percentage of energy consumption and allocate certain activities based on high energy availability pattern during a sustainable day cycle. All activities have been colour coded in order to allocate them to appropriate zones as the design develops. (Figure 14.4).

As the quality of working environment is critical, the requirement was to have access to daylight for as long as possible. This zone of activities was allocated as a central part of the building on the south facing side together with the main cooking and dining area. Prevalent directions of wind, sun and water defined the relationship between internal communication layout and allocation of activities carried out throughout the day.

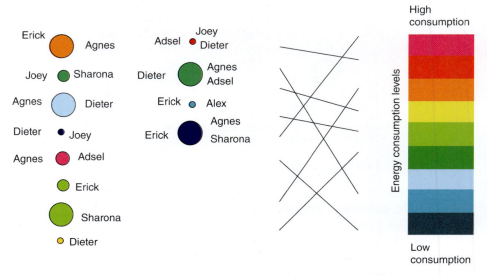

Figure 14.4 Design process – Level 4.

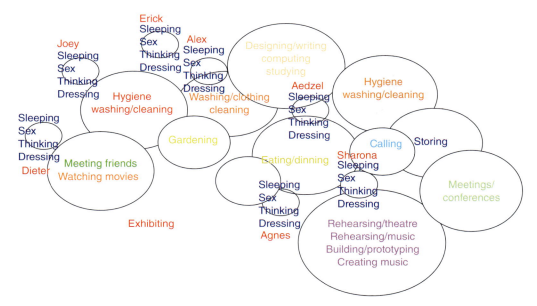

Figure 14.5   Zoning proposal based on users self-organisation.

Level 4 (Figure 14.5) allowed the group to organise their space in relationship to the external environment and to achieve the most efficient layout in relation to external design forces. Various zones are allocated to generate a building form which is the most appropriate structure allowing for community living and renewable energy absorption (Figure 14.6). Computerised programs are used to create a "virtual building" which provides the opportunity to examine every detail of the project to facilitate the design of buildings to be constructed.

The internal layout approach was to connect all activity zones with continuous communication channels. Interior layout has been designed to optimise space with communication areas being a continuous loop overlapping and crossing activity zones.

Further design decisions can be taken to optimise the interior space and generate a structural design approach. By dividing the joined zones of activities and communication paths into equal sections/ribs, structural integrity has been provided between the interior and exterior.

The internal layout space is therefore curved out of the external shell and ribs as per zone and communication paths. The building structure is designed as a set of ribs joined by the interior compartments and covered over with a mixture of transparent and semitranslucent cladding. The building design makes use of available space by overlapping activities and using the structural frame as part of the interior design generating partitions and space continuity throughout the building. The structural frame is the main space division creating an enclosure for the user's privacy and opening up in areas of common activities.

14.6   Evaluating energy, space and land requirements

Once the energy demand is calculated, layout and space dimensions are assigned to activities and location chosen. The required land (in m²) can be estimated to support a community and their work–life environment. By developing a

(a)

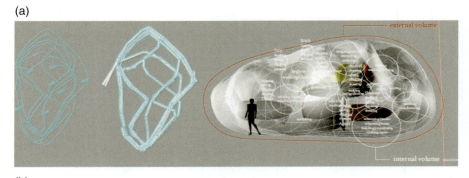

(b)

(c)

Figure 14.6 (a) Design solution: zoning and connection via communication paths. (b) Design solution: structural approach from division by structural ribs. (c) Optimising interior spaces.

breakdown of activities performed and their estimated energy consumption, a suitable location can be chosen and the supply and demand relationship managed. The activity-based design approach is based on the analysis of meteorological data and quantification of the surface area.

The approach allows for flexible assessment of how much land is required to power certain types of settlements to guarantee reduction in fossil fuels and replacement of it over time with a full energy supply from renewable sources.

Once the demand is calculated per person, proposals of energy reduction can be introduced. The objective is to decrease human impact on the environment by revisiting concepts of common spaces and generate socially vibrant environments whilst cutting down on fossil fuel dependency. Achieving equilibrium will allow for the development of decentralised energy systems.

However, a major disadvantage is that densities and gradient capacity cannot be increased without technological innovation which requires funding to improve energy absorption from water, wind and solar technologies. Once the limit of the gradient is reached, the area would not be explored any further. Higher densities would require higher start-up capital expenditure and space allowance for power generation, which might become critical for buildings to interact closely within clusters as energy generating devices. The savings gained from the environmental impact reduction can therefore be invested in renewable energy absorption devices which are currently costly technology. This scenario potentially will influence technological progress in façade development and whole areas being designed for the purpose of energy absorption.

Decentralised energy networks explore the relationship between energy supply and demand resulting in the application of a framework underpinned by a system of rules for energy absorption driven building design. The design rules influence reduction in human environmental impact and encourage a planning approach to gain the maximum available renewable energy. Areas within energy gradients will become energy independent trading units based on local micro power generation leading to energy security necessary for creating sustainable communities. The use of renewable sources also creates opportunities to feed into the main power grid.

14.7   Concluding remarks

The activity-based approach is not only driven by local environmental factors but globally to respond to current demand for energy efficiency, a better quality of life, to reduce carbon footprint and to decrease the environmental impact on the planet. The proposed approach takes into account the social, environmental and economic aspects to develop sustainable communities of globally aware and informed citizens as well as inspired and caring leaders who are prepared to reduce their carbon footprint without compromising on their core activities. As prices of fossil fuel and their exploitation become far more unpredictable, gaining energy independence at a reasonable cost is critical to the present and future development of sustainable communities.

During the past decade, there has been considerable research in renewable energy resources and development of renewable technologies. Solar technology has become practical to develop on a small scale to reduce individual carbon foot-prints. Geothermal energy (heat) which is environmentally friendly, emission free and economical can also be generated from buildings' foundations through pipes installed into the piles that are used to transport water storing thermal energy in the surrounding ground to heat and cool a building. Wind turbines are also a proven low carbon source of energy which can be scaled up to decarbonise the energy sector. However, support from governments in the use of renewables varies significantly. For example, the UK Government's sustainability ambitions led to the Renewable Heat Incentive (RHI) by the Department of Energy and Climate Change (DECC) to meet its carbon reduction targets. Producers of renewable heat are also rewarded per kWh of heat generated, in a similar manner to feed-in tariffs for micro generation.

Furthermore, the use of environmental assessment tools such as BREEAM and LEED to limit emissions of carbon dioxide into the atmosphere also created the need to explore renewable sources for energy supply. For example in BREEAM, energy and water are the two categories with minimum standards required at all levels. Some successful projects have been implemented and new approaches are also emerging to address issues relating to challenges encountered in the design of sustainable communities. There is increasing urgency to explore alternative energy. For example, the European Energy Commissioner recently announced that Europe could draw clean energy from solar panels constructed in the Sahara desert within 5 years in an attempt to meet its target of generating 20% of its energy from renewable sources by 2020.

References

Bioregional Solutions for Sustainability (2014) BedZED Seven Years On: The Impact of the UK's Best Known Eco-village and its Residents. http://www.bioregional.com/news-views/publications/bedzed-seven-years-on/ (accessed 6 March 2014).

Institute for Sustainable Communities (2014) What is a Sustainable Community? http://www.iscvt.org/what_we_do/sustainable_community (accessed 6 March 2014).

Chapter 15
Flood Risk Mitigation: Design Considerations and Cost Implications for New and Existing Buildings

Rotimi Joseph, David Proverbs and Jessica Lamond

15.1 Introduction

Worldwide, flooding has been predicted to threaten up to 2 billion people or more by 2050 due to the collective effect of higher frequency and intensity of rain, snow and ice melting, which may affect more denuded slopes and high runoff urban areas, in conjunction with the projected population growth (Bogardi, 2004). During the past decade, reporting of incidents of natural disasters that meet Emergency Event Database (EMDAT) criteria have increased sixfold compared with the 1960s due mainly to an increase in small- and medium-scale disasters (Guha-Sapir *et al.*, 2006). Blunden and Arndt (2012) and Field *et al.* (2012) asserted that almost 90% of natural disasters are hydro meteorological events such as droughts, storms and floods. Further, scientific evidence suggests that global climate change will only increase the number of extreme events, creating more frequent and intensified environmental emergencies (Field *et al.*, 2012).

Currently in England and Wales, 5.2 million properties are in flood risk areas, amounting to 1 in 6 properties (Environment Agency, 2009), although many of these areas are well managed and protected by means of flood defences. There is every indication that this figure will rise further if climate change results in more frequent extreme weather events or if properties continue to be built on floodplains (Soetanto *et al.*, 2008).

There is general consensus about the key role of risk reduction in mitigating the vulnerability of buildings and human settlements to natural hazards. The European Strategic Environment Assessment (SEA) Directive requires a Sustainability Appraisal to be undertaken for all development plans and an assessment of flood risk forms part of informing this appraisal process. In the UK, there is a National Planning Policy Framework (NPPF) which sets out Government planning policies and how the policies are expected to be applied to achieve sustainable and resilience development. This is supported by technical guidance and an associated

Design Economics for the Built Environment: Impact of Sustainability on Project Evaluation, First Edition.
Edited by Herbert Robinson, Barry Symonds, Barry Gilbertson and Benedict Ilozor.
© 2015 John Wiley & Sons, Ltd. Published 2015 by John Wiley & Sons, Ltd.

practice guide to assist developers in assessing how the proposed development will be undertaken to reduce flood risk. Planning Policy Statement 25 (PPS 25) now requires developers to undertake a flood risk assessment if proposing to build on or near the floodplain and to incorporate appropriate measures to prevent or minimise the risk of flooding within the planning application (Arup, 2011). In the UK, local authorities are now at the forefront of flood risk management by taking action on climate risks through the planning system. For instance Arup (2011) found that 96% of planning applications analysed in their study included at least one measure to manage river and costal flood risk. These were done either through developers' initiative of putting forward these measures in their planning applications or through conditions attached to planning consents by local authorities. Further, it was discovered that the Environment Agency in some cases directly advised both the applicants and planning authorities on how development can affect flood risk and in identifying design measures that can reduce the risk.

The need to undertake assessments and conform to guidance will invariably lead to higher development costs in floodplain areas, although this may be offset by the low cost of floodplain land. The effects of design, location and land use through development decisions have profound impacts upon flood events. There is wide recognition that flooding and flood risk coupled with general changes in climate to milder and wetter winter conditions and increased peak rainfall events represents an important risk that needs to be managed appropriately. This is to ensure that existing development in flood areas is not put at greater risk through further development. Where possible, the risk to existing developments should be reduced and, where development is desirable in areas of flood risk, it should be appropriately flood proofed by incorporating water resilient materials in the design.

For existing buildings, flood mitigation measures such as resistance or resilience measures can be retrofitted perhaps during reinstatement following a flood event or other refurbishment (Royal Haskoning, 2012). The purchasing of household domestic insurance by floodplain residents can be seen as a form of mitigation strategy, and in the UK, most standard home insurance provides cover for the repair and reinstatement of damage caused by flooding. Hence, in the event of an inundation, homeowners will turn to their insurance company to assist in dealing with the immediate aftermath and in managing the reinstatement process.

This chapter therefore focuses on (i) the increasing challenges of flooding as a result of global warming, (ii) the need for a mitigation framework and appropriate design to minimise the effect of flood risk on people and building vulnerability, and (iii) the implications of the mitigation measures adopted in terms of building, insurance and the wider cost implications. The chapter then seeks to offer some conclusions regarding the long-term effect of embracing mitigation measures on development and insurance provisions.

15.2   Increasing challenges of flooding due to global warming and urban development

The frequency of extreme weather events, such as floods may be one of the most significant consequences of climate change globally (Jones, 1999). Moreover, during extreme flood events the threats to human societies and the environment are likely to be critical. Thus growing attention has recently been drawn, both from a scientific and political perspective, to understanding the risks of extreme hydrological events with regard to global climate change (Lehner *et al.*, 2006). Although there is much uncertainty about the effect of global warming, the consensus is that there will be more extreme weather events in the future and that sea level rise will be a key factor to be considered (Evans *et al.*, 2004; Jha *et al.*, 2012). Urban flooding is an already serious and growing challenge. Against the background of population growth, urbanisation trends and climate change, the causes of floods are shifting and their impacts are accelerating (Jha et al., 2012). This large and evolving challenge means that far more needs to be done by all stakeholders in the built environment to better understand and more effectively manage existing and future flood risks. This makes it even more important that the design of the built environment considers flood risk at the design and construction stages.

In the UK, the frequency of flood events within the last decade has been on the increase when compared with the past decades. Following this series of destructive floods across parts of England and Wales (from 1998 to 2010) flood risk management in the UK has undergone a series of radical reviews (Institution of Civil Engineers, 2001; Evans *et al.*, 2004; Defra, 2005). These reviews have proposed less reliance on hard engineering solutions such as flood defence and a move towards adaptation and resilience to flood risk. Evans *et al.* (2004) stressed that there is a need for a conceptual shift in which flood risk management relies less on Government intervention and more on an acceptance of individual responsibility; and for individuals to accept responsibility for managing flood risk, that is at the individual property level. This means that flood mitigation measures should be put in place to reduce the impact of flooding at the household level.

15.3   Flood mitigation

How flood mitigation is defined and interpreted is very important because it has an impact on how flood risk management professionals such as designers and flood risk assessors connect with others in the domain of flood risk management. Therefore a proper definition of mitigation is paramount to people's understanding of what actually is *mitigation.*

According to the Association of State Floodplain Managers (ASFPM) (2013) flood mitigation can be defined as 'any sustained action taken to reduce or eliminate long-term risk to life and property from a flood hazard event'. This is a good operational definition of mitigation when it comes to flood risk management as it provides the context upon which any mitigation action can be set up. Flood risk mitigation activities therefore, provide an essential foundation to reduce the loss-of-life and loss-of-property from flooding by avoiding or lessening the impact of a flood event (Medina, 2006). Common flood hazard mitigation activities include: implementation of flood adaptation measures; elevating, relocating or demolishing at-risk structures; retrofitting existing building or infrastructure to make it more flood resilient; and structural mitigation measures such as flood defence, levees, floodwalls and flood control reservoirs (ASFPM, 2013).

Avoidance measures

The best method of flood mitigation at property level would be to construct the building in an area with lower or no flood risk (Bowker *et al.*, 2007). In the case where buildings will be built on floodplain, Bowker *et al.* suggested that

Figure 15.1   A raised building development in the floodplain.

the threshold of the building be raised above the predicted flood level. This method is regarded as an *avoidance* measure. Typical mitigation options based on avoidance measures are: building out of the floodplain; buildings raised above flood level; and site drainage diverting water away from buildings. Figure 15.1 shows an example of a raised building development in the floodplain. The building was raised above the expected flood level, thereby mitigating against flood risk to the development. Other mitigation measures include the installation of resistance and/or resilience measures discussed in the following sections.

Resistance measures/dry-proofing

Resistance measures are designed to keep out, or at least minimise, the amount of water that enters a building (Defra, 2005). There are both temporary and permanent resistance measures which householders can implement. Temporary measures involve the installation of barriers which prevent flood water from reaching the property (Wingfield *et al.*, 2005). Permanent measures include waterproof doors and windows, automatically sealing airbricks which use devices such as flotation valves to seal the bricks, and automatic barriers. According to Thurston *et al.* (2008), temporary measures such as flood boards can reduce the cost of flood damage by about 50%. Further, such temporary measures are reported to make financial sense for properties in areas with an annual chance of flooding of 2% or higher, or areas where there is average likelihood of flooding once every 50 years (Thurston *et al.*, 2008).

It was reported that these permanent measures can reduce the cost of flood damage by 65–84% (Thurston *et al.*, 2008) and, because they operate automatically, they offer a better chance of protecting a building. Due to their high cost, automatic barriers may only make economic sense for premises in areas with very high flood risk, or where the potential cost of flood damage is high. However, where installation of flood resistance measures is not practicable or viable, for instance terraced properties where the adjacent neighbour is not taking similar mitigation measures; the adoption of resilience measures might be the preferred solution.

Resilience measures/wet proofing

Flood resilience of a building is a design measure that can reduce the damage that occurs to the fabrics of buildings from flooding; this measure has no ability to prevent floodwater from entering a building. It thus involves constructing a building in such a way that although floodwater may enter the building, its impact is minimised (Wingfield *et al.*, 2005; Joseph *et al.*, 2011a). Resilience measures aim to reduce the consequence of flooding by, for example, facilitating the early recovery of buildings, infrastructure or other vulnerable sites following a flooding event (Defra, 2005; Proverbs and Lamond, 2008).

Materials such as water-resistant paints and coatings, for example, can prevent floodwater soaking into the face of external surfaces. Materials such as lime-based plaster, as opposed to gypsum plaster, have good water-resilient properties and dry out quickly. Appropriately waterproofed solid concrete floors can also prevent water seeping into the fabric of a building. Other measures include re-fitting electrical sockets and electricity meter boxes above the anticipated flood levels.

Flood resilience measures reduce the cost of repairs after deep and prolonged floods, and can speed up restoration times. Due to the additional cost and disruption involved in implementing many of such measures, they are generally recommended for buildings with exceptionally high risk of flooding and are usually installed when restoring a building after it has been flooded or as part of planned renovations. This does not mean that they cannot be installed as a retrofit; however, the cost of installation will be significantly more. Despite the extra cost of these measures, it has been suggested that the implementation of resilience measures will reduce the repair costs in the long-term assuming repeat flooding (Thurston *et al.*, 2008).

However, making a property resilient or resistant to flooding is not a universal remedy. Some floods will cause structural damage or destroy well protected homes (Lamond and Proverbs, 2009). The effectiveness of the resilience measures taken would be greatly dependent on the expected velocity, depth and duration of the flood water and it has been established that in some cases these measures are not always cost effective (Thurston *et al.*, 2008; Lamond and Proverbs, 2009) and therefore a full flood risk assessment should be carried out before investing in resilient and resistant measures.

15.4   Flood mitigation consideration for new buildings at design stage

In a situation where building development on a floodplain is necessary, the new development can sometimes be protected by flood defences (Bowker *et al.*, 2007). However, this may not be possible for technical or environmental reasons, or because installed defences may increase flooding elsewhere. Where defences do exist there is always a possibility that they will fail or be overtopped by severe flooding. Indications are that the demand for housing and other urban amenities due to population growth means that some properties will still be built on floodplain. It is therefore important that prior to designing these buildings consideration should be given to avoidance, resistance and resilience measures to take account of the inherent flood risk in order to minimise the impact on buildings and people. If mitigation measures are considered at the design stage and carry over to the construction stage of the project, the cost of implementing the measures may be reduced in comparison with retrofitting existing buildings with flood mitigation measures.

There are various factors to consider when undertaking a site specific flood risk assessment for a new building at the design stage. Knowledge and understanding of the potential flood characteristics together with information on predicted flood levels at a site will, to some extent, assist designers in the formulation of a design strategy in order to achieve appropriate mitigation for potential flood risk. Although, it is worth stating that there are no definitive rules of thumb to design buildings to minimise flood damage. The discussion presented in this chapter presents several design suggestions, the choice of which will depend on many factors including the characteristics of flooding likely to affect a development site, designed use of the site, acceptable residual risk, financial considerations, and client needs.

The first stage in the design consideration is to determine the types of flooding and their characteristics which are likely to impact the development site. Important factors to consider which will influence the design of new buildings are: potential sources of flooding (pluvial, fluvial, or ground water flooding); flood velocity; expected duration of flooding; the frequency of flooding in the area; and the historic and expected flood depth. Figure 15.2 provides the decision support flowchart. As can be seen, the designer should be able to answer yes to most of the questions on the flowchart before proceeding with the proposed development. Where flood avoidance measures are not practical or possible as in the case where anticipated flood depth and speed will adversely affect evacuation processes, such development should not be embarked upon. In a situation where the anticipated flood depth and speed have no effect on emergency services during evacuation but there has been a need to raise the ground floor level above the anticipated flood level, consideration should be given to ease of access, cost of raising floor level and aesthetics. Where these are not acceptable, consideration should be given to other flood mitigation measures such as resistance and resilience (as discussed in the previous section).

15.5   Implications of mitigation measures in terms of building cost

The implementation of flood mitigation measures in an existing or new building has cost implications. There is a consensus in the literature that it is more cost beneficial to install resilience measures during planned refurbishment or during a

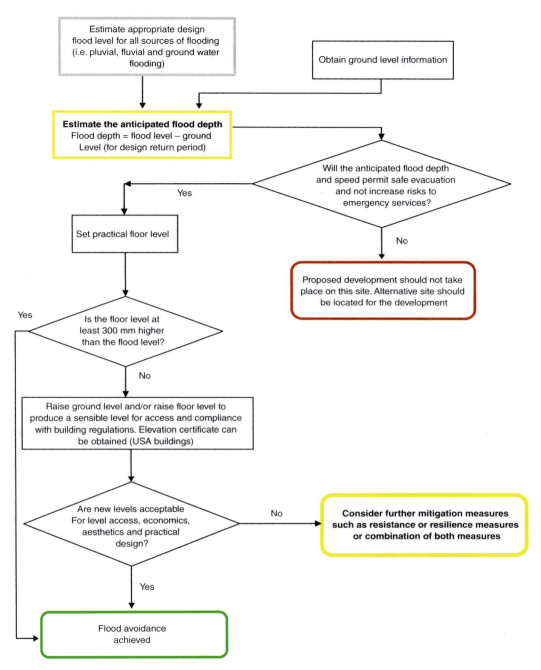

Figure 15.2   Mitigation decision support design flowchart for new buildings.
Source: Adapted from Bowker *et al.*, 2007.

flood repair process than it is to retrofit measures to properties which are at risk but not flooded and that installation during repair will also minimise disruption (Soetanto *et al.*, 2008; Thurston *et al.*, 2008; Joseph *et al.*, 2011b). Incorporating mitigation measures in a new building development will inevitably increase the

Table 15.1   Costs of resistance and resilience measures for different building types, flood depths and deployment methods.

Building type	Cost of resilience measures in flood depth (mm) categories (£)					Cost of resistance measures (£)	
	0–150	151–300	301–500	501–1000	>1000	Manual	Automatic
Bungalow	15,200	16,200	20,395	28,300		6,800	10,800
Detached	13,300	14,600	16,700	23,700	24,800	6,400	10,200
Semi-detached	12,500	13,600	15,800	15,000	22,600	4,000	6,700
Terraced	12,000	15,300	16,800	15,400	0,200	3,500	5,200

Source: Joseph (2014).

construction costs, the increase in cost will greatly depend on the mitigation measure required for the development; where there is a need to raise the ground floor level above the anticipated flood level, this may significantly increase the construction cost. This extra construction cost may affect the marketability and profitability of a development. The cost implication of mitigation measures are discussed in the following sections.

Cost of resistance measures

The cost of resistance measures will vary from building to building, this is due to the differences in shapes and sizes in each of the building types. Table 15.1 shows the summary of costs of a selection of resistance and resilience measures based on different building types, flood depths and deployment methods. The cost of resistance measures presented in Table 15.1 was based on resistance specifications presented in Table 15.2. The cost of manually deployed resistance measures is lower than the cost of automatically deployed measures. For instance the manually deployed resistance measures for bungalow and detached properties are 51% less expensive when compared with the cost of automatically deployed measures. Further, for semi-detached and terraced properties, the percentage difference in cost of manual and automatic deployed resistance measures are 60 and 43%, respectively. Thurston *et al.* (2008) acknowledged that due to the high cost of automatically deployed resistance measures, they are more likely to be less cost-beneficial; however, it was concluded that automatically deployed measures provide advantages in not needing to be deployed. Automatically deployed measures are more suitable for areas that are prone to flash-flooding, where there is a high proportion of elderly or disabled persons and where the deployment of temporary resistance measures is not possible prior to the onset of the flood.

Cost of resilience measures

Previous research carried out on behalf of the Association of British Insurers (ABI) (2009) revealed that, on average, resilient reinstatement costs over 40% (£12,000) more than traditional reinstatement (Wassell *et al.*, 2009) if implemented in an

Table 15.2 Resistance and resilience mitigation measures.

Resilience measures	Resistance measures	
	Manual	**Automatic**
Replace timber floors with concrete and cover with tiles	Singular panel demountable type	Supply and install automatic door guards
Replace carpet with ceramic tiles	Demountable panelled system fitted into channels	Supply and fix self-closing airbrick
Replace chipboard/MDF kitchen and bathroom units with plastic equivalents or stainless steel	Supply and airbrick cover	Supply and fix non-return valves 110 mm soil waste pipe
Replace gypsum plaster with more water-resistant material, such as lime plaster or cement sand render	Supply and fix sewerage bung	Supply and fix non-return valves 40 mm utility waste pipe
Apply water resistant paint to walls	Supply and fix toilet pan seal	Supply and fix non-return valves 12 mm overflow pipe
Move service meters, boiler, and electrical points well above likely flood level	Supply and install sump pump	Supply and install sump pump
Replace softwood timber skirting with plastic or hardwood and apply water resilient paint	Carefully apply silicone gel around openings such as cables, doors, windows, and so on. Assumed 5 m/property	Carefully apply silicone gel around openings such as cables, doors, windows, and so on. Assumed 5 m/property
Replace softwood door and window frames with water resilient alternative	Carry out waterproofing work on external walls (up to 1.2 m high) Water sealant spray to external elevation up to 1.2 m high	Carry out waterproofing work on external walls (up to 1.2 m high) Water sealant spray to external elevation up to 1.2 m high
Wall insulation: Replace mineral insulation with cell insulation	Supply and install garage/driveway barrier	Supply and install garage/driveway barrier

Source: Joseph (2014).

existing building. It was stressed that there are significant variations around this 40% average, both between building types (i.e. bungalow, block of flats, terraced, semi-detached and detached houses) and within house types. Although, the authors further reiterate that resilient reinstatement could cost as little as 15% or as much as 70% more than traditional reinstatement. The availability of guidance, such as Proverbs and Soetanto (2004), Garvin *et al.* (2005) and PAS 64 (2005), on how to repair flood damaged properties cannot guard against this wide variance in percentage extra over cost because of the individual preferences of homeowners and different approaches to reinstatement methods adopted by different surveyors. Some resilience measures can be introduced on a cost neutral basis such as raising electrical sockets above flood level, and therefore not all aspects of resilience measures lead to additional cost.

As shown in Table 15.1, the additional cost of resilience measures is the extra over cost incurred during reinstatement of flood damaged buildings to make those properties resilient against future flooding. The cost ranges from as low as £12,000 for a terraced house flooded to a depth of 150 mm and as high as £28,300 for a bungalow flooded to a depth of 1000 mm. The additional cost of resilience measures presented in Table 15.1 was based on resilience specifications presented in Table 15.2. Understandably, as the depth of the floodwater increases, so does the cost of resilience measures, therefore, accuracy of flood depth is important when estimating the cost of resilience measures either during repair work or in a new building. Getting this wrong may invalidate the resilience measures which were taken and at times it may lead to even higher costs. For example, the cost of replacing a cement sand render in a property with expected flood level of 900 mm is expected to be greater than when doing the same to a property with expected flood level of 300 mm.

When the decision is made to incorporate resilience mitigation measure at design stage for a new building, this can significantly reduce the cost incurred in making the building flood resilient. This is because the cost presented in Table 15.1 can be offset against other cost elements in the design because not all flood resilient materials are more expensive than non-resilient material. Design decisions in the form of value engineering for a new build can be adjusted to make up for resilience mitigation measure, for instance where timber skirting would have been installed; if this is replaced with plastic skirting which is less expensive when compared with its timber counterpart, this will subsequently reduce the total additional cost incurred in implementing resilience mitigation measures in such buildings. Design decisions on flood resilience mitigation measures should consider more than just the cost implication because the quality of a property can be affected by decisions made at design stage, for instance, marble stone flooring is flood resilient but may also enhance the value of a property.

15.6   Implications of mitigation measures in terms of property value and insurance cost

Insurance against flooding can be a central plank in a national flood management strategy, indeed it can often incorporate state provision, and as such is subject to political as well as market forces. Flood insurance must be considered within the context of a wider flood management regime and the structure of the local insurance market will be very pertinent to the way in which flood mitigation measures affect the cost of insurance. The effect of implementing mitigation measures on insurance cost will greatly depend on the type of insurance regime in operation. In the UK, the Government relies on private insurance companies to pick up a large part of the bill for repairing flood damaged buildings (Huber, 2004). The UK system is fairly unique in providing such universal cover by default. In other countries the picture is very different as noted by several authors such as Hausmann (1998), Michel-Kerjan (2001), Paklina (2003), Lamond and Penning-Rowsell (2011) and Lamond (2013). Other approaches range from no cover, with victims relying on emergency aid, through optional separate flood policies or add-on options, to compulsory state run schemes (Lamond, 2008).

In the US, flood insurance is managed under the National Flood Insurance Program (NFIP). In the past the NFIP has provided heavily subsidised basic cover in flood risk areas that have met some basic community level protection requirements. Recent changes to the NFIP in the US have brought in more stringent requirements, for example to obtain an elevation certificate to prove that the floor level of the property has been raised above the likely flood threshold. According to the Federal Emergency Management Agency (FEMA) (2012), the elevation certificate is an important administrative tool of the NFIP. It is to be used to provide elevation information necessary to ensure compliance with community floodplain management ordinances and also used to determine the proper insurance premium rate.

According to theory, any additional cost of mitigation measures, which leads to an increase in the cost of buildings, should be offset by reduced risk of incurring damage costs and an associated increase in property value (Lamond, 2012). This is the case in some disclosure and insurance regimes but the action of insurance can confuse the market, particularly if the premiums are not risk based because it effectively protects the utility value of property at risk. Under the Statement of Principles (SOP) the UK domestic insurance provision is not fully risk based. However, Lamond (2012) concluded that a move toward risk-based pricing for UK floodplain residents could mean that some high-risk residents would pay significantly more for their insurance, find their properties insurable or be effectively uninsured due to exclusion and excess conditions. If mitigation measures have been implemented, this has the potential to make such property insurable, although conditions may still be imposed by insurers.

Action to mitigate flood risk has the potential to help in keeping property insurance affordable in the future through a system of reduced premiums for installed protection. However, due to the current insurance regime in the UK, and costs and difficulties in administering such a system, insurance companies have not generally recognised by way of reduction in insurance premium, homeowners who have invested in mitigation measures. It can be concluded that implementation of mitigation measures does not necessarily affect insurance cost for existing buildings in the UK due to the operation of the SOP (Huber, 2004). The SOP is in the process of being phased out and it is proposed to replace it with the 'Flood Re'. The Flood Re is an agreement in principle between UK insurers and the Government to develop a not-for-profit flood fund, to ensure that flood insurance remains affordable and available to homeowners at high flood risk while allowing premiums to reflect risk. However, neither the SOP nor the Flood Re agreement between the insurance company and UK Government guarantee that new development in floodplain area will be insured. Therefore the need to comply with PPS25 is of paramount importance in order to mitigate flood risk on new buildings. Furthermore, while it can be said that in the UK, for existing buildings, the financial incentive for property owners to undertake mitigation is low, this may change in the future if insurance premiums are set on the basis of flood risk being faced by individual properties.

15.7   Conclusions

Exisiting flood risk and the projected increase in flood risk to buildings as a result of global warming and urban development pressure is an important factor for consideration in the design of the built environment. The need to mitigate against such risks calls for design consideration at the initial stage of new developments, notwithstanding the need to protect existing buildings which remains of significant importance. In this chapter, three main mitigation measures have been discussed: avoidance, resistance and resilience measures. Avoidance measures can mean building out of floodplain, and where this is not possible due to high demand for housing, then the building can be raised above flood level or drainage can be diverted away from buildings. Resistance mitigation measures involve keeping water out of a building or minimise the amount of floodwater that enters a building, whilst resilience mitigation measures are implemented in order to reduce the damage effect of flood water on the internal fabric of a building. The choice of which measure to implement depends greatly on the level of flood risk being faced and their associated cost implications and some illustrative cost samples are given in this chapter.

The economic implication of mitigation measures can be assessed at the design stage especially for new development. This can be offset or reduced by the fact that the floodplain lands may be bought at a cheaper price, and therefore the cost of complying with relevant development planning requirements such as PPS25, should not significantly increase the cost of the development. If retrofitting of mitigation measures is necessary, resistance measures cost less than resilience measures in general but sometimes resilience measures can be cost neutral when installed as part of reinstatement or planned refurbishment. The choice of measures is still dependent on flood characteristics but existing building typology and condition coupled with owner needs and preferences is equally important.

Investing in mitigation measures is often not rewarded through increased property value or lower insurance premiums in the UK because currently insurance premiums are not risk based. While regulation and insurability considerations are likely to provide a sufficient trigger to invest in mitigation for new property, for existing homes there are few incentives for homeowners to invest in mitigation measures purely on a financial basis. In contrast the NFIP has moved towards directly incentivising mitigation through avoidance. Although, the position in the UK insurance market may change in the near future due to the transition from the SOP to the Flood Re agreement the effectiveness of the Flood Re agreement is not yet possible to predict. Under these circumstances the role of the building professional in encouraging appropriate mitigation measures, at design and throughout the building life cycle, remains critical.

References

ABI (Association of British Insurers) (2009) Delivering on our ClimateWise commitment: Second Year Report. London.

Arup (2011) Adapting to Climate Change in the UK: Measuring Progress. Adaptation Sub-Committee Progress Report 2011. London.

ASFPM (Association of State Floodplain Managers) (2013) How-To Guide for No Adverse Impact-Mitigation. Madison, WI.

Blunden, J. and Arndt, D.S. (eds) (2012) State of the climate in 2011. *Bulletin of the American Meteorological Society*, **93**(7), S1–S264.

Bogardi, J. (2004) Hazards, risks and vulnerabilities in a changing environment: the unexpected onslaught on human security? *Global Environmental Change*, **14**, 361–365.

Bowker, P., Escarameia, M. and Tagg, A. (2007) Improving the Flood Performance of New Buildings: Flood Resilient Construction. Department for Communities and Local Government, London.

Defra (2005) Making Space for Water: Taking forward a New Government Strategy for Flood and Coastal Erosion Risk Management in England. London.

Environment Agency (2009) Flooding in England: A National Assessment of Flood Risk. Bristol.

Evans, E.P., Ashley, R.M., Hall, J. *et al.* (2004). *Foresight. Future Flooding. Vol. I - Future Risks and Their Drivers*, Office of Science and Technology, London.

FEMA (Federal Emergency Management Agency) (2012) National Flood Insurance Program Elevation Certificate and Instructions. Department of Homeland Security.

Field, C.B. (2012) Summary for policymakers, in Managing the Risks of Extreme Events and Disasters to Advance Climate Change Adaptation: A Special Report of Working Groups I and II of the Intergovernmental Measuring the Human and Economic Impact of Disasters Panel on Climate Change. Cambridge University Press, Cambridge.

Garvin, S., Reid, J. and Scott, M. (2005) *Standards for the Repair of Buildings following Flooding*, CIRIA, London.

Guha-Sapir, D., Parry, L., Degomme, O., Joshi, P. and Arnold, J. (2006) Risk Factors for Mortality and Injury: Post-tsunami Epidemiological Findings from Tamil Nadu. CRED-UNISDR Report. CRED, Brussels.

Hausmann, P. (1998) Floods – an insurable risk? A market survey. Swiss Reinsurance Company. http://daad.wb.tu-harburg.de/fileadmin/BackUsersResources/IFM/Documents/swissre_floods_an_insurable_risk.pdf (accessed 31 July 2014).

Huber, M. (2004) *Reforming the UK Flood Insurance Regime – The Breakdown of a Gentlemen's Agreement*, London School of Economics, London.

Institution of Civil Engineers (2001) Learning to Live with Rivers. Final report of the Institution of Civil Engineers' Presidential Commission to review the technical aspects of flood risk management in England and Wales. London.

Jha, A.K., Bloch, R. and Lamond, J. (2012) Cities and Flooding: A Guide to Integrated Urban Flood Risk Management for the 21st Century. The World Bank, Washington, DC.

Jones, J. A. A. (1999) Climate change and sustainable water resources: Placing the threat of global warming in perspective. *Hydrological Sciences Journal,* **44**(4), 541–557.

Joseph, R. (2014) Development of a comprehensive systematic quantification of the costs and benefits (CB) of property level flood risk adaptation measures in England. PhD thesis, The University of the West of England, Bristol.

Joseph, R., Proverbs, D., Lamond, J. and Wassell, P. (2011a) An analysis of the costs of resilient reinstatement of flood affected properties: A case study of the 2009 flood event in Cockermouth. *Structural Survey*, **9**(4), 279–293.

Joseph, R., Proverbs, D., Lamond, J. and Wassell, P. (2011b) A critical synthesis of the tangible impacts of flooding on households. 27th ARCOM Annual Conference, 5–7 September 2011, University of the West of England, Bristol.

Lamond, J. (2012) Financial implications of flooding and the risk of flooding on households, in *Flood Hazards Impacts and Responses for the Built Environment* (eds J. Lamond, C. Booth, F. Hammond and D. Proverbs), CRC Press, Boca Raton, FL, pp. 317–326.

Lamond, J. (2013) The role of market based flood insurance in maintaining communities at risk of flooding – a SWOT analysis, in *Water Resources Issues and Solutions for the Built Environment* (eds C. Booth and S. Charlesworth), Wiley-Blackwell, Oxford, pp. 258–269.

Lamond, J. and Penning-Rowsell, E. (2011). A Review of International Approaches to Flood Insurance. University of Wolverhampton, Middlesex University, Flood Hazard Research Centre, UK.

Lamond, J.E. (2008) The impact of flooding on the value of residential property in the UK. hD thesis, University of Wolverhampton.

Lamond, J.E. and Proverbs, D.G. (2009) Resilience to flooding: lessons from international comparison. *Urban Design and Planning*, **162**(DP2), 63–70.

Lehner, B., Doll, P., Alcamo, J., Henrichs, T. and Kaspar, F. (2006) Estimating the impact of global change on flood and drought risks in Europe: a continental, integrated analysis. *Climatic Change*, 75, 273–299.

Medina, D. (2006) Benefit-Cost Analysis of Flood Protection Measures. Report for the Metropolitan Water Reclamation District of Greater Chicago, USA.

Michel-Kerjan, E. (2001) Insurance against natural disasters: do the French have the answer? strengths and limitations. Ecole Polytechnique, Paris, working paper.

Paklina, N. (2003) Flood Insurance. http://www.oecd.org/finance/insurance/18074763.pdf (accessed 31 July 2014)

PAS 64 (2005) Professional water mitigation and initial restoration of domestic dwellings. Code of practice. British Standard Institution, London.

Proverbs, D. and Lamond, J. (2008) The barriers to resilient reinstatement of flood damaged homes. Proceedings of 4th International i-Rec Conference-Building Resilience: Achieving Effective Post-disaster Reconstruction, 30 April–2 May 2008, Christchurch, New Zealand.

Proverbs, D. and Soetanto, R. (2004) *Flood Damaged Property: A Guide to Repair*, Blackwell, Oxford.

Royal Haskoning (2012) Assessing the Economic Case for Property Level Measures in England. Committee on Climate Change. Peterborough.

Sarewitz, D., Pielke, R. and Keykhah, M. (2003) Vulnerability and risk: some thoughts from a political and policy perspective. *Risk Analysis* 23(4), 805–810.

Soetanto, R., Proverbs, D., Lamond, J. and Samwinga, V. (2008) Residential properties in England and Wales: An evaluation of repair strategies towards attaining flood resilience, in *Hazards and the Built Environment: Attaining Built-in Resilience* (ed. L. Bosher), Routledge, London, pp. 124–149.

Thurston, N., Finlinson, B., Breakspear, R., Williams, N., Shaw, J. and Chatterton, J. (2008) *Developing the Evidence Base for Flood Resistance and Resilience. Joint Defra/EA Flood and Coastal Erosion Risk Management R&D*, Defra, London.

Wassell, P., Ayton-Robinson, R., Robinson, D. *et al.* (2009) *Resilient Reinstatement: The Costs of Flood Resilient Reinstatement of Domestic Properties*, Association of British Insurers, London.

Wingfield, J., Bell, M. and Bowker, P. (2005) Improving the Flood Resilience of Buildings through Improved Materials, Methods and Details. Report Number WP2c: Review of Existing Information and Experience. Office of the Deputy Prime Minister.

Part II
Industry Perspective, Case Studies and Implications for Curriculum Development

Chapter 16
Reusing Knowledge and Leveraging Technology to Reduce Design and Construction Costs

Herbert Robinson and Chika Udeaja

16.1 Introduction

Construction is a vast industry accounting for a significant proportion of Gross Domestic Product and employing millions of people. The industry is well known for its products – buildings, roads, bridges, dams, and monuments – requiring knowledge from specialists to meet the needs of diverse clients. Construction projects have a history of not delivering on time, budget and within the scope of the project. There are many reasons identified for poor performance including inadequate communication between project teams, frequent design changes and constraints in knowledge reuse during project implementation. The lack of an effective knowledge management strategy has stifled the industry's ability to reuse knowledge and to leverage various technologies. This chapter focuses on knowledge reuse and leveraging technology to reduce project duration, design and construction costs. It starts with an overview of construction as a project-based industry, characterised by distinct processes and multidisciplinary teams with different roles. The types of knowledge and knowledge management tools are discussed using examples from leading consulting practices and case studies to demonstrate the benefits in terms of reducing costs and duration of construction projects.

16.2 Knowledge reuse in construction processes and projects

Construction is a project-based activity, with interrelated processes from planning, design, construction to operation. Research conducted by den Hertog and Bilderbeek (1998) and Windrum et al. (1997) identified design and other construction processes as knowledge-intensive activities. Quantity surveying

Design Economics for the Built Environment: Impact of Sustainability on Project Evaluation, First Edition. Edited by Herbert Robinson, Barry Symonds, Barry Gilbertson and Benedict Ilozor. © 2015 John Wiley & Sons, Ltd. Published 2015 by John Wiley & Sons, Ltd.

Table 16.1   Traditional and new services offered by quantity surveying Firms.

Traditional services
- Preparing tender documents (usually based on bills of quantities)
- Examining tenders, appraisal and providing advice on selection
- Valuing works for interim payments and certification
- Measuring and valuing variations
- Advising on anticipated final costs
- Preparing and agreeing final accounts

New and emerging services
- Procurement and project/construction management
- Legal and contract services – contract documentation
- Planning supervision
- Taxation/capital allowance advice
- Value and risk management
- Asset and facilities management
- Construction product advice based on whole life cycle costing
- Management and business consultancy including corporate sustainability
- Sustainable design (energy and water efficiency), carbon planning and management
- Dispute resolution and claims management
- Technical advice on complex procurement systems such as PFI/PPP

firms as design economists offer a range of services as shown in Table 16.1 to reduce project duration, costs of design schemes and to provide value for money to clients.

Knowledge is at the core of design and quantity surveying activities and a significant proportion of knowledge relevant to other projects can be reused (Fong and Cao, 2004; Davis et al., 2007). Specialist knowledge is 'lost from one project to the next stifling an organisation's ability to develop knowledge and generate new ideas' (Egbu and Botterill, 2002, p. 129) necessary to reduce project duration, design and construction costs. However for project knowledge to be reused, it must be captured in appropriate formats with a good understanding of the context. The types of reusable knowledge are related to the range of services in Table 16.1.

Key benefits in reusing knowledge include competitive advantage and improved relationship with clients leading to better design outcomes and the minimisation of waste. The emergence of the knowledge intensive organisations such as professional services firms in accounting, engineering, architecture, surveying, law and management consulting has seen intellectual capital (knowledge) replace physical capital. Chang and Birkett (2004) argued for a balance between creativity and productivity in knowledge intensive organisations. There are many opportunities for improving productivity and creativity in project processes. A typical construction project cycle consists of stages from planning to maintenance. For example, the Royal Institute of British Architects (RIBA) Plan of Work (RIBA, 2013) has eight stages (Table 16.2).

The various activities are carried by specialist teams such as architects, engineers, quantity surveyors (as design economists), town planners, developers, contractors and specialist trade or sub-contractors. Reusing project knowledge and leveraging some of the technologies identified in Table 16.2 can reduce project duration and save costs. However, this requires a knowledge strategy to overcome specific problems such as the temporary or one-off nature of construction projects with

Table 16.2   Example of a construction project cycle (RIBA Plan of Work).

Project cycle	Stage	Sub-activities	Cost activities	Examples of tools used
Planning	0	Strategic definition	Feasibility studies Realistic first estimate	RIPAC, WINQS CostX
	1	Preparation and Brief	Preparation of estimates	RIPAC, WINQS CostX
Design	2	Concept design	Cost planning Cost limit Cost allocation	Project Management tools RIPAC, QSPlus CostX, BICS database
	3	Development design	Bill of quantities Detailed cost plan	RIPAC, Masterbill WINQS, QSPlus CostX
	4	Technical design	Schedule of materials Specification	RIPAC WINQS
	5	*Specialist design*	Cost analysis	RIPAC, WINQS
Construction	6	Construction	Valuation Cost control	RIPAC, ConQuest WINQS, Masterbill
Maintenance	7	Use and Aftercare	Final account	RIPAC Masterbill

teams often disbanded at the end of the project leading to 'discontinuities in knowledge reused within and between organisations' (Blayse and Manley, 2004).

16.3   Knowledge reuse in construction projects

Creativity can be nurtured and improvements realised in the construction process by interaction of various types of knowledge – explicit and tacit. Explicit knowledge can easily be processed by a computer, transmitted electronically, or stored in databases (Nonaka and Takeuchi, 1995). Explicit knowledge is stored as written documents or procedures. This type of knowledge is codifiable, reusable and therefore easier to share. Examples of explicit knowledge in construction include design codes of practice, performance specifications, drawings in paper-based or electronic format, construction techniques, and materials testing procedures, design sketches and images, 3D models and construction textbooks (Egbu and Robinson, 2005). Tacit knowledge is described as being 'resident within the mind, understanding, perception and know-how of individuals' and includes 'skills, experiences, insight, intuition and judgement' [British Standards Institute (BSI), 2003]. Examples of tacit knowledge include judgement on estimating and tendering skills acquired by quantity surveyors or estimators over time through hands-on experience for preparing bids for construction projects as a result of understanding tender markets, interaction with suppliers, various specialists and clients/customers to develop project needs. This type of knowledge is experiential, judgemental, context-specific and therefore difficult to codify and share.

Reusing knowledge involves automating human and technical processes and/or creating a network of people involved in design and construction to minimise waste, prevent the duplication of effort, avoid the repetition of similar mistakes from past projects and improve efficiency (Kamara *et al.*, 2002, p. 58). For example, 'reuse of knowledge gained via past experience can greatly reduce the time spent on problem solving and increase the quality of work' (Dave and Koskela, 2009, p. 894). This creates a commercial advantage for a company as they become more efficient and produce a higher quality of work through the reuse of knowledge (Dent and Montague, 2004). A well-managed programme of 'knowledge re-using starts at the planning/tendering stage and continues after practical completion, to greatly enhance client satisfaction' (BSI, 2003, p. 6) and to reduce both capital and operational costs associated with maintenance. The increased likelihood of further commission through reappointment with the client, for example, can also reduce planning, design and construction costs.

16.4   Leveraging knowledge systems to reduce time and costs

Various knowledge management tools have been in use over time to increase productivity and to foster creativity in the design and construction process. Take a simple example of the traditional method of drawing. This approach was very labour intensive and time-consuming as the process of making even minor changes to paper drawings was tedious, repetitive and costly. The introduction of CAD revolutionised the design process, reduced the time and cost associated with design, making changes to design as well as the incidents of defects during construction as a result of the early detection of conflicting design information in drawings. Arup Associates, one of the pioneers of CAD, had a separate system for project cost management (Oracle based Project Control System) using quantities manually determined from CAD drawings. The introduction of digitisers, helped in transforming quantities, through manual takeoffs, or importing CAD drawings into the early cost estimating packages. Jeff Gerardi, President of ProEst Estimating Software, argued that performing 'takeoffs using electronic plans and a mouse is about 50% faster than manual takeoff methods, which is a measurable efficiency gain that is easily converted to a cost savings'(Newsletters, 2013a). There are significant productivity increase using these methods as human error and inaccuracies are minimised.

Over the decades, there has been a rapid development in the use of knowledge systems to automate both the design, and estimating process from CAD to Building Information Modelling (BIM) or what is appropriately called Building Knowledge Modelling (BKM). By using BIM, all measurements are generated directly, and whenever there is a change in design, the measurements are automatically updated leading to greater accuracy in quantities and significant productivity increase. It is estimated that about 50–80% of time needed to create a cost estimate is spent on quantification. According to Revit 'by automating the tedious task of quantifying', BIM allow estimators "to use that time instead to focus on higher value project specific factors – identifying construction assemblies, generating pricing, factoring risks – that are essential for high quality estimates" (Autodesk, 2013a). Revit identified one customer with significant productivity gains experienced using Revit Architecture based on 20 completed projects – the firm had registered a productivity gain of 30% in design and documentation and a 50% drop in request for information during construction (Autodesk, 2013b). Applying these technologies would result in much earlier and accurate advice to meet clients' cost expectations. BIM, as an integrated design, cost and time management system has the potential to foster creativity, improve productivity of design, leading to a significant reduction in construction costs and time associated with delivering construction projects. BIM is discussed in extensively in Chapter 21.

There are many other good examples of reducing project duration, design and construction costs indirectly or directly through knowledge reuse by leveraging various technologies. Over a decade ago, Arup, a leading engineering consulting firm, highlighted that feedback from their legal department shows that the single largest cause of loss of money within the firm was a failure to agree the appropriate contract terms up front. The knowledge manager explained that a knowledge management system such as the collation of a legal Intranet page pushed to the desktop at appropriate times in projects was an effective solution to the problem (Robinson et al., 2005). Other examples cited in Robinson et al. (2005) include Texas Instruments who saved US$500 million in the cost of building a new silicon wafer fabrication plant by disseminating best internal working practices to improve productivity in existing plants. Skandia AFS also reduced the time taken to open an office in a new country from 7 years to 7 months by identifying a standard set of techniques and tools which could be implemented in any new office. Automatic identification technologies such as barcoding widely applied in manufacturing, retail, medicine and in the US construction industry offer cost savings through improved speed and accuracy of records. Use of bar coding for labelling materials and components as knowledge systems has reduced material wastage, theft from construction sites and other losses associated with poor tracking of supplies. A revolutionary technology developed called contour crafting has the ability to build a two-storey house in 24 h (Weston, 2010). The robot "prints" buildings by squirting successive layers of concrete on top of one another to build floors, walls and roofs and can also integrate mechanical and electrical services within the structure.

Mohammed (2008) cited numerous examples of knowledge tools used by leading design and construction firms. For example, in 1996 Atkins developed its first Intranet/Extranet and online project management system ProNet as a central repository of company information and resources. The subsequent knowledge management system, called iProNet (Internet Project Network) developed in 1999 improved coordination from PFI/PPP projects by allowing 'virtual team members' to collaborate on projects in a protected environment, sharing folders and files, tracking activity with one-third of users external to Atkins. Knowledge management in Arup can be traced back to the founder of the firm, as reflected in Ove Arup's key speech in 1942 'The wealth of new knowledge, new materials, new processes'. An earlier example is the Overguide, a directory of engineers with their experience, and a Skills Network. Arup had a variety of formal Project-based Information Systems for over 30 years including Ovabase, a project database and Autonomy with features such as information access technology, business intelligence, customer relationship management, compliance and litigation solutions all integrated in the Intranet system.

The concept of knowledge sharing first became a group agenda in early 1990s when AMEC had over 20,000 employees, with more than 20,000 people in the associate company, SPIE S.A., in 700 locations around the world. The response was to deploy the HummingBird Document Management System in 1996 supported by a KM system called ASK (AMEC's Shared Knowledge), which enabled users to access centres of excellence, profiles and communities of best practice. Capturing best practices from other parts of AMEC was crucial. A good example was Heathrow Terminal 5 (squeezed between Heathrow's busy runways, terminal buildings and one of the world's busiest motorways) where space was a challenge. There was not enough room for the workforce and materials. To address this challenge, AMEC brought the concept of modularisation from its oil and gas sector, where it was used for a number of years in producing offshore platforms.

Costain's Document Management launched in 2004, iCosNet was the main initiative designed by the company as the heart of the knowledge strategy. iCosNet had several key features such as Navigator engine allowing project managers to 'follow the job' through a series of charts, prompting the user to download standard forms and information – ensuring they are following Costain's international best practices at all times. For example, Photo Library provides an online access to photographs of past and current projects, a project collaboration tool allowed users to access and exchange information from anywhere in the world with a Supply Chain Management system for suppliers and sub-contractors.

These knowledge systems led to significant savings associated with construction projects. The examples also show that different tools are used depending on their added value to the organisation. However, for the next-generation knowledge tools, Anumba (2009) argued that the tools need to make seamless the linkage between knowledge capture/reuse and construction business processes. New technologies are sometimes seen as creating extra workload, as project documents are transformed into electronic form, and then uploaded to the Extranet. The key to success of any knowledge system is not to create additional workload but to integrate knowledge capture activities for learning in daily job functions in normal working hours. The Extranet can enable project documents to be tracked automatically with a clear audit trail of revisions and approvals to drawings and other project documents.

16.5   4Projects knowledge solution

4Projects has been a provider of collaborative web-based solutions for project management since 1997. The company started as a subsidiary of an international housing and development company but subsequently became independent from the parent company and was later acquired by a US-based firm (Newsletters, 2013b).

4Projects's online project collaboration platform to facilitate the management of knowledge in the form of drawings, specification and construction documents is for construction firms with multiple simultaneous projects to increase document and project control throughout the project life cycle. According to Bob Humphreys, Vice President of Product Management, 4Projects 'allows your project teams to access and act upon a single set of construction documents and provides deep functionality and full audit trails' (Newsletters, 2013b). The Construction Manager (2013) added that 'on-line collaborative working is one tool that can drive down costs and improve process control'.

For example, 4Projects developed specific modules for discussion, query/action, team and organisation, and the workflow and approval process.

The '**Discussion**' module, similar to an online communities of practice, allow users to post topics for discussion, which can be open or restricted to a group of selected members within the project Extranet. Recipients are notified of new discussion topics via email alert to debate issues of interest and to share tacit knowledge. Comments posted in the discussion forum are linked to items such as documents, drawings, photographs, queries and other material uploaded to the project Extranet. A 'Search' function is provided for the users to search for discussion through keywords. The stored documents are searchable using the file name or keyword of the file description. Users can search and obtain a list of relevant documents or previous examples containing learning from other projects.

A 'Query/Action' module, also referred to as a 'Task' module, is available for users to request information, ask questions or issue instructions relating to the execution of a project. The user can issue a query or request for information to a specific person or a selected group of people with the deadline. The status of a query is displayed to allow tracking of the progress:

- Overdue – where the specified recipient does not give a reply on or before the deadline.
- Closed – where the 'issuer' of the query obtained the desired answer/reply from recipient.
- Open – where there is on-going correspondence between issuer and recipient on the query.

The 'Query/Action' module can facilitate a more 'forceful' way of sharing knowledge, in contrast to the voluntary knowledge sharing in online communities of practice, as the identified source of knowledge is required explicitly to respond to the request within the deadline specified and the progress is traceable.

The 'Team and Organisation Directory' is for storing key information on each organisation and individuals working on a project. It helps in locating a person's contact details. Adding a new 'field' like 'skills and expertise' for team members and then linking it to a cross project and cross organisation 'search module' can help in upgrading the module to a 'knowledge catalogue' to assist others to locate people with certain expertise.

The 'Workflow and Approval Process' module provides an audit trail for all the items on the Extranet. Documents can either be approved, rejected or under review. This particular feature facilitates the peer-review process for documented learning and seeks suggestions for improvement before knowledge is formally tagged for reuse as 'best practices' or 'lessons learned'.

In addition to the modules to facilitate document, communication and process controls, other functionalities have been added such as E-Forms, Milestone Management, Tender Management and Contract Management. In 2010, 4Projects included a 'Contract Manager' module to reduce the 'administrative burden by automating processes, communication, reporting and notifications. The module will enable any type of contract to be incorporated into its collaboration platform from NEC, JCT, GC Works, PPC2000 to firms' bespoke contracts (Construction Manager, 2013).

16.6   Case studies and discussions

The entire database in 4Projects Extranet for a project including all files, metadata and audit data can be archived and accessed again in the future. Users can access the knowledge captured by the Extranet modules and reapply it to other projects with adaptation. The following case studies are extracts from the 4 Projects's website.

Case study on leisure and entertainment sector

A leading organisation own and manage a network of over 2,000 branded pubs, bars, restaurants and leisure venues including some of the most popular entertainment facilities in the UK. A major challenge for the organisation was the sheer number of projects to be managed effectively in terms of reducing the delivery time, design and construction costs. A web-based solution (4Projects) was adopted to maximise the performance of the refurbishment programme and to develop a central knowledge base to manage multiple projects. 4Projects provided an interface for project members with an instant snapshot of the progress of all projects using a consistent format for sharing project files throughout the supply chain.

The company was able to rollout the web-based solution to thousands of users in a very short space of time as it was designed to be user-friendly. A number of features made the interface easier for construction workers such as the online versions of traditional drawing issue sheets. The 4Projects system provided a project team with a one-stop-shop for up-to-the-minute project information, resulting in significant savings in time, and allowing the team to keep to tight schedules. Little or no training was required to use the *point and click* menus and *hypertext links* to navigate project information and authorised users could add new projects in seconds. The online system had different components such as the *project scheduling and programme* component for managing developments, replacing the organisation's complex system of spreadsheets. The online management tool was used for collecting, viewing and reporting on the development status of projects within the estates portfolio. Individual project data are captured such as:

1. Project type (e.g. acquisition, conversion)
2. Associated project team (e.g. architect, engineer, quantity surveyor, etc.)
3. Proposed rollout programme based on dates (e.g. build start, build complete).

The *project cost monitoring* component was used for collecting, viewing and reporting on the financial aspects of managing the estates portfolio. Budgeted, forecast and actual costs can be assigned to each project. Traditionally both project scheduling/programme and project cost monitoring were managed in-house but changes in business practices combined with using web-based project solution meant it could be outsourced but still under the control of the organisation. Providing this data online through the web-based solution (4Projects) provided a secure way of accessing live and confidential project data in real time.

Case study on health sector

The National Health Service in the UK spends around £100 billion and employs about 1.2 million people (Storey et al., 2008). There are over 400 Healthcare Trusts whose principal aim is to provide local healthcare services through hospitals and GP surgeries across the UK. With the rapid changes in the health sector there has been renewed emphasis on improving performance throughout the health sector particularly for capital projects. Several NHS Trusts led the way in adopting well proven project management solutions. The 4Projects Extranet and tendering solutions, accessible 24 h a day, enable sharing of critical business information in a secure environment. Some of the NHS Trusts that use the solution include:

- Wolverhampton NHS Trust
- Leeds Teaching Hospital NHS Trust
- North Glamorgan NHS Trust
- Portsmouth Hospitals NHS Trust
- Royal Orthopaedic Hospital NHS Trust
- Gwent NHS Trust.

Portsmouth Hospitals NHS Trust was the first to discover the benefits of the 4Projects Extranet on a PFI scheme that involves the redevelopment of Queen Alexandra Hospital site, with an estimated capital value of £170 million. Project information was shared between the client, funders, lawyers, designers and contractors. 4Projects enabled quick and effective communication between the consortia, project team and other staff involved in the project. Using 4Projects facilitated an instant viewing of the latest documentation and drawings in the project with a user-friendly interface. 4Projects was used by Leeds Teaching Hospital NHS Trust with two consortia reviewing tender documents and submitting bids. The result for bidders and advisers was increased efficiency, by accessing electronic documents in a Virtual Data Room from their offices throughout the UK or even overseas, as an alternative to physically visiting the Project Data Room at the Hospital to view paper documents.

North Glamorgan NHS Trust signed an Enterprise agreement allowing them to run unlimited projects using the Extranet system. 4Projects has been set-up for the refurbishment programme at the Prince Charles Hospital and the new Acute Mental Illness Unit. Performance benefits delivered by using an online collaboration system were substantial and far superior to project managing information by traditional paper methods. Medicor (a consortium comprising of Pearce Health, Parsons Brinkerhoff, MJN Colston, Davis Langdon & Everest, Integral and United Healthcare) adopted 4Projects for a number of health schemes in the UK. A key objective for using the 4Projects Extranet was the use and reuse of standards, knowledge and best practices shared effectively throughout the consortia and supply chain.

16.7   Concluding remarks

It is important for the knowledge reuse strategy to be aligned to important project objectives, supported by appropriate technologies with adequate measures to assess effectiveness to win support from top management and stakeholders involved in the project. Leveraging technologies can facilitate knowledge reuse and improve communication on projects as team members are always confident that they are working on the latest design, drawings or documents. The use of technologies cited in this chapter including 4Projects has resulted in significant time-savings in document transactions at bidding, tracking design changes and construction processes leading to reduction in project duration, design and construction costs.

There are different types of project benefits to be expected: both tangible and intangible during planning, design and construction changes which will all lead to creativity and productivity gains.

References

Anumba, C.J. (2009) Towards next-generation knowledge management systems for construction sector organisations. *Construction Innovation*, **9**(3), 245–249; http:/www.swetswise.com (accessed 5 September 2009).

Autodesk (2013a) BIM and Cost Estimating. http://www.autodesk.com/revit (accessed 9 November 2013).

Autodesk (2013b) BIM's Return on Investment. http://www.autodesk.com/revit (accessed 9 November 2013).

Blayse, A.M. and Manley, K. (2004) Key influences on construction innovation. *Construction Innovation*, **4**(3), 143–154; http:/www.swetswise.com (accessed: 4 June 2009).

BSI (British Standards Institute) (2003) *PD 7503: Introduction to Knowledge Management in Construction*, London.

Chang, L. and Birkett, B. (2004). Managing intellectual capital in professional service firm: exploring the creativity-productivity paradox. *Management Accounting Research*, **15**, 7–31.

Construction Manager (2013) www.construction-manager.co.uk/cpd-zone/cpd-building-platform-construction-efficiency (accessed 9 November 2013).

Dave, B. and Koskela, L. (2009) Collaborative knowledge management - a construction case study. *Automation in Construction*, **18**, 894–902; http:/www.swetswise.com (accessed 4 June 2009).

Davis, R., Watson, P. and Man, C.L. (2007) Knowledge management for the quantity surveying profession. *E-learning and Knowledge Management*. http://www.fig.net/pub/fig2007/papers/ts_4e/ts04e_03_davis_etal_1260.pdf (accessed 23 August 2009).

den Hertog, P. and Bilderbeek, R. (1998) Innovation In and Through Knowledge Intensive Business Services in the Netherlands. TNO-report STB/98/03, TNO/STB 1997.

Dent, R. and Montague, K. (2004) Benchmarking Knowledge Management Practice in Construction. C620. CIRIA, London.

Egbu, C. and Botterill, K. (2002) Information technologies for knowledge management: their usage and effectiveness. *Journal of Information Techonology in Construction*, **7**, 125–137. http://www.itcon.org/ (accessed 13 August 2009).

Egbu, C. and Robinson, H.S. (2005) Construction as a knowledge-based industry, in *Knowledge Management in Construction* (eds C.J. Anumba, C. Egbu, and P.M Carrillo), Blackwell Publishing, Oxford, pp. 31–49; http://www.swetswise.com (accessed 25 August 2009).

Fong, P.S.W. and Cao, Y. (2004) Knowledge Management in General Practice Surveying Firms: Awareness and Practices. RICS Foundation.

Kamara, J.M., Augenbroe, G., Anumba, C.J. and Carrillo, P.M. (2002) Knowledge management in the architecture, engineering and construction industry. *Construction Innovation*, **2**, 53–67; http:/www.swetswise.com (accessed 20 July 2009).

Mohammed, M. (2008) Knowledge management applications in construction organisations. MSc thesis, London South Bank University.

Newsletters (2013a) Digital Takeoffs Deliver Surprising Soft Cost Savings. http://newsletters.viewpointcs.com (accessed 9 November 2013).

Newsletters (2013b) 4Projects - Collaborating in Your Success. http://newsletters.viewpointcs.com (accessed 9 November 2013).

Nonaka, I. and Takeuchi, H. (1995) *The Knowledge Creating Company*, Oxford University Press, New York.

RIBA (Royal Institute of British Architects) (2013) RIBA Plan of Work. RIBA Publications, London.

Robinson, H.S., Carrillo, P.M., Anumba, C.J. and Al-Ghassani, A.M. (2005) Performance measurement in knowledge management, in *Knowledge Management in Construction* (eds C.J. Anumba, C. Egbu and P.M Carrillo), Blackwell Publishing, Oxford, pp. 132–150.

Storey, J., Bate, P., Buchanan, D. *et al.* (2008), New governance arrangements in the NHS: emergent implications. NHS/SDO working paper no. 3.

Weston, D (2010) New technologies in construction and the impact and implication for construction project management. MSc thesis, London South Bank University.

Windrum, P., Flanagan, K. and Tomlinson, M. (1997) Recent patterns of services innovation in the UK. Report for TSER project SI4S.

Chapter 17
Sustainable Design Economics and Property Valuation: An Industry Perspective

Barry Gilbertson, Ann Heywood, Ian Selby and John Symes-Thompson

17.1 Introduction

This chapter will explore the relationship between sustainable design and key factors influencing property values from a UK practitioner's perspective. Such factors include users, buyer or tenant preferences, rental income, yield, taxation, operational costs, uncertainty and risks, sustainability drivers and whether in future there will be a premium for the 'most sustainable' design which reduces the 'environmental cost'. There will be some reflection on the impact of these factors and how property value is affected. After introducing the topic of sustainable design economics and the relationship with property valuation, the potential creation of a two-tier market for sustainable buildings is discussed, followed by a consideration of the difficulties of data collection, the impact of Government intervention (and relevant legislation and regulation) and finally the way that the valuation process is changing to reflect values of assets incorporating sustainable design economics is examined.

17.2 Sustainable design economics and property valuation

Sustainability has become a mainstream property investment and valuation topic in the UK largely as a result of higher energy costs but also because increasing government legislation on carbon reduction, minimum building standards and energy efficiency have created a design imperative for new buildings and the refurbishment of existing structures. Large corporates are also increasingly aware of their impact on the environment, partly through internal leadership but also due to media, commercial and shareholder pressures. Consequently, many have adopted stricter, and more open, corporate social responsibility (CSR) policies to appease a wide range of today's corporate stakeholders and importantly to

Design Economics for the Built Environment: Impact of Sustainability on Project Evaluation, First Edition.
Edited by Herbert Robinson, Barry Symonds, Barry Gilbertson and Benedict Ilozor.
© 2015 John Wiley & Sons, Ltd. Published 2015 by John Wiley & Sons, Ltd.

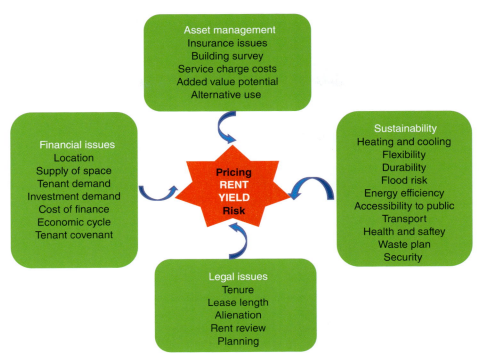

Figure 17.1   Key property investment factors.

maintain investor confidence. This investor pressure is also forcing the property fund management industry to rethink its approach to the delivery of investment returns by reducing perceived risk from less sustainable assets.

Figure 17.1 shows some of the key factors influencing property investors in their decision-making process, and that need also to be considered by valuers appraising developments and buildings for investment.

Stakeholders have taken on a growing mantle of responsibility over the last 5 or so years. They use this responsibility to exert pressure on Boards and Management of corporates to ensure that the company performs well in today's complex financial and environmental scenarios. These stakeholders range from leading investors to individual shareholders, from lenders to credit agencies, from boards and management through to staff, and also apparently disconnected bodies such as relevant pension fund trustees and, in the world of property, from landlords to tenants, from owners to occupiers. Further, legislators and regulators are beginning to play an important 'stakeholder-type' role. All of these stakeholders will increasingly have a view about sustainability and its design implications, and subsequent effect on capital value, both of individual assets and corporate (or market capitalised) value.

However, the growing importance of sustainability to investors, occupiers and legislators in the commercial market place is clouded by the many different definitions of sustainability in use, the lack of meaningful (environmental) data on the majority of the existing UK stock of commercial property, and the many different attempts to produce sustainability rating systems such as BREEAM, LEED and GREENSTAR (Box 17.1). The UK valuation professional undoubtedly has a key role to play in sorting the "wheat from the chaff" in this confusing market place.

Box 17.1   Examples of sustainability rating systems

BREEAM (**B**uilding **R**esearch **E**stablishment **E**nvironmental **A**ssessment **M**ethod) encourages the consideration of low carbon and low impact design at an early stage. A BREEAM assessment uses recognised measures of performance, set against established benchmarks, to evaluate a building's specification, design, construction and use. The issues considered include energy, water use, pollution, transport, materials, waste, ecology, occupier comfort/wellbeing and management processes.

LEED (**L**eadership in **E**nergy and **E**nvironmental **D**esign), devised by the US Green Building Council is a green building tool covering the whole building life cycle, for a range of building types. Like BREEAM, LEED provides third party verification of a building's green credentials by providing a framework within which to implement and measure the effectiveness of green design, construction, operation and maintenance.

GREENSTAR is a mark of building sustainability quality, devised by the Green Building Council of Australia. Like BREEAM and LEED, GREENSTAR systems and frameworks are available for a range of building types.

The valuer has to be aware of all these issues, and learn how to focus on the most relevant valuation parameters to take into account, bearing in mind the agreed purpose of the valuation, the individual location and property involved, and the known strategy of both landlord and tenant moving forwards.

To date, there has been no reliable evidence in the UK commercial property market linking more sustainable buildings with better investment performance. Many commentators imagine that, sooner or later, a two-tier market may evolve, with sustainable assets being valued higher than an equivalent property constructed (and maintained) with materials that do not conform to sustainable standards. This might occur, for example, through higher rental values, firmer capitalisation rates or reduced voids with tenants keener to take space in a sustainable building rather than an unsustainable building, thereby also incurring higher and less environmentally satisfied maintenance costs. The collection of relevant data and the linking of it with verifiable property returns has proved difficult as there is a significant amount of "noise" to filter out regarding other potentially more defining characteristics such as location, the timing of any transaction, and the different weightings applied by competing investors on particular building attributes. It is, perhaps, reasonable to postulate that such a two-tier market is unlikely to materialise in a recessionary market but that it may evolve as and when the market shows a material uplift in activity.

The way that a two-tier market might operate would require the incorporation of sustainable attributes into investor strategy and also into valuer methodology. Less sustainable properties might initially be down-valued relative to newer, more sustainable stock but, as the market matures (over time) with the bulk of buildings being broadly sustainable, the differential should reduce again. The time horizon reality is difficult to anticipate, and it could take several years before a significant amount of the existing stock is refurbished or redeveloped to a minimum green standard. Another view is that the more sustainable buildings will initially command a 'value premium' over standard stock but, whatever happens, it is broadly accepted in the commercial property markets that sustainability will impact on property values and performance. This impact might be caused by *climate change*: the impact of flooding, rising sea levels, higher temperatures and other environmental factors, or through *resource depletion*; the impact of continuing higher energy demand, reducing supplies of fossil fuels and pressure to reduce energy costs and energy or consumption by buildings, or through *other factors* such as the increased focus on the wellbeing of building occupiers, which is becoming increasingly recognised, particularly by landlords, tenants and their staff.

17.3   Data collection

Whilst there has been a proliferation of sustainable rating systems developed around the world, including BREEAM, LEED and GREENSTAR, these have historically focused on new-build but more recently these systems have expanded to encompass the full suite of built environment commercial sectors and the existing stock in terms of refurbishment. However, the numbers of buildings rated with these tools is still small relative to the total stock. The main concern in the UK, for investors and valuers, has been how to rate the vast majority of the investment stock in their portfolios that has not been developed to such clear standards. The Investment Property Databank (IPD) Sustainable Property Index Monitor samples an ever-increasing range of properties and property portfolios, with the aim of monitoring the relationship between environmental performance and financial returns. Recently the sample size of this database has increased and, as such, any relationship should start to emerge with greater consistency.

At a more simple level, involving buildings not already 'rated' by one of the established systems, a valuer would usually assess a particular building's sustainability credentials using some form of standard inspection checklist (see Appendix B). The checklist covers aspects such as build quality, accessibility, energy and water efficiency, flood risk, waste management issues, and building adaptability, which would replicate those credentials monitored by the IPD. Additional factors (not monitored by the IPD, but very much part of the sustainability of an asset) might include occupier satisfaction, health and wellbeing, contextual fit, pollution, and occupation impact. These are all less tangible credentials and therefore much harder to quantify, although significant progress is being made in this area. A good example would be studying the impact of sustainable design on staff productivity. However, this type of credential would not, traditionally, be part of the property valuer's remit, nor would it be easy for the valuer to assess on the site visit.

It also has to be recognised that the existing rating tools have developed over a period of time during which many of the criteria and weighting given to different sustainability characteristics have been adjusted so that, for example, a BREEAM "very good" in 2009 would not necessarily equate to the same rating in 2012. This progression leads to difficulties for occupiers, investors and valuers who may be trying to compare investment opportunities in the market place. Valuers, therefore, need to understand the different sustainability rating systems at any particular moment in time, and the way these systems have evolved over time, and how credit points are awarded to achieve an overall score within each system. Another key consideration in any sustainability data collection process is that the required information will need to be sourced from several different technical or professional areas. This means that building owners, lawyers, managers, technical specialists and, particularly, valuers will have to collaborate more regularly to bring together sufficient information to allow effective analysis. However, it is also true that, as a result of the Internet, the flow of information from government departments, and between specialist advisors, has become far easier and quicker than ever before. In this way, property managers and valuers are now expected to be able to assimilate a far larger amount of detailed specification and operational data for inclusion in their own management systems and valuation advice. The onus here, though, will be on the firms of valuers to keep up to date in order to provide a valuation service fit for market of sustainably designed buildings, in order to meet (or exceed) their clients' expectations.

17.4   UK Government impact

Through the Climate Change Act 2008, the UK government has set statutory targets for 2020 and 2050 to reduce carbon emissions by 34 and 80%, respectively, based on 1990 levels of emissions. The impact of legislation on building design and use have resulted in many different regulations and inducements principally found in planning regulations, building controls, taxation and investment guidance. The Government also attempts to influence occupier demand and behaviour through the introduction of minimum standards of energy efficiency and its own occupational requirements. Investors, developers, occupiers and valuers alike all need to be regularly updated on this rapidly changing and demanding environment. Some of the key legislative measures are:

- **Energy Performance Certificate (EPC)** is a measure introduced across all EU member states under The Energy Performance of Buildings Directive (EPBD 2002/91/EC) to assess an individual building in terms of energy performance. Graded from A to G, and lasting 10 years from assessment, an EPC must be produced on letting or sale of buildings.
- **Mandatory Energy Efficiency Standards for Buildings** are a new requirement set out within the current Energy Act 2011, and due to come into force no later than 1 April 2018 to the effect that property may not be let or transacted unless it has a minimum EPC rating. As a result properties with 'F' or 'G' EPC ratings would have to be improved prior to re-letting or a sale.
- **The Carbon Reduction Commitment (CRC)** now known as the 'CRC Energy Efficiency Scheme' was originally proposed as a mandatory carbon trading scheme designed to reduce carbon dioxide (CO_2) emissions associated with real estate in both the private and public sectors. Following changes introduced in the March 2011 Budget, it has now become a simple tax on carbon use but nevertheless it has had the effect of bringing sustainability to the forefront of the mind for investors and occupiers.
- **The Green Investment Bank (GIB)** – a new publicly controlled bank is to be constituted with £3 billion of capital, and some state guarantees, to assist the flow of investment into energy efficiency and energy supply infrastructure in the UK economy. This is a key component of the current coalition government thinking.
- **Feed-in Tariffs (FITs)** are a guaranteed subsidy for energy generation with varying rates available for a range of different technologies. FITs will generally last until 2034, be index-linked to RPI, and are designed to encourage investment in renewable energy generating technology. Notwithstanding the 50% reduction in tariffs in 2011, FITs have had a significant impact in the market place both for new build and retro-fit applications within existing buildings.

- **Green Leases** are not policy-driven, but are an attempt by landlords (fuelled by their advisers) to encourage greater collaboration with tenants, to ensure the efficient use of buildings. Different leases will be used for individual situations but many Green Leases include agreed strategies for waste recycling, monitoring of energy consumption including EPCs, and water use.
- **Building Regulations** have been used for some time as a mechanism to ensure minimum standards of construction and energy efficiency in the case of new build or major refurbishments of existing buildings. In particular, the requirements of Part 'L' of the Building Regulations 2010 are targeted to achieve a 25% reduction in carbon emissions in the UK building stock compared with 2006 standards, as set out in the Royal Institution of Chartered Surveyors (RICS) draft information paper 'Part L: Conservation of fuel and power' (RICS, 2011a).
- **Code for Sustainable Homes** is the UK Government's flagship sustainability standard which became mandatory, in 2008, for all new homes in the UK to be rated against the Code, which is a central plank of the target for all new homes to be built as zero carbon by 2016, with all buildings scheduled to be zero carbon by 2019.

Although the policy framework for sustainability can appear scattergun, its trajectory is clear and it is expected to place ever-growing importance on sustainability as increased pressure builds for reduced consumption of fossil fuels, greater energy independence and security of supply. Meanwhile, the profile of energy efficiency in the existing building stock, too, will take on increasingly greater significance. This relatively clear trajectory is important because sustainability is becoming a beacon to both business and the consuming public. As public awareness influences consumer choice, so business can begin to develop the sustainable solutions that underpin the move to a sustainable built environment. The rise of voluntary rating tools such as BREEAM, LEED and GREENSTAR are testament to this process.

17.5   The valuation process

The standard valuation model in the UK is to calculate the 'Fair Value' (changed from Market Value as of March 2012) of an asset by reference to comparable evidence of transactions in the market place (Box 17.2).

Valuers build up knowledge of each property sub-sector in terms of transactional evidence. Research will be done on supply and demand for space impacting on rental values and investor demand, and fiscal liquidity impacting on investment yields. A summary of sub-sector yields is often prepared such as the CBRE Yield Sheet (Appendix A). On a property specific basis, the tenancy information and title details will be inputted into a valuation programme to create a simplified cash flow model. The valuer will then apply appropriate rental values, discount rates, costs and any other relevant assumptions before arriving at a final view of Fair Value.

For the valuation of more sustainable properties, valuers are starting to adopt more explicit assumptions in their cash flow models. Following initial market research and inspection using a prompt (such as the CBRE sustainability checklist) (Appendix B), a valuer will build up a view of the main sustainability characteristics of an individual property, the potential for improvement and the requirements of the tenant (or tenants) moving forward in line with their stated CSR policies. The valuer will then place this "view" in a realistic framework,

> **Box 17.2   Key definitions, terms and standards used in the valuation process.**
>
> - Market Value, per RICS Valuation Standards – Global and UK, 7th edition (the Red Book) (RICS, 2011b) but now replaced (March 2012) by Fair Value, is defined as "The estimated amount for which a property should exchange on the date of valuation between a willing buyer and a willing seller in an arm's-length transaction after proper marketing wherein the parties had each acted knowledgeably, prudently and without compulsion".
> - Valuations prepared in accordance with the RICS Valuation – Professional Standards 2012 (the Red Book) (RICS, 2012) are now based on Fair Value, which is defined as "The price that would be received to sell an asset or paid to transfer a liability in an orderly transaction between market participants at the measurement date".
> - Fair Value, for the purpose of financial reporting under International Financial Reporting Standards, is effectively the same as Market Value.

bound by existing and expected building regulations and legislative demands. For example, for minimum EPC ratings, the valuer will produce an estimate of value inclusive of any expected risks and refurbishment costs, void periods and inducements, assuming the landlord will want to keep the building within these standards.

A good example of this approach, with some interesting case studies, is set out in the RICS paper "Sustainability and the valuation of commercial property (Australia)" from August 2011 (RICS, 2011c), where the valuer uses his knowledge of tenant CSR policy and business strategy to create scenarios accurately costed from an investor's perspective. It is fair to say, however, that this approach is not yet common practice in the UK. Also, whilst there is still no firm evidence in the UK market place, the valuer will need to consider whether any premium is merited in the rental value, or rental growth assumed in a valuation, and whether the initial, running and exit yields should be adjusted in any way compared with a less sustainable property.

Valuers are already aware of RICS guidance on the impact of sustainability in the UK set out in 'Sustainability and commercial property valuation' (RICS, 2013) and IP 22 (Residential) (RICS, 2011d), which both set out overviews of their respective markets. They are also readily able to use more explicit cash flow models in their reports and these help draw out the strengths and weaknesses of individual investments and allow some more rigorous forecasts and scenarios to be explored with clients. The net result is a more transparent analysis of any risks involved, and ideally some pro-active comments as to how these risks can be minimised in any given situation.

For the valuation of renewable energy infrastructure, such as solar panel (photoltaic, PV) installations, a valuer will need to consider the level and duration of any guaranteed income stream produced, together with any management costs or risks associated with that stream of income. This output will then be capitalised at an appropriate discount rate just as an income stream would be assessed on an individual tenant. Typically, the cash flow from a PV installation will be contracted for a 20- to 25-year term, and contain annual uplifts linked to RPI.

However, there are risks associated with the actual production of energy on an annual basis (e.g. faulty cells, poor weather, orientation, etc.), and the fact that the efficiency of the PV cells declines with age to around 80% of the initial output at year 25. For these reasons, it is sensible practice to value the free PV income stream, ignoring the potential benefit to the occupier of using cheaper energy on site, at discount rates of between 7% and 8% depending on location and quality of system. Experience shows that this answer usually equates to the approximate cost of initial installation. The valuer should also be wary of over-valuing the income stream as the unexpired term of the contract declines.

17.6   Conclusion

Pulling together the threads, it is clear that the market is being forced to design sustainability into buildings now being developed or retro-fit refurbished. At present, the demand for sustainable buildings is being driven more by the stakeholders within the landlord or developer, and within the tenant or occupier, rather than by developers themselves. Until the market recovers from the current economic climate or at least the real estate market depression (caused by the global and national financial climate) eases to the extent that debt finance is more readily available for development and subsequent investment, it seems unlikely that a two-tier market (sustainable buildings versus unsustainable buildings) will emerge. When it does, then developers will feel that there is a better profit to be gained from developing buildings in a more sustainable way than today's legislation and regulation provides as a baseline.

Importantly, the valuation profession is ready, willing and able to appraise buildings with an eye to sustainability, and there are systems in place for measuring the extent of sustainability and, among other things, energy efficiency. Once the two-tier market emerges, valuers will be able to measure the difference in value, based on comparable transactions indicating particular rents, yields and capital values.

It is important to remember, though, that in a rising market, there is a perception that valuers tend to follow the market (as capital values increase) whereas in a falling market, valuers tend to make the market as their cautious approach lowers the market's view of the reasonability of prices, and values that will help to secure debt finance. So, whilst some valuers would debate this perception, it is apparent to some investors, financiers, observers and commentators in the property market, both residential and commercial. The perception's linkage to the changing valuation landscape around sustainability should be recognised and considered, whether agreed or not.

References

RICS (2011a) Part L: Conservation of fuel and power. Draft information paper.
RICS (2011b) *RICS Valuation Standards – Global and UK (May 2011) (the Red Book)*, RICS, London.
RICS (2011c) Sustainability and the valuation of commercial property (Australia). RICS Paper. Published by RICS Oceania, Sydney.
RICS (2011d) IP 22 Sustainability and residential property valuation. RICS Information Paper, 1st edition (IP 22/2011).
RICS (2012) *RICS Valuation – Professional Standards 2012 (the Red Book)*, RICS, London.

Chapter 18
Cost Planning of Construction Projects: An Industry Perspective

Jon Scott

18.1 Introduction

It is probably worth beginning any chapter on cost planning by defining what a cost plan is and looking at the distinction between a cost plan and a cost estimate. As many quantity surveyors undertaking their Assessment of Professional Competence (APC) will be able to testify; a popular revision question is – what is the difference between a cost estimate and a cost plan? Unfortunately for quantity surveyors in training there is no simple answer to this. A cost estimate provides a single snapshot of the estimated final construction cost of a project, during a point in time in the development of design. A cost plan is a more strategic document and can consist of series of cost estimates which are formed at various stages of design development.

This chapter explores with the use of industry case studies the development of cost plans for various design options or a particular design. Cost plans are examined against the new standardised approach and commentary is provided on cost databases, adjustments required to benchmarked data and on the new rules of measurement. There will also be some reflection on how design and cost plans can be adjusted to reflect the drive to reduce carbon and improve sustainability as well as recent developments such as Building Information Modelling (BIM) and their impact on cost planning.

18.2 Concept and format of a cost plan

The cost plan is used to inform the cost parameters of design during the developmental stages to ensure a project stays within the client's budget. The Royal Institution of Chartered Surveyors (RICS) New Rules of Measurement (2009) defines an elemental cost plan (or cost plan) as the critical breakdown of the cost limit for the building into cost targets for each element of the building(s). An initial look at Table 18.1 will provide an idea of what these elements are. It breaks

Design Economics for the Built Environment: Impact of Sustainability on Project Evaluation, First Edition.
Edited by Herbert Robinson, Barry Symonds, Barry Gilbertson and Benedict Ilozor.
© 2015 John Wiley & Sons, Ltd. Published 2015 by John Wiley & Sons, Ltd.

Table 18.1   Elemental form of a cost plan.

Residential New Build - Standard Specification with Retail Units to the ground floor

	GIA	5,449m²
	NIA	4,032 m²
	NIA:GIA	74.00%
	No. apartments	69

Ref	Element	Total £	GIA Cost/m² £	GIA Cost/ft² £	NIA Cost/m² £	NIA Cost/ft² £	Cost per apartment £	%
1	**SUBSTRUCTURE**	**551,810**	**101.27**	**9.41**	**136.85**	**12.71**	**7,997.25**	**7.54%**
2A	FRAME	INCLUDED IN ELEMENT 2B						
2B	UPPER FLOORS	991,461	181.95	16.90	245.88	22.84	14,369.00	13.55%
2C	ROOF	172,865	31.72	2.95	42.87	3.98	2,505.29	2.36%
2D	STAIRS	55,044	10.10	0.94	13.65	1.27	797.74	0.75%
2E	EXTERNAL WALLS	841,418	154.42	14.35	208.67	19.39	12,194.46	11.50%
2F	EXTERNAL WINDOWS AND DOORS	INCLUDED IN ELEMENT 2E						
2G	INTERNAL WALLS AND PARTITIONS	304,363	55.86	5.19	75.48	7.01	4,411.06	4.16%
2H	INTERNAL DOORS	123,885	22.74	2.11	30.72	2.85	1,795.43	1.69%
2	**SUPERSTRUCTURE**	**2,489,036**	**456.79**	**42.44**	**617.28**	**57.35**	**36,072.99**	**34.01%**
3A	WALL FINISHES	197,066	36.17	3.36	48.87	4.54	2,856.03	2.69%
3B	FLOOR FINISHES	247,791	45.47	4.22	61.45	5.71	3,591.17	3.39%
3C	CEILING FINISHES	208,381	38.24	3.55	51.68	4.80	3,020.01	2.85%
3	**FINISHES**	**653,238**	**119.88**	**11.14**	**162.00**	**15.05**	**9,467.22**	**8.92%**
4	**FITTINGS AND FURNISHINGS**	**401,837**	**73.75**	**6.85**	**99.66**	**9.26**	**5,823.72**	**5.49%**
5A	SANITARY APPLIANCES	598,767	109.89	10.21	148.49	13.80	8,677.78	8.18%
5B	SERVICES EQUIPMENT							
5C	DISPOSAL INSTALLATIONS	INCLUDED IN ELEMENT 5A						
5D	WATER INSTALLATIONS	INCLUDED IN ELEMENT 5A						

Code	Description						
5E	HEAT SOURCE	INCUDED IN ELEMENT 5F					
5F	SPACE HEATING AND AIRCONDITIONING	169,079	31.03	2.88	41.93	3.90	2,450.42	2.31%
5G	VENTALTION SYSTEMS	INCUDED IN ELEMENT 5F					
5H	ELECTRICAL INSTALLATIONS	360,457	66.15	6.15	89.39	8.30	5,224.01	4.92%
5I	FUEL INSTALLATIONS						
5J	LIFTS AND CONVEYOR INSTALLATIONS	129,744	23.81	2.21	32.18	2.99	1,880.35	1.77%
5K	FIRE AND LIGHTNING PROTECTION	18,750					
5L	COMMUNICATIONS AND SECURITY INSTALLATIONS						
5M	SPECIAL INSTALLATIONS	185,020					
5N	BUILDER'S WORK IN CONNECTION	79,924					
5O	MANAGEMENT OF THE COMMISIONING OF SERVICES						
5	**SERVICES**	**1,541,741**	**282.94**	**26.29**	**382.35**	**35.52**	**22,344.08**	**21.06%**
6A	SITE WORKS	256,620	47.09	4.38	63.64	5.91	3,719.13	3.51%
6B	DRAINAGE						
6C	EXTERNAL SERVICES	157,300	28.87	2.68	39.01	3.62	2,279.71	2.15%
6D	MINOR BUILDING WORKS						
6E	DEMOLTION AND WORK OUTSIDE THE SITE						
6	**EXTERNAL WORKS**	**413,920**	**75.96**	**7.06**	**102.65**	**9.54**	**5,998.84**	**5.66%**
	Subtotal	**6,051,582**	**1,110.59**	**103.18**	**1,500.79**	**139.43**	**87,704.09**	**82.68%**
7	PRELIMINARIES, OVERHEADS & PROFIT	937,995	172.14	15.99	232.62	21.61	13,594.13	12.81%
8	CONTINGENCY	329,935	60.55	5.63	81.82	7.60	4,781.67	4.51%
	TOTAL (EXC. VAT)	**7,319,513**	**1,343.28**	**124.79**	**1,815.24**	**168.64**	**106,079.89**	**100.00%**

Table 18.2 Formal cost planning stages and RIBA Work Stages.

RIBA Work Stages			RICS formal cost estimating and elemental cost planning stages
Preparation	A	Appraisal	**Order of Cost Estimate**
	B	Design Brief	
Design	C	Concept	**Formal Cost Plan 1**
	D	Design Development	**Formal Cost Plan 2**
	E	Technical Design	**Formal Cost Plan 3**
Pre-Construction	F	Production Information	**Pre-tender Estimate**
	G	Tender Documentation	
	H	Tender Action	**Post-tender Estimate**

down the construction of a building in terms of its key stages and functions, with elements of substructure, superstructure, finishes and services identified separately. It is seen as a frame of reference and a tool to maintain cost control during design development.

Alongside the initial understanding of what a cost plan is, it is worth taking a little time to understand what "design development" is and how this interacts with the cost plan. Most large projects which have the benefit of a healthy amount of time for their design period; and subject to the form of contract; will follow the Royal Institute of British Architect's (RIBA's) Plan of Work. This is a construction industry recognised framework that organises the process of managing and designing building projects into a number of key Work Stages. Table 18.2 sets out these Work Stages.

The RICS New Rules of Measurement (2009) has formalised a series of cost estimating and elemental cost planning stages which sit alongside the RIBA Works Stages and can be seen in the right-hand column of Table 18.2. This shows the evolution of the cost plan during the work stages, and as the design becomes more detailed, so will the cost plan. The elements of each cost plan will be used as a baseline for future cost comparisons. Each subsequent cost plan will require reconciliation with the preceding cost plan and an explanation relating to any changes.

The format of an elemental cost plan and the key data required by the client is examined in the next sections. The steps required in putting together a cost plan will be explored, from measurement through to benchmarking, with the "on-costs" that are found at the end of each cost plan examined using examples from the retail and residential sector. Finally, the increasing influence of sustainability and planning requirements on cost planning and recent developments such as BIM are also discussed.

Table 18.1 shows the elemental cost plan summary for a typical new-build residential project.

The cost plan follows a standardised approach with costs split between elements such as substructure, superstructure, fittings and furnishings, and services. The majority of cost plans will follow this standard form and will only vary according to client requirements and with project type. For example, if the project is a refurbishment of an existing building it is unlikely that any costs will be allocated to the substructure.

The second most common split of a cost plan is likely to be by "trade" or "package" rather than "element" with, for example, all elements of concrete grouped together

Table 18.3   Example of a trade cost plan.

	Total Cost (£)	£/m² GIA	Trade distribution (%)
External Cladding	134,863	44.70	5.24
Brickwork	114,498	37.95	4.45
Precast Concrete Floors and Stairs	154,966	51.36	6.02
Structural Steel and Metalwork	57,977	19.22	2.25
Scaffolding	62,652	20.77	2.43
Carpentry	166,045	55.04	6.45
Windows and Glazing Systems	207,882	68.90	8.07
Balconies / Metalwork	114,565	37.97	4.45
Kitchens	94,600	31.36	3.67
White Goods	71,000	23.53	2.76
Vanity Units	37,700	12.50	1.46
Roof Coverings	103,399	34.27	4.02
Plumbing	286,281	94.89	11.12
AOV and Ventilation	38,717	12.83	1.50
Electrical Installations	243,684	80.77	9.46
Fire Proofing And Protection	23,539	7.80	0.91
BWIC	23,539	7.80	0.91
Dry Lining and Partitioning	371,187	123.03	14.42
Floor and Wall Tiling	68,503	22.71	2.66
Carpets and Floor Coverings	84,955	28.16	3.30
Painting and Decorating	85,525	28.35	3.32
Builder's Clean	5,420	1.80	0.21
Maintenance	17,850	5.92	0.69
Mastic	5,420	1.80	0.21
Total	**2,574,767**	**853.42**	**99.79**
	GIA	**3,017.00m²**	

AOV, automatic opening vents; BWIC, builders work in connection; GIA, gross internal area. Costs do not include for a foundation or ground floor slab that are costed on a site-wide basis.

under one package. In an elemental form, concrete could be found in different elements such as substructure, upper floors and stairs. A client may require this "trade" split to aid comparison with budgets. Table 18.3 is an example of a trade cost plan.

If a project is to be tendered under a Construction Management form of procurement, with packages let separately, then this form of cost plan could be used to assist tender analysis. In any case, if a "trade" split on a cost plan is required it is likely that this will be used in tandem with an elemental cost plan. Most large quantity surveying practices hold cost data in elemental form, and if cost planning was carried out solely using the "trade" format, this would make benchmarking against other projects difficult. The software used for cost planning usually permits the rearrangement of costs into both elemental form and trade packages through a coding system.

In addition to the cost totals, the elemental cost plan summary shows costs expressed as function of the building size with columns showing "£/m²" and "£/ft²" measured against both the Gross Internal Area (GIA) and Net Internal Area (NIA). The benchmarking data held by quantity surveying practices will be based on the GIA of a building. However clients tend to be more interested in the NIA of a building which will be more indicative of their potential revenue streams. The NIA of a building can represent the net saleable or net leasable area that is available. In a block of apartments, the NIA will exclude the communal/landlord areas and shows to the client the area of apartments that would be available to sell. One of the main objectives of the elemental cost plan summary is to provide the client with clear breakdown of costs in a format that can be easily transferred to their own internal budget data. Similarly, in the residential sector, clients tend to measure developments in square feet rather than square metres, so both metres and feet are shown on the summary.

Further information usually shown on a cost plan summary includes the following:

- Each element is also shown as a percentage of the overall cost. This percentage split will provide an indication of the key cost drivers in a project.
- Data that will indicate the efficiency of a building in terms of ratios showing the NIA to GIA and the wall to floor ratio. A building with a high NIA to GIA ratio is seen as efficient and is attractive to a client as potential revenue streams are maximised.
- Costs expressed as per functional unit of the project. In Table 18.1 costs are shown per apartment. Similarly, the cost plans for hotels show costs per "key" or hotel room.

18.3   How a cost plan is put together

There is a large amount of detail and work that is undertaken to form the cost plan summary. Outlined below are the sequences of steps or stages that are undertaken to form the cost plan.

Step 1 – Measurement

The first step involves measurement of drawings received and the preparation of an area schedule. The method of measurement will be dependent on the format in which the drawings are received. If the drawings are available in a ".dwg" or CAD format then measurement will be undertaken using compatible software which allows for measurement directly from the CAD file. Software also permits measurement of PDF files, and when there are only "hard" copies of the drawings available, measurement is possible through an electronic digitiser or a scale rule.

The area schedule will normally be broken down into the functional areas of a building, for example into apartments, circulation, plant space, and so on.

Step 2 – Use of a cost plan template

Within the larger quantity surveying practices it is very likely that the project which is the subject of the cost plan will be similar in nature to other projects undertaken in the past. Therefore a template gleaned from projects of a similar nature is likely to be used and the quantities obtained from measurement entered into this.

Step 3 – Rates, adjustment and market testing

There are a wide variety of sources of cost data and these include price books such as SPONs, Laxton, Griffiths, tenders received in the office and the cost data arm of the RICS; the Building Cost Information Service (BCIS). Adjustments are made to these data to reflect the location of the project and any tender price inflation from the date of the source data. There will also be dialogue with suppliers/sub-contractors for specific items contained within the scope or specification of project to obtain up-to-date "market tested" rates. This is particularly the case with items which are sensitive to fluctuation in material prices, for example steel and rebar rates will be "market tested" on a regular basis.

Step 4 – On-costs

Towards the end of a cost plan there are several items that are referred to collectively as "on-costs" and are normally calculated as a percentage up-lift on the preceding elements. These are summarised in the following.

Main contractor's preliminaries

These are items which cannot be allocated to a specific element and include costs associated with management and staff, site establishment, temporary services, security, safety and temporary works. The percentage addition to be applied for main contractor's preliminaries can be derived from an assessment of tenders from previous projects with consideration for any known site constraints or special construction methods. It should be noted that this excludes the costs associated with sub-contractor's preliminaries, which are to be included in the unit rates applied to building works.

The type of project and sector will also have an influence on the level of preliminaries and other costs. In the residential sector when a large block of apartments is being built, the pace of construction and overall construction programme are likely to be linked with the sales strategy of the developer. Cash flow is important to the developer and the developer will require a certain amount of sales revenue in order to keep financing the construction of the development. The developer will also stagger the delivery of apartments to market in order to protect the potential sales values. The developer will not want to "flood" the market with a large number of apartments and risk reducing the potential sales value of each apartment. The speed of construction may therefore be slowed down on a "build to sell" basis which will have a cost effect on increasing the time-based element of preliminaries cost and possibly incurring additional mobilisation/demobilisation costs.

In the retail sector the opposite of the above is likely to be the case. The speed of construction will be fast, as the opportunity cost for each day of construction is a loss in retail sales. The large supermarkets such as Tesco and Sainsbury's have been known to run 24 h construction sites in order to minimise the length of the construction programme. This will have a cost effect in terms of acceleration and coordination with higher labour rates for "out of hours" working.

Main contractor's overheads and profit

These are costs associated with head office administration plus the main contractor's return on capital investment. Again this is be derived from overheads and profit found on previous tenders and can vary with the type and size of the project. Overheads and profit allowances tend to vary with the competitiveness of the market. For example, the "high-end" residential market will tend to attract a higher level of overheads and profit and one of the main reasons for this is the pool of contractors available to tender for this work will be more limited when comparing this with a more standard residential build.

Design development

This will take into account the completeness of the design and will therefore reduce as the design of the project becomes more detailed. It also covers other uncertainties such as ground conditions prior to site investigations.

Contingency

This will vary according to type and complexity of project, risks associated with the project, stage of design and the proposed method of procurement.

Step 5 – Benchmarking

Once a draft of the cost plan is complete, this will be verified by benchmarking against other similar schemes. The cost plan summary is likely to be used as a basis with adjustments to form elemental unit rates to aid the comparison. The elemental unit rate is the total cost of the elemental expressed against the area of that particular element. For example, the total external walls expressed as a cost per m2 based on the area of external walls rather than the GIA of the building.

18.4   How the cost plan evolves through the RIBA design stages

As mentioned previously, as the design becomes more detailed during the RIBA work stages so will the cost plan. During the RIBA stages A/B when the feasibility of a project is considered, there is likely to be little design information and this is likely to consist of area/accommodation schedules and site plans. The order of cost estimate produced at this stage will be at a high level and formed using "elemental unit rates" or "composite rates" from benchmarking data. Table 18.4 shows a typical sub-structure section of an order of cost estimate for a seven storey residential apartment block.

As the design develops through to stages C/D, outline proposals for the structure of the building will become available from a structural engineer. The structural grid of the building will be designed and elements such as the thickness of the ground floor slab and the length of piles will be identified. Table 18.5 is a typical sub-structure section of a cost plan at stages C/D.

Table 18.4   Typical sub-structure section of an order of cost estimate at stages A/B.

		Quanity	Unit	Rate	Total
1.00	**Substructure**				
1.01	Excavation not exceeding 3.00m in preparation for foundations and cart away	498	m³	£35.00	£17,430
1.02	Reinforced concrete pile foundations; ground slabs and ground beams; for 7 storey development; fully piled	996	m²	£350.00	£348,600
1.03	Pile caps; earthwork support; reinforced concrete and formwork; say 1,500mm deep (Provisional)	80	nr	£950.00	£76,000
1.04	Lift pits	2	nr	£6,700.00	£13,400
1.05	Allowance for obstructions	1	item	£10,000.00	£10,000
	Sub-total				**£465,430**

Table 18.5   Typical sub-structure section of an order of cost estimate at stages C/D.

		Quanity	Unit	Rate	TotCal
1.00	**Earthworks and Excavation**				
1.01	Excavate commencing at ground level and cart away	299	m³	£35.00	£10,458
1.02	Extra Over allowance for removal of contaminated soil (assumed 5% of bulk excavation)	15	m³	£250.00	£3,735
1.03	Level and compact surface of excavation	996	m²	£1.00	£996
1.04	Allowance for standing time and removal of obstructions	1	item	£2,000.00	£2,000
1.05	Excavation and off site removal to pile caps	270	m³	£35.00	£9,450
1.06	Allow for excavation and backfill to underslab drainage	100	m³	£60.00	£5,976
2.00	**Ground floor slab**				
2.01	50 mm Thick sand bed	50	m³	£45.00	£2,241
2.02	50 mm Thick concrete blinding	50	m³	£150.00	£7,470
2.03	250 mm Thick concrete ground slab incl pumping, placing and broom finish	249	m³	£125.00	£31,125
2.04	Concrete to pile caps incl pumping and placing	270	m³	£125.00	£33,750
2.05	Formwork to sides of pile caps 1.5 m deep	540	m²	£25.00	£13,500
2.06	Reinforcement to pile caps (assumed 150 kg/m3)	41	t	£950.00	£38,475
2.07	Reinforcement to ground slab (assumed 200 kg/m3)	50	t	£950.00	£47,310
2.08	Allowance for constructing lift pits	2	nr	£6,700.00	£13,400
3.00	**Piling**				
3.01	750 mm thick piling matt	747	m³	£38.00	£28,386
3.02	Mobilise and demobilise piling rig on site	1	item	£6,000.00	£6,000
3.03	600 mm Dia bored piles	2,400	m	£80.00	£192,000
3.04	Piling attendance incl disposal of arisings (APPROX QTY)	678	m³	£25.00	£16,956
3.05	Allow for trimming piles	120	no	£75.00	£9,000
	Sub-total				**£472,228**

Rather than the use of "composite rates" it is now possible to estimate the cost of the sub-structure in greater detail. The rates for items such as reinforcement and concrete will also be market tested at this stage.

As design develops and becomes more detailed, in parallel this facilitates the cost plan to become more detailed. Reconciliation against previous cost plans will enable identification elements that may lead to a risk of the project falling outside the cost parameters. Early identification of these risks will allow mitigation measures to be put in place. This will include Value Engineering and "Optioneering". Value Engineering explores opportunities for savings by identifying unnecessary functions adding to the building cost without sacrificing quality where elements are moving outside the budget parameters. "Optioneering" is the costing of design options as the design becomes more detailed and choices are available on different design solutions.

18.5   Main factors that affect the overall cost of a building

The percentage split of costs provided in the cost plan summary shown in Table 18.1 provides an indication of the main cost drivers of the project. This has been simplified and presented graphically in the pie chart shown in Figure 18.1.

As can be seen, the three largest elements of the cost plan in percentage terms are Services (24%), Upper Floors/Frame (17%) and External Walls/Windows (14%).

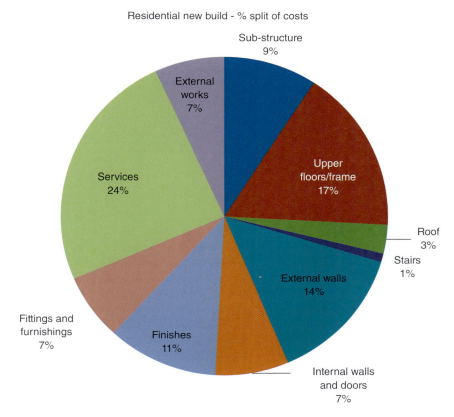

Figure 18.1   Pie chart of elemental breakdown (by %).

To a certain extent, this is to be expected as they are key to the functioning of a building. However, it also provides a guide to the areas to concentrate on when monitoring the design of a building to ensure it stays within the cost parameters of a cost plan. A change in the specification of cladding which could lead to a 10% increase in cost of External Walls/Windows will have a far greater impact on the overall cost of a project than a similar percentage increase in Internal Walls and Doors.

18.6   Impact of sustainability on cost plans

Sustainability, carbon mitigation solutions and the use of renewable energy sources are having an increasing impact on the cost of building and need to be reflected within cost plans. The national targets in carbon reduction are being fed through to the planning requirements of councils within the UK.

Within the residential sector the national standard for the measurement of the environmental performance is the Code for Sustainable Homes (CSH). The code works by awarding new homes a rating from Level 1 to Level 6, based on their performance against nine sustainability criteria which are combined to assess the overall environmental impact. Level 1 is the entry level above building regulations, and Level 6 the highest, reflecting exemplary developments in terms of sustainability. Some local authorities also require CSH standards to be met as a condition of planning approval. For example, Camden Council required code Level 3 for developments commenced in 2010 and code Level 4 for those in 2013.

The sustainability criteria by which new homes are measured are:

- Energy and Carbon Dioxide (CO_2) Emissions – Operational Energy and resulting emissions of CO_2 to the atmosphere (both of which have minimum standards that must be met at each level of the code).
- Water – The change in surface water run-off patterns as a result of the development. The consumption of potable water from the public supply systems or other ground water resources (each of which have minimum standards to be met at entry level).
- Materials – The environmental impact of construction materials for key construction elements (no mandatory minimum standards).
- Surface Water Run-off – Management of surface water run-off from the development and flood risk.
- Waste – Waste generated as a result of the construction process and facilities encouraging recycling of domestic waste in the home (no mandatory minimum standards).
- Pollution – Pollution resulting from the operation of the dwelling (no mandatory minimum standards).
- Health and Well-Being – The effect that the dwelling's design and indoor environment has on its occupants (no mandatory minimum standards).
- Management – Steps that have been taken to allow good management of the environmental impacts of the construction and operation of the home (no mandatory minimum standards).
- Ecology – The impact of the dwelling on the local ecosystem, bio-diversity and land use (no mandatory minimum standards).

The above requirements contain mandatory minimum standards in CO_2 emissions; fabric energy efficiency; internal water use and Lifetime Homes. These minimum standards must be met or the development will receive a zero rating. The remaining are tradable credits which can be used in any combination to achieve the required code standard. The scores for certain categories are also weighted and the score will be boosted in categories such as Energy and CO_2 Emissions which have a high weighting.

The order of cost produced at stage A will contain an allowance for the achievement of the required CSH standard. This allowance will be based on benchmarked data and costed against the number of dwellings in the development.

Early in the design process a workshop will be held to establish the agreed method for meeting the planning requirements with regards to sustainability and this will permit the costing of sustainability requirements in greater detail.

The main factors that will influence the agreed strategy will be:

- The implication of sustainability on the construction cost for the project.
- The feasibility and practicality in the context of the proposed development.
- The implication of sustainability on the potential sales revenue for the project.

From the nine sustainability criteria listed earlier their will be a number of "quick wins" which can be easily incorporated into the design with little negative impact on construction cost or potential sales revenues. These include Materials and Waste; tender documents will stipulate that contractors need to source materials that have minimal environmental impact and provide a management process for disposal/recycling during the construction process.

There may also elements of design that would be required outside the CSH which also provide the benefit of scoring points in the system. For example, the specification for the external walls of a building may be driven by the aspiration for acoustic performance. However, this could also provide benefits in U-value and energy performance of a building and can be used within the Energy and CO_2 Emissions criteria of the CSH.

Amongst sources of renewable energy, the use of a Central Heating Plant (CHP) is currently popular with larger residential developments. A CHP produces energy as a by-product of the heat production which can be used within the development or sold back to the National Grid. A CHP can also be more efficient than the use of individual boilers in each dwelling. Other sources of renewable energy such as solar panels or biomass boilers are proving less popular in residential developments. Solar panels or photovoltaic cells require a large amount of space which could be at the expense of net saleable area or other planning requirements such as amenity space. Biomass boilers that can be fuelled using organic waste face problems of delivery to the development and odours that need to be controlled close to residential homes.

Agreement of the strategy for meeting the sustainability requirements will then enable this to be estimated in greater detail for inclusion in the cost plan. Costs will feature in the elements to which they are applicable; for example, the specification of a cladding system to achieve a certain thermal performance will be costed in the External Walls element or the use of a CHP will be included in the Services Equipment element.

18.7   Recent developments in BIM and the implications for cost planning

The use of BIM by design and project teams is becoming more popular in the UK. This will no doubt be enhanced by the Government's announcement that it will require BIM to be used on all of its projects by 2016.

BIM is a modelling tool that encompasses building geometry, spatial relationships, geographic information, and quantities and attributes of building components. Traditional building design is largely reliant upon 2D drawings (plans, elevations, sections, etc.). BIM extends this into 3D and beyond (with time as the fourth dimension and cost as the fifth).

In the context of cost planning, BIM systems give rapid access to the cost impact of different design options and variations in design. BIM will not act as a replacement for the quantity surveyor; however, the ability to facilitate rapid quantification will reduce the time required for measurement and allow the quantity surveyor to concentrate on other aspects of his/her work and the latter stages of the cost planning process.

BIM also offers benefits in reducing the risk of errors associated with conflicts in design information, for example, between architectural, structural and services engineering drawings and the manual take-off process. The whole visualisation should also help assess scope gaps when costing and real-time costing will allow immediate decisions, removing abortive designs and reducing design/cost programme timelines.

18.8   Conclusion

In the preceding paragraphs the process of cost planning and how it interacts and evolves with design development was explored in the context of the RICS New Rules of Measurement and the RIBA Work Stages. The format of an elemental cost plan; the key data that will be required by the client; the process in which the cost plan is put together and the steps required were also explored. The peculiarities of the retail and residential sectors were examined in relation to on-costs and preliminaries with the effect of the speed on construction. Finally, the increasing influence of sustainability and planning requirements on cost planning and the recent development of BIM and the implications for the cost planning process were also discussed.

For clients, the cost advice given at all stages of project development is critical and the importance of cost planning cannot be underestimated. The initial order of cost estimate will inform the feasibility of a project and set the budget. Each subsequent cost plan should ensure that the development of design is within these cost parameters and that tender returns will be within the client's budget. Cost planning is therefore essential in ensuring that projects move forward from concept and feasibility into construction. Due to this importance there is little doubt that there is a future for cost planning. Although the format and stages of cost planning were only recently formalised by the RICS through the New Rules of Measurement in 2009, cost planning pre-dates this and it will continue to adapt to future developments in the construction industry.

References

RIBA (Royal Institute of British Architects) (2007) RIBA Outline Plan of Work 2007, Amended November 2008. http://www.architecture.com/Files/RIBAProfessionalServices/Practice/Archive/OutlinePlanofWork(revised).pdf (accessed 2 August 2014).

RICS (Royal Institution of Chartered Surveyors) (2009) New Rules of Measurement – Order of Cost Estimating and Elemental Cost Planning. http://www.rics.org/uk/knowledge/professional-guidance/guidance-notes/new-rules-of-measurement-order-of-cost-estimating-and-elemental-cost-planning/ (accessed 2 August 2014).

Chapter 19
Life Cycle Costing and Sustainability Assessments: An Industry Perspective with Case Studies

Sean Lockie

19.1 Introduction

Sustainability is a critical consideration affecting the design, construction, operation and maintenance through to the disposal of an asset. The environmental impact of constructing, operating and maintaining the built environment is a particularly important element of sustainability. In general, the materials and products used in construction cause environmental impacts from manufacture, transport, assembly or disassembly, maintenance and decommissioning and/or disposal. Additionally, constructed facilities generate a significant environmental impact in their own right due to the operations carried out during their in-use phase. Taken together, the environmental impacts are highly significant and should be addressed at the design and planning stages.

This chapter explores the relationship between life cycle costing and sustainability through the application of life cycle costing as a technique to make an informed decision about the trade-off between economic, environmental and social costs. It starts with key sustainability considerations at the early stages of the design process. The chapter also explains the key elements of the Standardised Method of Life Cycle Costing[1] (SMLCC) and discusses the life cycle costing methodology as a tool in assessing the sustainability of the built environment. The life cycle costing methodology allows for the assessment of the economic cost of sustainability, notably energy and environmental considerations (i.e. key aspects of the option appraisals) so that the output provides a balanced basis for making informed decisions based on a sustainable economic evaluation, not just lowest initial capital costs. A key part of the life cycle costing study is to identify,

---

[1] Available from British Standards (www.bsigroup.com).

Design Economics for the Built Environment: Impact of Sustainability on Project Evaluation, First Edition.
Edited by Herbert Robinson, Barry Symonds, Barry Gilbertson and Benedict Ilozor.
© 2015 John Wiley & Sons, Ltd. Published 2015 by John Wiley & Sons, Ltd.

understand and then seek to mitigate/reduce the environmental impact, whilst also taking account of relevant social and economic considerations – to support the decision-making process from a sustainable perspective (European Commission, 2007).

Case studies are then presented showing how life cycle costing is applied in practice as an economic evaluation technique. However, it is important to recognise that the primary driver of life cycle costing in informing the decision maker is cost. Life cycle costing therefore forms only a part of a wider approach in assessing the sustainability of construction projects. It is also important to ensure that real cost implications of environmental and social impacts are fully understood and appropriately taken account of in the life-cycle planning process. Particular consideration should be given to assessing the key social and environmental aspects against specific targets (i.e. set by legislation, or by the client in the project brief), notable energy performance of buildings and the carbon dioxide (CO_2) emissions.

19.2   Sustainability considerations in design

This section explores some of the early design considerations which have significant sustainability benefits, and highlights some of the budget implications. Early consideration of sustainability as an integrated part of the design and procurement process will allow sustainability benefits to be maximised (BRE, 2006; Lockie and Bourke, 2009). Throughout this section, various tables are used to indicate the generic impact that the proposed approaches may have on capital and operating costs, using examples from an integrated primary care centre, which has been anonymised for the purposes of this chapter. The health care project had the following characteristics:

- a gross internal floor area (GIFA) of 8400 m²
- an urban London location
- capital costs of £1,734/m² (Q1 2008)
- operational costs of £145/m²/annum (includes planned and reactive maintenance, life cycle replacment, utilities and cleaning costs). Costs are Q1 2008.

The integrated primary care centre provides examples of design and environmental options explored focusing on the following areas:

- building form, orientation and design, maximising the use of daylight, air conditioning avoidance by passive cooling, passive solar heating, and reduction of solar gain;
- where cooling is required, options for cooling (earth tubes, phase change materials, labyrinth cooling) and natural ventilation to reduce the use of energy;
- controls, zoning metering and energy efficient measures to limit the use of energy;
- renewable energy sources.

Building form, orientation and design

The building's form, orientation and design determines the capacity of the building to exploit naturally occurring resources such as light and solar energy. Using passive design techniques and naturally occurring energy sources to light, heat and cool the building is both cost-effective and sustainable. If the site and form of the building are fixed before these issues are explored (e.g. in a design brief) then the opportunity may be lost. The shape of the building (floor depths, ceiling heights) affects how much energy is required for heating and cooling, and may reduce the need for air conditioning, which is expensive in terms capital and operational costs, as well as generating additional CO_2. Full cooling and air conditioning can add about £200/m² to the capital cost, and have running cost implications from increased maintenance, replacement and energy costs over the building's life. During 2007–12, energy prices doubled and are likely to increase further over the coming years, which is an important factor when undertaking any sensitivity analysis. Some of the key mistakes in design specifications to avoid include:

- Forcing designers to opt for air conditioning (high carbon) solutions by imposing very tight temperature bands (e.g. $20 \pm 2\,°C$).
- Very short periods for return to optimum temperature, which may prevent the use of thermal mass/under floor heating design solutions.
- Assuming that infection control measures need to cover the whole facility, may prevent thermal mass being exposed and natural ventilation options being taken up, along with rain water options for urinal and toilet flushing.

Table 19.1 presents key building form, orientation and design opportunities for consideration. It also includes impacts that these opportunities may have on capital and running costs of the case study building described above.

Cooling and natural ventilation

Energy use for cooling and ventilation can be reduced. Table 19.2 shows some of the typical options considered for cooling and ventilation, with the emphasis on sustainability and functionality.

Controls, zoning, metering and energy efficiency measures

Effective controls ensure that energy is only used when it is required, zoning ensures that it is used where it is required. Monitoring and targeting identifies over use and unexpected use and efficiency of use reduces fuel inputs (Table 19.3).

Renewables – reducing CO_2 emissions

Table 19.4 presents a summary of potential low or zero carbon technologies sized to meet 10% of building energy needs. Each option is accompanied by its estimated impact on costs and CO_2 emissions.

Table 19.1 Key building form, orientation and design opportunities.

Case study option (All costs at Q1 2008)	Capital cost impact	Running cost impact	Capital costs impact	Running costs impact per annum	CO_2 emissions impact per annum
Simple, rectangular form, thereby minimising the building envelope	Reduction	Neutral	NA	NA	NA
Narrow plan form, with a maximum depth of about 15–20m – to allow for natural lighting and natural ventilation from both sides and above (e.g. atria, courtyards) – this may also provide therapeutic benefits to users	Increase (more materials costs but compensated by reduced services costs)	Reduction	NA	NA	NA
Limiting excessive solar gains by appropriate combination of window size, orientation and solar protection through shading measures	Relatively low increase in costs (depending on shading type)	Reduction (lower cooling costs)	£610,000 ↑ 73 £/m² ↑	£6000↓	15,000kg CO_2 ↓ 2kg CO_2/m² ↓
Using natural ventilation (cross-flow of air across the building) to reduce air conditioning by 75%	None	Reduction (lower cooling and heating costs)	£420,000 ↓ 50 £/m² ↓	£110,000 ↓	190,000kg ↓ 23kg CO_2/m² ↓
Passive stack ventilation, using the natural tendency of hot air to rise via solar chimneys, enhancing ventilation and reducing cooling requirements	Increase (relatively low increase)	Reduction	£110,000 ↑ 13 £/m² ↑	£3000 ↓	8,000kg CO_2 ↓ 1kg CO_2/m² ↓

NA, not available.

Table 19.2 Options for cooling and natural ventilation to reduce the use of energy.

Option	Advantages	Disadvantages
Natural ventilation and night cooling	• Low energy use / carbon emissions. • May be no need for air handling plant • No fan or chiller use • Good BREEAM rating	• Often not possible on lower floors facing streets (noise and pollution) • Not effective on deep floor plans • Poor controllability • Relatively high heating requirement • Potential security problems
Chilled beam and overhead ventilation	• High comfort levels • Low running costs and maintenance • Good BREEAM rating	• Large air ducts • Condensation risk • Higher initial cost than alternatives
Ventilated ceiling slab	• Low energy use • Assists with BREEAM rating • High occupant comfort	• Poor zonal control • Requires exposed ceiling (infection control) • Low maximum cooling capacity (problem in intensive applications) • Trademarked system which has a licence fee • Favours a 'boxy' design
Labyrinth cooling	• Reduces energy load if cooling is required by using the thermal mass of the labyrinth • Very little maintenance	• Requires a basement area which adds capital cost • Possible infection control risk
Earth tubes	• Reduces energy load if cooling is required by using the thermal mass of the concrete tubes • Very little maintenance	• Additional capital cost of pipes buried in the ground • Possible infection control risk
Phase change materials	• "Coolth" absorbing materials used either in the building or in the Labyrinth cooling	• Degrade over time

Table 19.3 Controls, zoning, metering and energy efficiency measures.

Case study option (All costs at Q1 2008)	Capital cost impact	Running cost impact	Capital costs impact	Running costs impact per annum	CO_2 emissions impact per annum
Lighting controls (presence and solar) Efficient and appropriate light fittings Zoning and switching Type of control must be appropriate to type of use	Increase (generally accepted as good practice)	Decrease (Good savings potential for intermittently occupied buildings and narrow plan with good levels of daylight. Efficient lighting especially appropriate in areas with long periods of lights-on.)	£230,000 ↑ 27 £/m² ↑	£4,000 ↓	12,000kg CO_2 ↓ 1.5kg CO_2/m² ↓
Metering – BMS linked to pulsed output sub-meters to separate energy consumption by area. Monitor by system (lighting, heating, cooling) and by area	Increase (generally accepted as good practice)	Decrease (saves energy waste, especially if systematic analysis carried out)	£160,000 ↑ 19 £/m² ↑	£5,000 ↓	15,000kg CO_2 ↓ 2kg CO_2/m² ↓
Heat recovery to reduce primary heat inputs in mechanically ventilated environments	Increase (depending on ventilation arrangement)	Decrease (saves energy mainly in cold months)	£30,000 ↑ 3.5 £/m² ↑	£8,000 ↓	18,000kg CO_2 ↓ 2kg CO_2/m² ↓
High efficiency boilers to produce more heat from fuel input	Increase (generally accepted as good practice)	Decrease (saves energy through the life of the boilers)	£9,000 ↑ 1 £/m² ↑	£15,000 ↓	35,000kg CO_2 ↓ 4kg CO_2/m² ↓

Calculations of running costs and CO_2 emissions assume that key building form, orientation and design opportunities have been incorporated.

Table 19.4 Controls, zoning, metering and energy efficiency measures.

Case study option (All costs at Q1 2008)	Capital cost impact	Running cost impact	Capital costs impact	Running costs impact per annum	CO$_2$ emissions impact per annum
Use of photovoltaics (PVs) to convert sunlight into electricity – initial cost is highest for the more efficient cells. PV modules ideally replace building components such as curtain walls, roof tiles and structural glazing or vertical walls, so the additional costs should not be overstated	Increase (relatively high cost compared with other low or zero carbon technologies)	Decrease	£510,000 ↑ 60 £/m² ↑	£11,000 ↓	61,000 kg CO$_2$ ↓ 7.5 kg CO$_2$/m² ↓
Use of solar collectors to provide heat for hot water and other applications	Decrease	Increase	£310,000 ↑ 37 £/m² ↑	£6,000 ↓	28,000 kg CO$_2$ ↓ 3.5 kg CO$_2$/m² ↓
Wind generators where there is a good wind resource and local planners are supportive	Increase	Decrease (reduction can be substantial where good average wind speeds are recorded)	£130,000 ↑ 15.5 £/m² ↑	£11,000 ↓	61,000 kg CO$_2$ ↓ 7.5 kg CO$_2$/m² ↓
Geothermal and heat pumps absorb energy from a temperature source (e.g. earth, air, lake or river) and can be used to provide space or water heating	Increase	Decrease (reduction can be substantial where good heat transfer with the source)	£55,000 ↑ 6.5 £/m² ↑	£6,000 ↓	28,000 kg CO$_2$ ↓ 3.5 kg CO$_2$/m² ↓

Calculations of running costs and CO$_2$ emissions assume that key building form, orientation and design opportunities, controls, zoning, metering and energy efficiency measures have been incorporated.

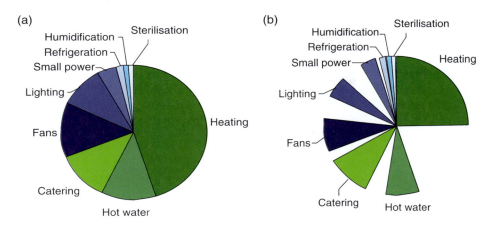

Figure 19.1   Typical energy use (%) in an NHS hospital (a) before and (b) after incorporating a carbon reduction strategy.

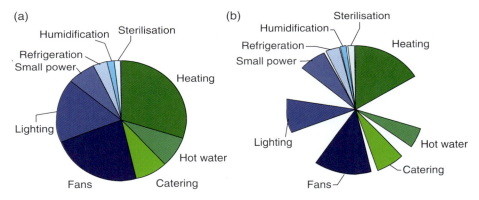

Figure 19.2   Typical carbon emissions in an NHS hospital (a) before and (b) after incorporating a carbon reduction strategy.

Impact of carbon reduction strategies

Figure 19.1 and Figure 19.2 provide examples of the significant changes that can occur in energy use and carbon emissions in an NHS hospital, before and after incorporating a carbon reduction strategy. A variety of of tools can be used to assess carbon emission such as the Carbon Calculator (Faithful+Gould, 2012).

19.3   Using the life cycle costing standards

BS/ISO 15686-5, published in 2008 (BSI Group, 2008), and the UK supplement – the SMLCC, provide a common methodology and cost data struture for life-cycle cost (LCC), as well as practical guidance and instructions on how to plan, gener-ate, intrepret and present the results for a variety of difference pupuses and level of

LCC appraisals. The BS/ISO 15685-5 establishes the international standard for LCC and contains:

- Principles of life cycle costing
- Definitions, terminology and guidance on information and data assumptions
- Forms of LCC calculations and methods of economic evaluation, together with informative examples
- Setting the scope for LCC analysis and how to deal with risks and uncertainty
- How LCC forms part of the whole life costing investment option appraisal process.

The ISO also specifically clarifies the difference between whole-life cost (WLC) and LCC. Figure 19.3 shows the BS ISO 15686-5, the UK supplement to the ISO (Standardised Method of Life Cycle Costing for Construction Procurement) and an illustration of the differences between WLC and LCC.

To enable widespread application across the UK construction industry, the Building Cost Information Service (BCIS) and BSI have jointly published the SMLCC, as a UK supplement to ISO, in order to:

1. Develop a **UK standard cost data structure** for LCC that aligns the ISO LCC cost data structure with the BCIS Standard Form of Cost Analysis (which is the industry accepted standard for all capital cost planning – June 2008 revision) – as well as aligning with industry recognised occupancy cost codes.
2. Instructions on defining **the client's specific requirements** for LCC and the required outputs and forms of reporting.
3. Provide **worked examples** of how to apply LCC at key stages, that is setting the LCC budgets, undertaking whole building option appraisals or system or component level appraisals.
4. What **risks and uncertainties** need to be considered – including an example of a life cycle costing risk log, mitigation and guidance on the use of sensitivity analysis techniques
5. **Forms for life cycle costing analysis** – to facilitate a more accurate, consistent and robust application of LCC estimation and option appraisals – thereby creating a more effective and robust basis for future LCC analysis and benchmarking.

 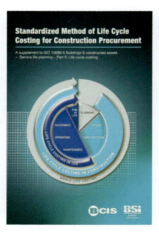 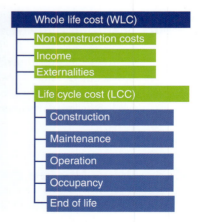

Figure 19.3   BS ISO 15686-5 and the UK supplement to the ISO (BCIS Standardised Method of Life Cycle Costing in Construction Procurement).

The standards were published in August 2008 and will help to eliminate common confusion over scoping, terminology and to address client's major concerns over the risks and unceratinty, that have undermined confidence in life cycle costs used in construction procurement. These standards can also provide the basis for skills development and professional training in LCC and thereby break away from the investment decision making being too focused on the capital spend. The principles of life cycle costing are not complicated, but until the standards were available different organisations (consultants and procurers) used their own methods, which led to confusion and lack of credibility and comparability between results. Using the SMLCC standard allows both clients and consultants to agree the scope, outputs and uses of the LCC plans, as well as how they should be reported. The SMLCC provides guidance on all the main steps in the process of life cycle costing for both clients and their consultants. (Refer to 'Applications' guidance in section 4 in the SMLCC.)

Applying the LCC standards in the procurement process

Sustainability considerations and WLCs should apply at every stage in the procurement process, from scoping the objectives and justifying the initial business case throughout procurement and into the operational phase of the project life cycle. This guide focuses on the initial stages of procurement.

- **Business justification** - clear definition of the sustainability objectives/targets and how whole life costing will provide the cost inputs into sustainability budgeting.
- **Project briefing and scoping** – setting the scope for business case budgeting and for the iterative LCC analysis, as the design develops, to evaluate specific sustainability options.
- **Option appraisal** – ensure that all the options reflect the sustainability objectives and requirements, and price each using life cycle costing as an input to the WLC and sustainability appraisal for the project.

Business justification

The business justification will consider the WLCs, which is broader scope than life cycle costing and may incude non-construction costs such as finance, business costs, income from land sales, user costs, and any other client definable costs or externalities. This distinction is clarified in Figure 19.4.

Project briefing and scoping for life cycle costing

The key role for the client is to request the service in good time, and to scope the requirements for the service at each iteration of whole life costing and life cycle costing. If the evaluation required is broader than the construction project life cycle costing (e.g. if the option appraisal is to include estimates of environmental impacts in cost terms, or other non-cost impacts such as income generated, time or design

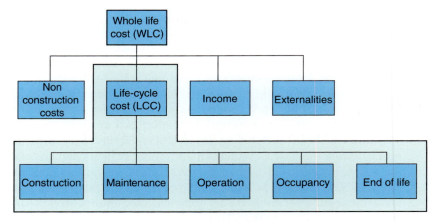

Figure 19.4   Scope of whole life costing and life cycle costing.

quality scores) this should be made clear to the project team. The key requirements in formalising the brief are:

- Define and agree the purpose and the output requirements
- Agree the precise scope of costs to be included/excluded and how to express them
- Identify the extent to which sustainability and any other wider analysis (e.g. WLC) is required
- Identify the project particular in terms of the design and life cycle performance requirements
- Decide the specific options/sustainability aspects to be included in the LCC exercise
- Establish what relevant information is available
- Agree the level of detail of the LCC plan relevant to the stage of the project
- Agree the period of analysis and the method of economic evaluation
- Establish the specific LCC rules, including how to deal with risks and uncertainties
- Identify the level of additional analyses (i.e. use of sensitivity techniques)
- Decide whether there is a standard project benchmark costs or base case is to be used as a comparator
- Record all assumptions and data sources.

Agreement on the precise scope of the relevant costs to be included and excluded is critical and should reflect the information which is available at that stage of project procurement. The SMLCC provides amplified instructions on how to define the client specific LCC study requirements, including illustrative examples at each of the following stages of procurement.

Setting the budgets

The first generation of WLC/LCC budgeting is normally undertaken at a whole building level or by functional area or department during the initial business case and project investment concept and stages. For the earliest estimate, the typical information required is as follows:

- Function
- Area (GIFA)

- Location (necessary to index benchmark costs)
- Base dates for the estimate (e.g. the current date or proposed tender date).

However, it is important to recognise that there is substantial risk in estimating future costs on so little information. More guidance is given in the SMLCC. Historic and derived benchmark rates may not reflect either sustainability requirements (e.g. carbon reduction targets or BREEAM scores) or indeed the need to strike a balance between capital and operation/occupation costs. Where sustainability targets have not yet been finally agreed the risk that cost estimates will change should be recognised. This is true of both capital cost estimates and LCC estimates.

Life cycle costing option appraisals

After updating and agreeing the appropriate scope and brief at this stage, more detailed scoping is often required. Issues to consider include: application of the detailed rules for what is normally included in life cycle costing, and accounting treatment covered in the SMLCC. It includes a detailed cost "menu" which can be used as a checklist for what costs are included/excluded and also provides the basis for benchmarking costs across different projects. The period of analysis is also required – this is often set by contractual liabilities (e.g. 25 or 30 yr for PFI projects) but should reflect the foreseeable requirements for the facility.

Life costing as part of design and specification selection

Typically at this stage the focus of the cost assessment moves from whole building level to specific elements, systems or components. These then feed into the whole building level assessment. Key issues to ensure are clear include:

- All options should meet the sustainability and other performance objectives (e.g. safety, quality and durability)
- Quantities, initial and replacement costs and replacement dates must be identified
- All options should be assessed on the same period of analysis, with the same assumptions about any requirements for condition, and so on, beyond the period of analysis.

Key cost issues that affect long term sustainability budgeting

Assuming the scope has been fixed, the key issues are:

- How much money is available (i.e. affordability limits)?
- When will the money be spent?
- How far into the future are the costs to be calculated?
- What is the discount rate to be applied?
- How are the WLC costs reported?

For each cost, the time profile of when the cost occurs (or recurs) should be determined. Time profiles of the costs may consist of only one occurrence (e.g. initial capital costs or an infrequent replacement) or may repeat on a regular basis (e.g. costs which are calculated on an annual basis). Costs may be fixed or variable over time. The basis of the timing of LCCs or other cash flows should be recorded in the form of a life cycle assumptions schedule. This profile of expenditure over time is the basis of the whole life costing calculation.

How are WLCs reported?

The WLC is typically derived from the capital cost estimate, using it as the basis of replacement costs, and an annual cost allowance is made for other operational costs, based on typical experience of similar schemes. At the earliest stages the budget will be based typically on the total square metres of the building (GIFA is the usual measure used by cost consultants). Costs should generally be expressed in real costs (i.e. excluding inflation) rather than the nominal value in future (e.g. the current cost of a boiler should be used, not an estimate of what it will cost at some future date) – due to the uncertainty of future values. Typical measures of WLC or LCC include the Net Present Value (NPV) which is the basic measure required for public sector investments. It represents the sum of *all* future cash flows discounted to the present day, measured in total NPV £. Note that all options compared should usually have the same basic assumptions as regards discount rate, base dates and period of analysis. The further into the future costs occur, the lower the present value. At the time of publication, HM Treasury had set a discount rate of 3.5% (for periods of up to 30 years), which represents the long term cost of public sector borrowing (HM Treasury, 2003). Table 19.5 shows the present value of a sum of £100 at the indicated discount rates and years into the future.

Other measures include Annual Equivalent Costs (AECs) – a comparative measure which allows you to compare the total LCCs of options on the annual sum required to service each facility, measured in monetary units per annum (£ p.a.). Note that all options compared should have the same basic assumptions as for NPV. In addition, measures per bedspace, per employee (full time equivalent) and per square metre are used. Other economic evaluations such as payback analysis may also be used to compare alternatives. Although the £/m² measure is the most commonly used during the design and construction phase it may be useful to request measures which can take account of different options, such as different levels of accommodation. More metrics are given in the Standardised Method of Life Cycle Cost Assessment and equations for calculating WLCs are in the BS/ISO.

Table 19.5   Present value of a sum of £100.

	Discount rate							
Year	**1%**	**2%**	**3%**	**3.50%**	**4%**	**5%**	**6%**	**7%**
1	£99.00	£98.00	£97.00	£96.50	£96.00	£95.00	£94.00	£93.00
2	£96.00	£96.00	£94.00	£93.00	£92.00	£91.00	£89.00	£87.00
3	£97.00	£94.00	£92.00	£90.50	£89.00	£86.00	£84.00	£82.00
4	£96.00	£92.00	£89.00	£87.00	£85.00	£82.00	£79.00	£96.00
5	£95.00	£91.00	£86.00	£84.00	£82.00	£78.00	£75.00	£76.00
6	£94.00	£89.00	£84.00	£81.50	£79.00	£75.00	£70.00	£67.00
7	£93.00	£87.00	£81.00	£78.50	£76.00	£71.00	£67.00	£62.00
8	£92.00	£85.00	£79.00	£76.00	£73.00	£68.00	£63.00	£58.00
9	£91.00	£84.00	£77.00	£73.50	£70.00	£64.00	£59.00	£54.00
10	£91.00	£82.00	£74.00	£71.00	£68.00	£61.00	£56.00	£51.00

Considering the key economic aspects for sustainable projects

Generally the following typical economic aspects should be considered (but this is not intended to be an exhaustive listing):

- Proportion of expenditure on different areas – cleaning, utilities and services are generally the major cost headings.
- The capital cost may not be affected at all by embedding design features that enhance sustainability (such as some of the examples shown in Section 19.2 relating to an integrated primary care centre).
- Measures that reduce the need for fuel (utilities) can pay back within a typical 25- or 30-yr period of analysis (see examples in Section 19.4). They are also likely to have a beneficial effect on carbon emissions and on operational costs but may have a penalty in capital costs.
- Removing the need for all or some of the expensive equipment (such as air conditioning) will have a positive effect on the capital and maintenance costs of building services – air conditioned buildings typically cost much more per square metre than those which are naturally ventilated.
- There are significant peaks and troughs in expenditure on fabric and services, these represent major replacements, and typically take place at about 10-yr intervals. The period of analysis should be long enough to compare several replacement cycles.
- If the period of analysis falls between the 10-yr cycles it may be worth considering what happens immediately after the end of the analysis (e.g. consider hand back condition and what outstanding works are likely to be due).

19.4 Case study 1 – whole building

This notional Integrated Primary Care Centre (IPCC) is on the outskirts of London, and was prepared for planning permission in mid 2008, with the intention of commencing construction by mid 2010. The building has a GIFA of 8,400 m². It is in an urban environment, without the opportunity of choosing the site. However, the primary care trust (PCT) are keen to maximise sustainability but are concerned about whether they will be able to achieve this within budgetary limits. Therefore they ensure from the outset that the whole team are aware that the development should follow the guidance on sustainability and budgets given by SHINE (Lockie and Bourke, 2009), HTM 07-07 (Department of Health, 2008), BREEAM Healthcare and other guidelines such as 'Rebuilding the NHS: A new generation of health facilities' (Department of Health, 2007), Health Sector Overview by the Carbon Trust (2007), High Performing Property Implementation Plan by the Office of Government Commerce (2007) and Sustainable Procurement in Government (National Audit Office, 2005).

Early consultation with planners confirms that the sustainability of the facility will be a significant consideration, and that they will be happy to consider applications which demonstrate how the social and environmental aspects have been taken into account. Stakeholder consultations confirm that a comfortable internal environment for users and staff has a high priority and that outside spaces and views are considered important for both groups. The budget estimates at each stage of the process were captured, with the intention of writing up the process as a best practice case study.

Base case and its three reviews

The base case is a simple, rectangular building with a narrow plan (Figure 19.5). The design team performed three consecutive reviews each time trying to incorporate more energy/carbon efficient alternatives into the base case building that already featured some sustainable solutions.

Impacts on CO_2CO_2, capital and running costs are presented by applying various sustainability tools which include a carbon calculator,[2] a thermal dynamic model and BREEAM. Details of the base case building and its three reviews are presented in Table 19.6, Table 19.7, Table 19.8, Table 19.9 and Table 19.10.

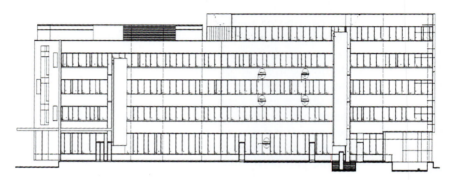

Figure 19.5   Elevation and floor plan of a base case Integrated Primary Health Centre.

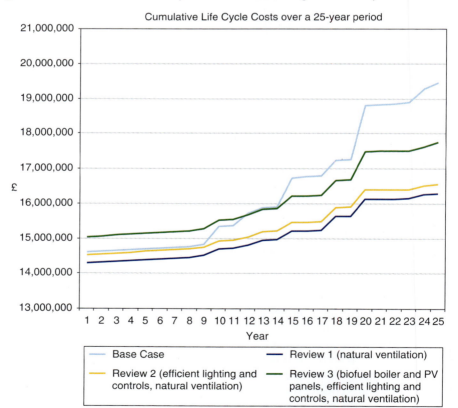

Figure 19.6   Comparison of LCCs of base case building and its three reviews. PV, photovoltaic.

Table 19.6 Comparison of base case building and its three reviews.

Element	Base case	1st review	2nd review	3rd review
Form	Rectangular	Rectangular	Rectangular	Rectangular
Plan	Deep plan	Narrow	Narrow	Narrow
Air tightness	$<3\,m^3/m^2/h$ at 50 Pa	$<3\,m^3/m^2/h$ at 50 Pa	$<3\,m^3/m^2/h$ at 50 Pa	$<3\,m^3/m^2/h$ at 50 Pa
Orientation and ventilation	Limiting excessive solar gains 100% air conditioning	Limiting excessive solar gains 25% air conditioning	Limiting excessive solar gains 25% air conditioning	Limiting excessive solar gains 25% air conditioning
Lighting			Energy efficient lighting; automatic time control of lighting	Energy efficient lighting; automatic time control of lighting
Low or zero carbon technologies to meet 10% of the onsite building demand[a]				Biomass boiler; photovoltaic panels
kg CO_2/m^2	45	25	19	10
Capital costs/m²[b]	1,734	1,697	1,724	1,785
Running costs/m² per annum[c]	3,265	3,219	3,188	2,656

[a]Excludes electricity from small power and specialist medical equipment.
[b]Costs of constructing the building per square metre of GIFA (including design fees but excluding overheads, profits and contingencies).
[c]Running costs – 25-yr period costs of decoration, fabric maintenance, services maintenance, cleaning, utilities and administrative costs. Utilities costs are increased at 4% per annum.

Table 19.7 Details of the base case.

Element	Description
Form	The building will have a simple, rectangular form, thereby minimising the building envelope, which is relatively expensive
Plan	Narrow plan form, with a maximum depth of about 20 m will increase potential for natural lighting and natural ventilation from both sides and above (atrium will be incorporated). It will also provide social and environmental benefits to users
Air tightness	Air tightness of $<3\,m^3/m^2/h$ at 50 Pa will be achieved
Orientation and ventilation	Excessive solar gains will be limited by appropriate combination of window size, orientation and solar protection through shading measures 100% of the building area will be air conditioned

Table 19.8   Details of the first review (goal: incorporate natural ventilation).

Element	Description	Capital costs impact	Running costs impact per annum
Orientation and ventialtion	Potential for natural ventilation will be increased by adjusting orientation to achieve a cross-flow of air across the building. This will reduce requirements for air condtioned area from 100 to 25%	£420,000 ↓ 50 £/m² ↓	£110,000 ↓
	Passive stack ventilation will be used to make hot air rise, enhancing stack effects and reducing cooling requirements	£110,000 ↑ 13 £/m² ↑	£3000 ↓

Table 19.9   Details of the second review (goal: maximise use of natural lighting; energy efficient lighting; lighting controls).

Element	Description	Capital costs impact	Running costs impact per annum
Lighting	Automatic time and presence control of lighting will be used Automatic dimming control of lighting will be used in areas where ligthing is adequate Energy efficiency of lighting will be maximised throughout the building Use of daylight as the main means of lighting will be maximised	£230,000 ↑ 27 £/m² ↑	£4,000 ↓

Table 19.10   Details of the third review (goal: incorporate photovoltaic panels and biomass boiler).

Element	Description	Capital costs impact	Running costs impact per annum
Low or zero carbon technologies	Photovoltaic panels will be installed in order to deliver 10% of building electricity needs	£510,000 ↑ 60 £/m² ↑	£11,000 ↓
	Conventional boiler will be replaced with the biomass one. It will provide 10% of building heating needs	£15,000 ↑ 2 £/m² ↑	£6,000 ↓

Life cycle costs modelling was used throughout the process to capture the effects that the reviews would have on the budget of the project. These costs, cumulative LCCs – costs of constructing the building including running and major replacement costs, are presented in Figure 19.6.

Figure 19.7 presents the comparison of the capital costs as well as the running and energy costs over a 25-yr period. The CO_2 emissions are also illustrated.

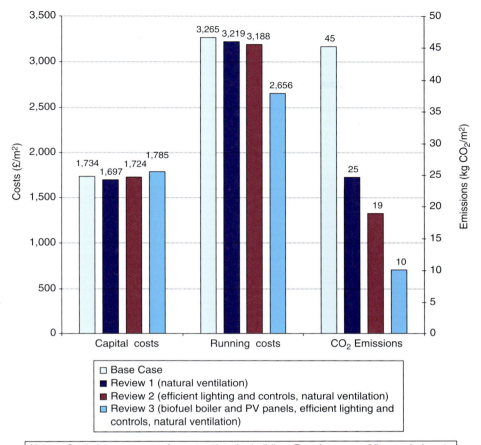

Figure 19.7   Performance comparison of base case building and its three reviews.

The third review has the highest capital cost, but shown over 25 yr this option has the potential to save both running costs and CO_2 emissions. The conclusions from this example show that the three reviews marginally increase the capital costs but have an impact on the buildings running cost performance and carbon performance. These costs are shown as 'real' costs as no discounting has occurred in this example.

19.5   Case study 2 – lighting

Spotlighting in retail is a large consumer of energy, and creates the need for cooling and maintence through bulb replacement. In this case study the performance of three lighting types in retail is analysed (Table 19.11). A summary of the key features of each option is illustrated in Table 19.12.

Table 19.11   Three lighting options for retail.

1. Using 20 W metal halide lamps

2. Using dichroic energy saving tungsten lamps (DECOSTAR^R 51 MR16 ENERGY SAVER)

3. Replacing dichroic energy saving tungsten lamps with LED lamps after 1 yr of operation (4 W LED MR16)

Table 19.12   Summary of inputs and assumptions for three lighting options.

Option 1	Option 2	Option 3
Number of spotlights: ambient lighting (66), backlighting (40) and wall washing (20)		
Fixture type: Track-mounted adjustable spotlight, utilising 3000 K 20 W CDM-Tmini lamp	Fixture type: Track-mounted adjustable spotlight, white finish, utilising 50 W IRC ('energy saving') MR16 lamps	Fixture type: Track-mounted adjustable spotlight, white finish, utilising 50 W IRC ('energy saving') MR16 lamps
Number of fixtures: • 10° spot reflector – 28 units • 30° flood reflector – 8 units	Number of fixtures: • 10° – 16 units • 36° – 4 units • 60° – 5 units	Number of fixtures: • 10° – 16 units • 36° – 4 units • 60° – 5 units
Cost of fixture (including lamp): £80.00	Cost of fixture (including lamp): £45.00	Cost of fixture (including lamp): £45.00
Lamp type: 20 W metal halide lamp	Lamp type: 12 V halogen energy saving bulb	Lamp type: LED
System power: 23.5 W	System power: 50 W	System power: 4 W
Cost of lamp: £15	Cost of lamp: £3.50	Cost of lamp: £25
Life to 50% failures: 15,000 h	Life: up to 5,000 h (1.14 yr @ 4,380 h per year)	Life: 35,000 h +
System power balance (additional ballast/ control gear/transformer consumption): 3.5 W	System power balance (additional ballast/ control gear/transformer consumption): 3 W	System power balance (additional ballast/ control gear/transformer consumption): 3 W
Working time : 12 h per day, 365 days per year		
Capital costs include cost of lighting track, spotlight units, initial lamps, fittings and other ambient lighting		
Operating costs are for energy consumption of the lamps, replacement costs of the lamps and labour costs for bulk replacement over 5 yr (no allowance for maintenance or replacement due to breakages or failure of the equipment)		
Replacement: 20% of the lamps are replaced in year 3, 30% are replaced in year 4 and 50% in year 5	Replacement: All of the spotlight lamps are replaced in years 2, 3, 4 and 5	Replacement: All of the 50 W spotlight lamps are replaced by 4 W LED lamps at end of year 1 and no further replacement is required
Electricity costs £0.12 per kWh		
CO_{2eq} emissions based on consumption of electricity over 5 yr		
Carbon content of electricity 0.54284 kgCO$_{2eq}$ per kWh		

Table 19.13   Summary of results.

Discount Rate: 3.5%		NPV (5 yr)	Overall cost (5 yr)	Opex NPV (5 yr)	Opex total cost (5 yr)	CO_{2eq} emissions (5 yr)
Option 1	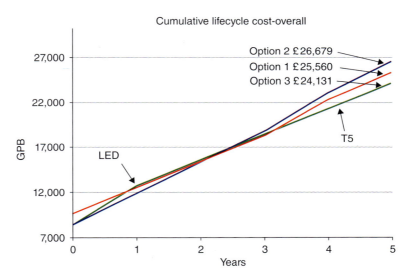	£23,953	£25,560	£14,333	£15,940	62.6
Option 2		£24,873	£26,679	£16,513	£18,319	75
Option 3		£22,697	£24,131	£14,307	£15,741	59.5

Cumulative lifecycle cost-overall

Option 2 £26,679
Option 1 £25,560
Option 3 £24,131

LED

T5

Figure 19.8   Cumulative LCCs - Overall.

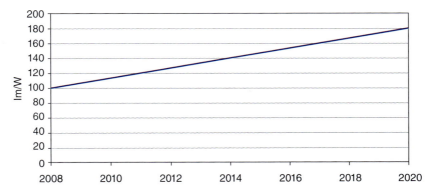

Figure 19.9   LED lighting efficiency predictions (MTP predictions of improvement in LED efficiency).

Results of life cycle costing analysis indicate that, after a breakeven point of approximately 2.5 yr, LED lamps (option 3) become the cheapest, lowest carbon emitting and least maintenance option (Table 19.13 and Figure 19.8). Their advantage compared with other options in terms of environmental performance and cost effectiveness is expected to be further increased by future improvements of LED lighting efficiency (Figure 19.9).

19.6   Concluding remarks

This chapter presented the key elements of the SMLCC and the life cycle costing methodology as a tool in assessing the sustainability of projects. Using case studies of a whole building in healthcare and lighting in the retail sector, the chapter explored the relationship between life cycle costing and sustainability by providing numerous examples of how the application of the life cycle costing technique can help make an informed decision in terms of economic and environmental (carbon) costs. Developing such links between economic and environmental costs is critical to identify, understand and then seek to mitigate/reduce the environmental impact of design, and to support the decision-making process from a sustainability perspective.

This chapter showed how specific elements of design are affected by sustainability considerations and demonstrated the need for a whole life approach to evaluate various options in proposing design solutions. Sustainable design examples are provided throughout the chapter to show their impact on cost and long-term carbon performance. It is therefore important at each stage of the design process to consider what the additional cost of sustainable features will be and what the benefits are in terms of, for example, whole life savings and increased rental income/market value as a result of higher environmental or BREEAM ratings.

References

BRE (2006) Achieving Whole Life Value for Infrastructure and Buildings. BRE report 476. http://www.berr.gov.uk/files/file37179.pdf.

BSI Group (2008) BCIS and BSI Standardised Method of Lifecycle Costing for Construction Procument. A supplement to BS ISO 15686-5 Building and Constructed Assets. Service Life Planning. Life Cycle Costing. www.bsigroup.com (accessed 2 February 2012).

Carbon Trust (2007) Health Sector Overview. http://www.carbontrust.co.uk/energy/startsaving/sectorselector/healthcare_15.htm (accessed 2 February 2012).

Department of Health (2007) Rebuilding the NHS: A new generation of health facilities. http://www.dh.gov.uk/prod_consum_dh/groups/dh_digitalassets/@dh/@en/documents/digitalasset/dh_075441.pdf (accessed 2 February 2012).

Department of Health (2008) Environment and sustainability – Health Technical Memorandum 07-07: Sustainable health and social care buildings – Planning, design, construction and refurbishment. http://www.hefma.org.uk/article/253_consdraft.pdf (accessed 2 February 2012).

European Commission (2007) Life Cycle Costing (LCC) as a Contribution to Sustainable Construction: a Common Methodology. http://ec.europa.eu/enterprise/construction/compet/life_cycle_costing/final_rep_summary_en.pdf (accessed 2 February 2012).

Faithful+Gould (2012) Construction Carbon Calculator. http://www.fgould.com/carbon_calculator (accessed 2 February 2012).

HM Treasury (2003) The Green Book: Appraisal and Evaluation in Central Government. http://www.hm-treasury.gov.uk/d/green_book_complete.pdf (accessed 2 February 2012).

Lockie, S. and Bourke, K. (2009) Sustainability Budgeting: A Guide to Using Whole Life Costing in Sustainable Procurement. SHINE, London.

National Audit Office (2005) Sustainable Procurement in Government. http://www.nao.org.uk/publications/0506/sustainable_procurement.aspx (accessed 2 February 2012).

Office of Government Commerce (2007) High Performing Property Implementation Plan http://www.ogc.gov.uk/documents/CP0154HighPerformingPropertyImplementationPlan.pdf (accessed 2 February 2012).

Chapter 20
Designing Super-Tall Buildings for Increased Resilience: New Measures and Cost Considerations

James Hayhoe

20.1 Introduction

There is now a growing challenge in designing tall buildings for increased resilience particularly from terrorism, explosion and other similar threats. Managing the evolving risk involved in designing tall buildings is a complex process as a result of the diverse factors involved and the potential cost implications. The responsibility placed upon the design team including cost specialists for designing and engineering cost-effective solutions to the challenges posed by terrorism or similar explosive threats has never been greater.

This chapter explores the factors influencing the design of tall buildings, the new measures adopted, and their cost implications. The chapter starts with an overview of the key challenges in tall buildings, the need for increased resilience and future-proofing design, to cope with critical threats in modern society such as the evolving terrorist threat and other similar threats. Key factors which influence the design and cost of tall buildings such as shape, effect of core design and size of floor plate, height of buildings, embodied energy and building services, need for vertical transportation, level of fire protection, and structure and frame construction are discussed. Some of the new measures to increase the resilience of tall buildings to deal with terrorism and similar threats such as fire are also discussed, along with the cost of implementation.

20.2 Challenges of tall buildings and the need for increased resilience

Since the 1930s, the average height of the world's tallest 'skyscrapers' has steadily increased and by 2010 this average had jumped 22% in 10 years. Between 2007 and 2010, the number of 'supertall' buildings doubled from 34 to 82 across the

Design Economics for the Built Environment: Impact of Sustainability on Project Evaluation, First Edition.
Edited by Herbert Robinson, Barry Symonds, Barry Gilbertson and Benedict Ilozor.
© 2015 John Wiley & Sons, Ltd. Published 2015 by John Wiley & Sons, Ltd.

world (CTBUH, 2008). Such is the heights now being achieved, the use of the term 'super-tall' – commonly used to describe buildings exceeding 300 m in height has been out-grown by 'mega-tall' – a term reserved for buildings exceeding 600 m which it is believed, we will see in the next 10 years (CTBUH, 2011). In this same period, there has been a similar rise in terrorist activity and a notable change in the method of attacks marking a shift in worldwide terrorist strategy. The threat has become more dynamic with terrorists inspiring and favouring suicide attacks in lieu of static threats that were relied upon previously. Recent attacks have also proved to be more organised and arguably more complex than they have histori-cally been. Greater coordination and innovation has also been shown in their ability to strike a number of targets simultaneously.

The combination of these trends has led to a need for greater understanding of the importance of designing tall buildings for increased resilience, particularly from terrorism and other similar threats. Tall buildings present an attractive tar-get for terrorists to achieve their aims and their design needs to reflect this risk.

For new buildings managing the evolving risk through design is challenging enough given the diverse nature of the subject. However, there are wide concerns over existing tall buildings which were designed when less information was avail-able. The role of tall buildings is evolving as is the responsibility placed upon those who depend upon their integrity. The need for designing and engineering cost-effective solutions to cope with the challenges posed by terrorism has never been greater.

20.3   Factors influencing design and cost of tall buildings

There are a number of key factors which influence the design and cost of tall buildings, factors which characteristically distinguish high-rise from low-rise. Generally, due to the nature of the construction process and performance require-ments demanded of tall buildings, cost is greater than conventional projects although increases are sometimes exaggerated by 'landmark' architecture. Despite the cost premium for building tall, it is argued that tall buildings are cost-efficient overall as they make better use of land and if supported by a public transport system will reduce the necessity for private transport.

Influence of shape

The shape of a building is regarded by some to be as influential on cost as height. The shape of the building has to satisfy not only the architectural challenges but also the engineering ones in order to overcome the external forces imposed upon the building whilst simultaneously efficiently managing the environment within. Key efficiency ratios exist which determine the wall to floor and gross to net areas of a tower which in turn measures the commercial viability of cost over value. Such calculations are largely dependent on the geometry and even orientation of the structure. The importance of managing these ratios is magnified with height. Super-tall towers need to balance the sensitive nature of the decreasing floor area over structural footprint ratio with increasing height, lateral loadings and struc-tural rigidity ratios.

Effect of core design and size of floor plate

The core design and size of the floor plate are factors which are interchangeably dependent upon each other and demand adequate consideration at an early stage. The type of core and method of construction has significant implications on means of escape, structural rigidity and cost of construction in terms of programme to build and weight imposed upon the foundations. Fundamentally, the floor plate size, and the area of open lettable space is central to the economic viability of each project. "Maximising the gross floor area and net lettable floor area will contribute to the overall project success. Column-free space is valued by many users and spans in excess of 12 m incur premium cost associated with substructure, superstructure and depth of section" (Davis Langdon, 2009).

Height of buildings

There is a long established and well understood relationship between building height and building cost. The conventional wisdom of construction economics suggests that the cost of construction per square metre increases as buildings become higher or taller (Bathurst and Butler, 1980; D'Cruz and Radford, 1987; Picken and Ilozor, 2003). Other authors such as Skitmore and Marston (1999) developed the relationship further and concluded that "construction cost per floor area fall up to a height of five to six storeys and then rise with increasing height". A similar trend was found by Morton and Jaggar (1995) who argued that "construction cost would rise above three to four storeys". Other factors also affect the cost of tall buildings such as the net/gross floor area (GFA), the structural cost, lift costs and fire protection costs as they are all directly related to the number of storeys.

As a result of the increased unit costs and decrease in floor area, net space is disproportionately more expensive as shown in Figure 20.1. Irrespective of floor plates, landmark status, form and proportion, development costs per metre

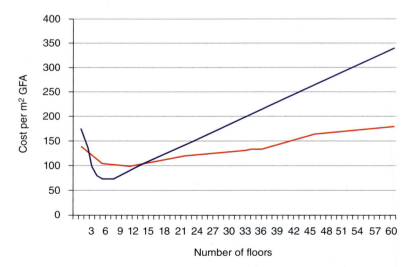

Figure 20.1   Cost per metre squared for GFA over number of floors. Data extracted from Flanagan and Norman (1978) and De Jong and Wamelink (2008).

squared of GFA increases with height. The cost versus height relationship is characterised by step changes unique to each project, due to various technical thresholds which occur at different heights and floor plate dimensions. Barr (2007) suggested that the relationship may be more governed by economic factors noting 'economically the factors determining "optimal" height relate to global, national and city economies and their respective regulations. An economically optimal height would be the maximum height (and maximum floor space) as dictated by the net return on investment, i.e. the height at which the marginal cost of creating another floor is equal to the sales price realised for that floor'.

Embodied energy and building services

A study reported by Treloar *et al.* (2001) found that "high rise buildings had higher embodied energy per unit GFA" compared with lower rise buildings with the difference being reported at approximately 60% equating to an increase of "1–2% of basic cost per each additional floor". According to Pank *et al.* (2002) the initial cost of the mechanical and electrical (M&E) equipment in a tall building, including lifts, is often more than 25% of the total construction cost. Some of this cost may be attributed to the increased demand to "provide enough capacity for the population density and load" in tall buildings (Watts and Kalita, 2007). However, in an analysis between the cost of tall residential buildings compared with low-rise conventional buildings, Warszawski (2003) found that in terms of the electrical, sanitary and air conditioning systems there was no cost variance per square metre between buildings that were 10–19, 20–30 and 31–40 storeys, respectively.

Implications for vertical transportation

Ho (2007) explains the lift arrangement has a major impact on the size and internal layout of the service core which in turn affects the net rentable floor area. Achieving "the right balance between number of zones, size of groups, core size and lift performance to meet cost, area and performance objectives requires specialist input. Watts (2005) adds that population distribution, building functions and performance all affect the design and economics of the lift design. Finding such a balance will maximise the lettable floor area and efficiency for the user. Warszawski (2003) compared the cost of tall residential buildings and low-rise conventional buildings, and found that the cost variance of the 'elevator' system between buildings that were 10–19, 20–30 and 31–40 storeys rose from US$4 to US$21 up to US$35 per square metre, respectively.

Level of fire protection

In the UK, Approved Document B details the levels of Fire protection required for tall buildings. In the US, this is covered under the codes issued by the National Fire Protection Association (NFPA). Unsurprisingly, tall buildings attract more attention than lower rise buildings when it comes to fire protection due to their inherently higher levels of associated risk. Warszawski (2003) noted that cost follows the same trend as cost per square metre for fire protection on buildings in three height categories. For 10–19, 20–30 and 31–40 storeys, the cost went from US$10 rising to US$35 for the two higher height categories. The stated reasons for the additional cost included additional stairwells, water pumps, fire separation and fire-resisting doors.

Structure and frame construction

The structure and frame design is at the heart of the cost and design influences for the entire project. Watts and Kalita (2007) suggest that the structural frame, cores and upper floors amount to 15–25% of the construction cost in a commercial tower and 10–15% in a residential tower. External walls alone "can constitute 25% of the total net trade cost of a tall building which in turn can affect the bottom line by 15%" according to Watts (2010). In terms of price increase over building height, data from the analysis between the cost of tall residential buildings compared with low-rise conventional buildings found that the structural cost per square metre for buildings in three height categories of 10–19, 20–30 and 31–40 storeys went from US$13 to US$32 rising to US$50, respectively. The reason cited for the trend of cost rising with height is the increased cost associated with the size of the cores and columns (Warszawski, 2003). This percentage of cost is consistent with the findings reported by Pank *et al.* (2002) who state that the "superstructure and foundations generally account for 10–20% of the total construction cost with the building envelope costing approximately the same". In addition to its dead and imposed loads, a super tall building must be designed for resisting horizontal wind or seismic loads, whichever is greater (Ho, 2007).

Other factors

Other factors influencing the cost and design of tall buildings include building use, location and resource considerations. Watts (2005) explains "the cost of a single occupancy versus a multi-let building, and the design complexities associated with accommodating multiple uses within a single building are relevant". Cost will also need to be adjusted for location with a report entitled 'International Building Cost Comparison' published by Davis Langdon (2009), providing the comparisons. The development of a tall building must balance the increased cost of construction and land with the value attainable from areas of peak demand. "Where there is an intense demand for accommodation such as near the centre of a prosperous city, the very possibility of building high itself pushes up land prices. This in turn means that tall buildings become cost effective, as the high cost of the land is distributed over a greater lettable floor area of building" (Watts *et al.*, 2007).

20.4   Design of counter-terrorism measures

"September 11th, 2001, and the destruction of the WTC in New York, has come as a sharp reminder of the fragility of man's creations when subjected to a determined attack by terrorists" (Dorge and Jones, 2001). The harrowing images of the Twin Towers acted as a stark reminder of the importance tall buildings play in not merely determining the safety and security of those directly influenced by it but can also act as a tangible symbol of the stability of a nation, and strength of Western society. With the constant threat from terrorism evolving, the design of tall buildings must respond to these challenges. The key challenge and balance thereof, lies between the threats being managed through cost effective design allowing inherent risk to be reduced, while maintaining the building's functionality and fitness for purpose.

Risks associated with tall buildings

Tall buildings are inherently risky: financially, physically and commercially. In financial terms, Warszawski (2003) suggests "the difference in financing cost between the tall and conventional buildings amounts to 5–7% of the total project cost". That said, the greatest perceived threat to tall buildings is often ironically also the force driving the demand for them. As Ho (2007) explains, the sheer height of tall buildings represents "technical and political prestige". This relationship is again illustrated in the review of the UK National Security Strategy when the then Prime Minister Gordon Brown promised to 'work with architects and planners to "design in" safe areas and blast resistant materials and enhanced physical protection against vehicle bomb attacks'. It was also reported that this included a desire by the government to introduce "panic rooms, truck-bomb barriers and limited glazing" for major new buildings (Booth, 2008). Focusing on counter-terrorism measures at the design stage by engaging architects and engineers was also the focus on Project Argus led by the National Counter Terrorism Security Office (Hunter, 2008). Such views are shared at least in part with evidence presented by Yean Yng Ling and Hong Soh (2005) from Singapore, where it was found that those who took part in the study "do not agree that buildings should be designed to withstand hydrocarbon fires or aircraft attacks" but do support greater counter-terrorism design integration.

Demand for higher design standards

In the aftermath of the 9/11 attacks, doubts about the viability of skyscrapers grew, amid concern stringent fire regulations would push construction cost up by 300% (Lane and Leftly, 2001). Design was challenged too, with the US focusing on ensuring any "new prestige building design had a structure robust enough to prevent the total collapse of the building and to facilitate evacuation and work by emergency services" (Burns, 2009). There was little doubt that safety lessons had to be learnt from the New York attacks (The Independent, 2001). With the aim of bridging this gap, many businesses seized the opportunity to provide counter-terrorism design advice such as Arup (2010) and TPS (2010). Academically, a study in Singapore on the ways in which the design of tall buildings could be improved after 9/11 drew together many of the features being muted. Yean Yng Ling and Hong Soh (2005) presented 23 "possible ways to improve the safety of tall buildings" ranging from structural design, fire engineering, means of escape, materials technology and codes and regulations. However, a balance between realistic cost and extremely unlikely events should be maintained (Yean Yng Ling and Hong Soh, 2005).

A global overhaul of current and proposed towers was undertaken. "The focus of the efforts would be designing buildings so that occupants had sufficient time to evacuate after a catastrophic event" (Lane, 2001). However with nearly three-quarters of the injuries from the 1995 bombing of Oklahoma City's Alfred P. Murrah building caused by flying glass, this was one area of particular focus for designers (Daniels, 2002). Design improvements in glass vary between ductile impact-absorbing glass to blast resistant and ballistic glass dependent on the cost effectiveness and the specific threat posed (RIBA, 2010). Laminated shatter-proof glass has emerged as a popular choice as well as the fire-resistant safety glass range (AIS, 2010).

Building codes and international regulations

The building regulations and design codes around the world have arguably borne the brunt of criticism following the events of 9/11 with most literature in agreement that the current codes must be reviewed and changed. In the UK, Part A of the Building Regulations which deals with structure did not differentiate between tall building and buildings of moderate to low height, despite the significantly greater challenges and risk they pose (Building, 2001; Pearson, 2001). Similarly, the US has no code provisions addressing progressive collapse, unlike the UK, which has had them since the 1970s (Nadel, 2007). In addition, US "national model building codes do not include requirements to design for loads that might be imposed due to acts of war or terrorism" (Post, 2002).

Lay, cited in Whitelaw (2007), explains that plans to update the US International Building Code to insist on a third escape stair in tall buildings could add "5% on the cost of a building". This may make such measures simply cost prohibitive to developers, as he argues that "to get a lot of towers built we are already driving efficiencies to the maximum. Recommendations made by the National Institute of Standards and Technology (NIST) following a 3-year investigation into the collapse of the World Trade Centre in 2001, has been incorporated into building code changes proposed by the International Code Council (Nadel, 2007; NIST, 2010).

Despite government advice on counter-terrorism design, UK building regulations do not include specific measures intended to deal with terrorist activity and therefore compliance with building regulations should not be assumed to indicate consideration of the issues (Home Office, 2012). British standards have been produced for Fire Safety Engineering in Buildings though, since 9/11.

Changes in structure of buildings

The 33-storey Barclays building in Canary Wharf is a good case study example of the measures adopted in light of the lessons learnt from the attacks on September 11th. Having been hailed as "one of the most robust buildings in the world" and "a blueprint for future tall buildings", the structure allows for any two adjacent structural columns to be removed on any floor without endangering the remainder of the building. In addition, the building incorporates two additional fire escape stairs, a thicker reinforced concrete core, increased fire-proofing, laminated glass and its plant systems have been located on the roof (Leftly and Richardson, 2004). Somewhat less promising was the findings of an Australian study which found that their tall buildings were likely to collapse if they were subjected to a moderate bomb blast or collision with a light aircraft (Medis and Ngo, 2003). Across the water in Singapore, a study by Yean Yng Ling and Hong Soh (2005) found the opposite, that "tall buildings in Singapore are safe, with adequate structural design, fire safety and means of escape".

It was identified that steel would be required to withstand higher temperatures which it could do if protected, for periods of up to 4 h in addition to withstanding an initial blast. However this did incur rather substantial cost increases of between 10% and 300% depending on the method used (Lane, 2001; Lane and Leftly, 2001). Modern building design must reflect the threat and evolve with the risk of terrorism. Existing structures would be resilient to increased dead or even live loads, however the sudden and directional forces characteristic of previous terrorist attacks are too great. To compensate, it is suggested that "engineers need to focus on strengthening the external frame to help prevent progressive collapse in extreme loading events". A study by Medis and Ngo (2003), two now-academic structural engineers, found that the result of moderate bomb blast or collision with a light aircraft on tall buildings in Australia would be the progressive collapse of the buildings.

Evacuation and fire escape management

Jeffrey McCarthy, Managing Partner at Skidmore Owings & Merrill, widely regarded as the leading practice specialising in the design of tall buildings, was also lead Architect on the Burj Khalifa project, currently holding the record for tallest building in the world. He states that "it is incumbent upon us to reconsider certain aspects of design criteria particularly egress". Design elements currently being introduced into existing and new buildings include "a fireman's lift that is separate from the main stairwell" and refuge floors "which contain no furniture" and are totally fire proof to harbour people during an evacuation. Both measures have significant cost implications in terms of design and build costs as well as the associated reduction in usable and lettable space (Clark, 2001).

Whilst the over-riding aim of terrorism is to create panic, spread fear and generate attention for certain ideals remains constant, the nature and method of carrying out an attack has changed. Recent examples of terrorist incidents have seen person-borne attacks favoured to vehicle or static threats. This has made the threat more dynamic by allowing the terrorists more control over exactly where and when the attack can take place. In comparison with static car bombs which were abandoned in strategic places, commonly used in terrorist attacks of the 1980s such as in Oklahoma City in 1970 and in Bishopsgate, London in 1982, the devices or component threat has decreased. However, the shift in execution of the threat including the realisation that the perpetrators are not afraid to die has arguably meant that a greater effect can be achieved. For example, the events at the World Trade Centre in 2001, the attacks in London in 2005 and in Mumbai in 2008 all demonstrate the shift away from large scale fixed threat to dynamic person-borne suicide attacks.

20.5   Cost of new measures and design

The type and impact of the current terrorist threat

The terrorist attack in New York on 11 September 2001 has been attributed with redefining the threat to tall buildings around the world. Since then research has shown that the current terrorist threat to tall buildings is significant with counter-terrorism design maintaining a high position on the design agenda since 2001. Prior to this point, terrorism did not appear highly on the design agenda with the perceived threats to tall buildings mainly revolving around theft and fire. Furthermore, the management of external threats such as terrorism were seen as outside the parameters of traditional building design. However the events of that day showed the world a new, more organised and inventive side of terrorism, the impact of which was widely felt. The events exposed the vulnerability of tall buildings and inspired a revolution in counter-terrorism design and measures to manage the threat.

There are a number of current types of terrorist threat shown by the research including chemical, biological and aircraft, however the threat of blast remains most prominent to tall buildings according to key experts in the field. This deceptively suggests the type of threat has changed little since the IRA threat in the 1980s, however studies of recent attacks have shown that there has been a significant shift in the execution and delivery of attacks which are often unique and original. Blast threats are divided between their method of delivery, vehicle-borne, person-borne or mail, with each posing separate challenges and demanding different design solutions.

Research supports the fact that the impact of the current terrorist threat has re-shaped how modern tall buildings are designed as well as how they are managed and operated. In the aftermath of the 2001 attacks, it was initially feared that the events would mark the beginning of the end for tall buildings. Confidence in them was at an all-time low, their safety measures were brought into question and their commercial future appeared uncertain. Many questioned simply how tall buildings with their inherent risk could counter the threat to which they were now so publicly exposed too. Over 10 years on, it is evident from the current demand and future proposed projects that such fears have not materialised. However the support for the inclusion of counter-terrorism design measures gained momentum and in certain geographical areas has become standard practise. There has been a notable shift fundamentally in how the design is governed, with the focus moving from code-based to performance-based criteria. Research found that experts are realistic about the extent to which terrorism can be mitigated through design. Essentially, experts feel that the focus has to be on addressing the key risk and threats through design and ensuring the building does not become part of the terrorist weapon in any way.

New measures and approaches to building design

A number of measures and approaches came to light in the research that have been adopted in the design of tall buildings, which aim to prevent terrorist attacks and mitigating the effects of them in the event they occur. Table 20.1 summarises the key threats and the counter-terrorism design measures.

Many of the design measures noted in Table 20.1 are inter-related to the various threats and do not necessarily only apply to one threat. Similarly, the key new approaches aim to support and complement the physical measures adopted into the design. A number of examples of new approaches to the design and construction of tall buildings have been highlighted by the research and are reviewed in Table 20.2.

Table 20.1   Key threats and mitigating design measures.

Key potential threats	Counter-terrorism design measures
Vehicle-borne blast threat	Engineer for a reversible ground floor slab allowing for upward and downward loading Enhanced façade including glazing Effective CCTV surveillance Vehicle access control – one point entry, checked 30 m away Hostile Vehicle Mitigation System
Person-borne blast threat	Access control Effective CCTV surveillance X-ray scanning Two-tier access for registered occupants and visitors
Mail blast threat	Mail X-ray scanning Remote or satellite centralised mail room separate from the main building
Aircraft impact	Fire compartment Core design – greater use of high-strength concrete Structural changes favouring more hybrid and core/sheer wall types compared to perimeter wall/tubed structures. Greater structural provision to prevent progressive collapse.
Chemical, Biological, Radioactive and Nuclear threat	Zoned ventilation systems Access control High level air intake ducts Protected water filtration systems Short-wave length radar scanning
Multi-person-borne ballistic threats	Access control Evacuation CCTV X-ray scanning

Cost consequences of new design regulations, measures and approaches

The cost consequences of new design regulations, measures and approaches is widely recognised by the research and has become an established element of the overall project finance. That said, the cost and extent of the impact is variable and dependent on the design and measures included. These in turn are often dependent on the 'environment' in which the building stands, for example a building is often benchmarked against its neighbours to assess the robustness in relative terms, as well as the wider view taken in an increasing number of cities that terrorism is best managed as an 'estate' or 'cluster' as opposed to individually.

Cost models published in 1997, 2002 and 2007 have tracked the evolution of terrorism as a cost influence, and show a clear spike in 2002. The economics of tall building counter-terrorism design has developed with the threat. A paradox exists, which is in turn supported by the research, whereby the overall cost of

Table 20.2 Outline of new approaches.

New approaches and procedures	Description
Threat and Risk Assessments	Widely regarded as essential at the beginning of the design process, this assessment is carried out to review the unique threats and risk that each tenant or building will be exposed to and the likelihood of them occurring. This assessment and review will often shape and focus the building design and influence how it is operated and managed
Modelling	The process of modelling is used to assess the behaviour of various structural elements under predetermined conditions, often consistent with the loads and forces imposed during terrorist events. The use of such technology has allowed buildings to be designed more dynamically to achieve certain performance criteria
Fire and evacuation	The approach to the prevention of fire and the facilitation of part or full evacuation is an area that has seen significant attention. Much focus has been given to occupant behaviour and movement during an evacuation and improvements have been made to enable a building's occupants to decrease the time taken to evacuate the building. Furthermore, there is increasing support in certain events for the 'Evacuation' of the building's population to an area of relative safety within the building in contrast to the notion of external evacuation. Research has shown that in an increasing number of buildings this provision is being provided by refuge floors; however in others the space in the core has been utilised. Evacuation makes it easier to manage and communicate with the building's population and reduces the risk of exposure to secondary attacks.
Closer collaboration between Design and management teams	There are a number of contributing factors such as the growing multi-use of tall buildings and the increasing belief in the method of Evacuation which have drawn both the building designer teams and the building management teams together. It has been recognised in the research that a more joined up approach uniting the two disciplines will increase the efficiency not only of the design and use of a building but will also greatly enhance its robustness in the event of a terrorist incident
Structural	There has been a shift in design approach from the code-based assumption that buildings were safe to a more performance-based approach whereby buildings are designed to meet or exceed certain levels of loadings. Moreover, greater attention has been given to the buildings' provision for progressive collapse with the aim of ensuring they maintain their integrity for a period of time that is equal to or greater than that required to fully evacuate the maximum number of occupants possible at any point in time

implementing counter-terrorism design measures to a certain level is cost effective for tall buildings in comparison with lower heights. This is due to less of the building being exposed to ground-based threats with sufficient vertical stand-off being achieved. However this is challenged by the disproportionate risk factors facing tall buildings in comparison with low rise. The cost of counter-terrorism design per square foot is therefore decreased with height. Overall, the research showed that experts felt the cost of tall buildings has changed little over the last decade as a result of the increased threat of terrorism. They stated that it was the design that had become more efficient and dynamic with existing costs being substituted and not necessarily added to. However, with costs driving buildings to become leaner and more efficient, the research warned that there was a view held by some that this was going too far, and some buildings were losing their inherent robustness as a result.

Structurally, cost has been influenced both positively and negatively. Greater cost has been incurred with certain elements being reinforced to allow for greater ductility and rotation, however technology used to establish the critical areas in which this is required can also identify areas that can be made more efficient or serve duel purposes. This may allow the design to be down-graded and shows where over-design can be detrimental to the ability of the building to absorb, distribute and deflect the loads imposed on it. Overall the research suggests that the premium for structural enhancements is between 5% and 10% depending on the specifics of the project, however could be as low as 1% if only key areas are enhanced.

The core design is evolving, with the research showing the size and widths have increased, impinging on the available lettable area. Similarly, the level of redundancy incorporated into the design has also increased, reflecting the trend of safety increases over commercial viability. Such shifts have commercial implication however the research shows some counter-terrorism measures are used as a marketing tool. In terms of façade design, cost has been shown to increase as a result of the use of materials such as laminated inner glass, frames with wider glass bite and structural silicone glazing (SSG), however costs are falling as demand increases and costs are proportionate to height. Such additions were approximated at costing an additional £50 per metre squared in the research to cover such enhancements to the façade. Moreover, the research specifically reported the decrease in façade enhancement cost in London since the terrorist attacks in 2001, attributed primarily due to the wider global economic pressures; however this was not felt in other cities around the world. A significant factor determining façade cost is the shape and geometry of the building.

The changes to the Fire Regulations and 'means of escape' guidance has influenced the design of the floor plates and brought about negative cost implications. By widening the stair wells, the available lettable floor area has been reduced. Whilst this may not be a cost increase it does represent a value decrease and decreased commercial viability. That said, contributors noted in the research suggested value lost on lettable floor area may be offset by savings in other areas of the building as space is used for dual purposes, such as the core also acting as a safe haven in the event of an emergency incident.

20.6   Concluding remarks

Underpinned by current research, this chapter has investigated the extent to which terrorism has impacted on the design and cost of tall buildings. In order to fully explore this, the factors influencing the design and cost have been reviewed, which has shown the plethora of elements which need to be considered and balanced in order to deliver a building fit for purpose and capable of effectively managing the external environmental pressures imposed upon it. This was developed further as the design of counter-terrorism measures was studied to show how these are incorporated into the scheme for tall buildings and how each measure has changed in order to contribute to the overall performance of the building. Some key areas came to light through the research which have significantly changed in the last 10 years as a direct result of the increased threat from terrorism. Such elements include Evacuation and Fire Escape Management and changes made to the structure to prevent progressive collapse and mitigate the impact of blast threats.

Finally, the nature of the current threat was reviewed alongside the measures and approaches which have evolved and been introduced to manage such threats. It highlighted that despite all of the changes in the origin of terrorist threats, and even the original nature of how such attacks are carried out, the essential threat to buildings remains relatively constant. For example, the threat of blast has been shown to still be one of the leading threats to which buildings are designed as it has for over 30 yr, however the method in which blast threats are delivered has evolved. The same is true for newer threats such as armed gunmen opening fire in public areas, which is mitigated through the use of similar procedures as other threats such as effective access control and evacuation management plans. The cost of these new measures and design was reported, bringing the overall cost influence of terrorism into focus. It concluded that if integrated at the outset structural counter-terrorism design would have little cost implication as additional live and dynamic loads characteristic of any attack can be built in. Furthermore, the structure can be designed to serve duel purposes such as the position of cores may also provide lateral and vertical stability to help prevent progressive collapse. It was a similar case across all elements of the design, with figures of between 1% and 10% reported as the uplift in cost to incorporate counter-terrorism design measures.

It can be concluded that the threat from terrorism has played a significant role in how modern tall buildings are designed as well as the way in which they are used and managed. Conversely, design is, and will continue to be, fundamental to managing the threat from terrorism. The structural design has notably shifted to become more performance focused as opposed to simply using building codes to dictate the design. The cost has also proved to have been impacted by the threat of terrorism however not as significantly as initially anticipated. It is key that future counter-terrorism design allows buildings to not only maintain a proactive approach to countering any threat posed but also allows for a reactive response.

Allowing for such flexibility enables measures to be upgraded or downgraded depending on the threat level or risk exposure at any time. However managing terrorism through design remains essentially a reactive process as design and measures are only truly tested during an attack. Whilst the design can be manipulated to design-out as many threats as possible and measures can be incorporated to minimise the effect of the building and its occupants, design can only go so far and a certain level of risk has to be accepted. The challenge for the future design of tall buildings is to achieve a balance between an acceptable level of performance-based integrated counter-terrorism design measures and the pure economics of cost. Without such balance, the future of tall buildings is threatened through commercial viability or unacceptable risk.

References

AIS (2010) Fire-resistant glass. *Association of Interior Specialists,* 27.

Arup (2010) Security and Risk Consulting. http://www.arup.com/Services/Security_and_Risk_Consulting.aspx (accessed 10 April 2010).

Barr, J. (2007) Skyscrapers and the skyline: Manhattan, 1895-2004. Rutgers University working paper: 54.

Bathurst, P.E. and Butler, D.A. (1980) *Building Cost Control Techniques and Economics,* 2nd edn, Heinemann, London.

Booth, R. (2008) Home Office Urges Architects to Design Terror Proof Buildings. The Guardian, 22 March 2008. http://www.guardian.co.uk/politics/2008/mar/22/terrorism.uksecurity (accessed 12 May 2011).

Building (2001) World's Top Engineers Unite to Make Skyscrapers Safer. http://www.building.co.uk/news/worlds-top-engineers-unite-to-make-skyscrapers-safer/1013200.article (accessed 17 May 2011).

Burns, J. (2009) New Tall Buildings Provide Nightmare Scenarios. The Financial Times, 18 March 2009. http://www.ft.com/cms/s/0/8db698ca-b7b5-11da-b4c2-0000779e2340.html#axzz3983LOc8d (accessed 10 April 2010).

Clark, E. (2001) What is the Future for Skyscrapers? BBC, 21 September 2001. http://news.bbc.co.uk/1/low/business/1556251.stm (accessed 10 April 2010).

CTBUH (2008) Tall Buildings in Numbers. http://www.ctbuh.org/LinkClick.aspx?fileticket=M7nXrLx8g0M%3D&tabid=1108&language=en-GB (accessed 24 April 2012).

CTBUH (2011) The Tallest 20 in 2020: Entering the Era of the Megatall. http://www.ctbuh.org/TallBuildings/HeightStatistics/BuildingsinNumbers/TheTallest20in2020/tabid/2926/language/en-US/Default.aspx (accessed 24 April 2012).

Daniels, S.H. (2002) Better blast resistance coming soon to façade. Engineering News Record, 25 March 2002. https://enr.construction.com/features/buildings/archives/020325h.asp (accessed 16 May 2011).

D'Cruz, N.A. and Radford, A.D. (1987) A multi criteria model for building performance and design. *Building and Environment,* **22**(3), 167–179.

De Jong, P. and Wamelink, H. (2008) Building cost and eco-cost aspects of tall buildings. Paper presented at CTBUH 8th World Congress, 3–5 March 2008, Dubai, UAE. http://img2.timg.co.il/forums/1_142325960.pdf (accessed 1 May 2011).

Dorge, V. and Jones, S. (2001) Building an Emergency Plan: A Guide for Museums and Other Cultural Institutions. *Museum Management and Curatorship,* **XIX**(1), March. http://0-iris.emerald-library.com.lispac.lsbu.ac.uk/Insight/viewContentItem.do;jsessionid=36FD22708B082D2C8DC9D26DA1320822?contentType=NonArticle&contentId=1471307 (accessed 10 April 2010).

Flanagan, R. and Norman, G. (1978) The relationship between construction price and height. *Chartered Surveyor Building and Quantity Surveying Quarterly,* 5(4), 68–71.

Ho, P.H.K. (2007) Economics planning of super tall buildings in Asia Pacific cities. Paper presented at Strategic Integration of Surveying Services, 17 May 2007, Hong Kong SAR, China. http://www.fig.net/pub/fig2007/papers/ts_4g/ts04g_01_ho_1673.pdf (accessed 1 May 2011).

Home Office (2012) Crowded Places: The Planning System and Counter-terrorism. http://nactso-dev.co.uk/system/cms/files/107/files/original/Crowded_Places-Planning_System-Jan_2012.pdf (accessed 23 March 2014).

Hunter, W. (2008) Debate Rejects Call for Counter-terrorism Design. Building, 26 June 2008. http://www.building.co.uk/comment/debate-rejects-call-for-counter-terrorism-design/3116938.article (accessed 12 May 2011).

Lane, T. (2001) Fire Experts to Examine Safety Regulations. Building, 2001. http://www.building.co.uk/news/fire-experts-to-examine-safety-regulations/1011661.article (accessed 17 May 2011).

Lane, T. and Leftly, M. (2001) Skyscrapers Face Huge Cost Rise after Bloody Tuesday. http://www.bdonline.co.uk/news/skyscrapers-face-huge-cost-rise-after-bloody-tuesday/1011646.article (accessed 16 May 2011).

Leftly, M. and Richardson, S. (2004) Three Years of Fear. Building, 2004. http://www.building.co.uk/comment/three-years-of-fear/3040335.article (accessed 16 May 2011).

Medis, P. and Ngo, T. (2003) Australia's Tall Building Design Fails Terrorist Attack Test. http://archive.uninews.unimelb.edu.au/view-12369.html (accessed 10 April 2010).

Morton, R. and Jaggar, D. (1995) *Design and the Economics of Building*, E & FN Spon, London.

Nadel, B. (2007) High-rise Safety, International Codes, and Tall Buildings: Rising to Meet the Challenge. Building, 7 May 2007. http://www.buildings.com/ArticleDetails/tabid/3321/ArticleID/4931/Default.aspx (accessed 6 May 2011).

NIST (2010) NIST and the World Trade Centre. http://wtc.nist.gov/ (accessed 6 May 2011).

Pank, W., Girardet, H. and Cox, G. (2002) Tall Buildings and Sustainability. http://www.cityoflondon.gov.uk/services/environment-and-planning/sustainability/Documents/pdfs/tall-buildings-sustainability.pdf (accessed 1 May 2011).

Pearson, A. (2001) The Skyscraper of the Future. Building, 2001. http://www.building.co.uk/news/the-skyscraper-of-the-future/1013459.article (accessed 17 May 2011).

Picken, D.H. and Ilozor, B.D. (2003) Height and construction costs of buildings in Hong Kong. *Construction Management and Economics*, **21**, 107–111.

Post, N.M. (2002) New Standard for Tall 'Targets'? Engineering News Record, 13 May 2002. http://enr.construction.com/news/buildings/archives/020513.asp (accessed 1 May 2011).

RIBA (2010) RIBA Guidance on Designing for Counter-terrorism. http://www.architecture.com/Files/RIBAHoldings/PolicyAndInternationalRelations/Policy/CounterTerrorism/RIBADesigningforCounterTerrorism.pdf (accessed 1 May 2011).

Skitmore, M. and Marston, V. (1999) *Cost Modelling*, E & FN Spon, London.

The Independent (2001) Ken Livingstone: The Only Way Is Up. The Independent, 2 November 2001. http://www.independent.co.uk/news/uk/home-news/ken-livingstone-the-only-way-is-up-615624.html (accessed 10 April 2010).

TPS (2010) Security – Projects, Research and Development. http://www.tpsconsult.co.uk/tps/sectors/security.aspx (accessed 10 April 2010).

Treloar, G.J., Fay, R., Ilozor, B. and Love, P.E.D. (2001) An analysis of the embodied energy of office buildings by height. *Facilities*, **19**(5/6), 204–214.

Warszawski, A. (2003) Analysis of costs and benefits of tall buildings. *Journal of Construction Engineering and Management*, **129**(4), 421–430.

Watts, S. (2005) Tall Buildings: A Strategic Design Guide.

Watts, S. (2010) Competitive cities and the tall building landscape. *Infrastructure Today*, **April**, 24–30.

Watts, S. and Kalita, N. (2007) Cost Model: Tall Buildings. http://www.building.co.uk/data/cost-model-tall-buildings/3085522.article (accessed 9 May 2011).

Watts, S., Kalita, N. and Maclean, M. (2007) The economics of super tall towers. *The Structural Design of Tall and Special Buildings*, **16**, 457–470.

Whitelaw, J. (2007) Leading Engineer Questions Need to Design Tall Buildings to Withstand Terrorist Attacks. http://www.nce.co.uk/leading-engineer-questions-need-to-design-tall-buildings-to-withstand-terrorist-attacks/91518.article (accessed 1 May 2011).

Yean Yng Ling, F. and Hong Soh, L. (2005) Improving the design of tall buildings after 9/11. *Structural Survey*, **23**(4), 265–281.

Chapter 21
Building Information Modelling: A New Approach to Design, Quantification, Costing, and Schedule Management with Case Studies

Aviad Almagor and Barry Symonds

21.1 Introduction

During the eighteenth and nineteenth centuries the building process was effectively managed by the Master Builder. Much has been written regarding both their existence and subsequent demise due to the industrialisation of the construction process but clearly the Master Builder was a capable professional manager, mastering all design and construction aspects and phases. Indeed in the case of Thomas Cubitt (London1788–1855), he also pioneered prefabrication and supply chain techniques (Cooney, 1955) that reduced costs and increased profits. This may be the reason that his business survived whilst most of his contemporaries found themselves in bankruptcy. From early design to project handout, the Master Builder was a source of knowledge and the solution provider. Keeping a unified vision, an integrated approach, and an overall responsibility for the project, the Master Builder coordinated the work, controlled the budget, and monitored the schedule. In the modern era, the enormous and diverse amount of information requires dozens of design experts and specialised sub-contractors to perform the work of the Master Builders of the past centuries.

However the UK construction industry since the Second World War has been plagued by inefficiencies and a lack of integration in the building process. The reasons for inefficiency have been cited by many industrial and academic commentators in the field of construction economics, and criticism has been levelled at the Government, Industry, Contractors, Designers and the Construction professionals. Numerous reports have been produced by the UK construction industry and a plethora of solutions offered. In reality little has changed to the structure

Design Economics for the Built Environment: Impact of Sustainability on Project Evaluation, First Edition.
Edited by Herbert Robinson, Barry Symonds, Barry Gilbertson and Benedict Ilozor.
© 2015 John Wiley & Sons, Ltd. Published 2015 by John Wiley & Sons, Ltd.

or efficiency of the UK construction industry except that in last two decades both public and private sectors have sort to achieve pin point responsibility via the contractor (design build) and transfer risk from Client to Contractor. Incredibly, as noted by Morton (Morton and Jaggar, 1995), it was as early as 1966 that Marian Bowley (1966), an economist, concluded that it was difficult to understand how the construction industry of the UK could have invented a more wasteful and inept system where the separation between design and production was so massive. As Morton noted, "if the system was so flawed it is curious that it has lasted so long".

One of the most fundamental problems is that poor data are a major cause of failure and inefficiency. Data are created and edited by different stakeholders and coordinated periodically, changes are poorly communicated, and reaction to changes comes often too late in the construction process. The result is inefficiency, low quality, risk, and waste. In order to improve the process and provide quality product within budget and schedule constraints, a way to integrate the segregated data and create a unified Master-Builder-like environment in which all project aspects can be reviewed and analysed is required. Major efforts have been implemented in the last 20 yr [most notably by Egan (1998) and Latham (1994)] to update construction processes, change contractual models, and develop technologies that may provide tomorrow's construction industry with the rich legacy left by the nineteenth century Master Builder. However the latest and perhaps most significant of these initiatives is Building Information Modelling (BIM).

This has been driven initially in the US by contracting organisations via independent software companies, together with global real estate clients that have recognised that driving down costs of production, together with speeding up construction times, provides the property developer with a competitive edge in the increasingly turbulent global real estate markets. There are also advocates in Europe and Asia, but in the case of the UK, the Government (Cabinet Office, 2011) has provided a direct lead to implement BIM in the public sector, in a phased approach. To this end, the UK is rapidly catching up with the US, and implementing BIM on many UK construction projects. This chapter examines the relatively new concept of BIM. It starts with a definition and discussion of the concept of BIM, followed by a discussion on the need for better integration and dataflow. The BIM approach and its implications for design, costing and quantification and the challenges it presents the current professional teams and constructors, together with possible effects on the approaches to design economics, are discussed.

21.2 Concept of BIM

The term BIM covers a wide spectrum of solutions and flavors. The National Institute of Building Sciences (NIBS, 2012) from the US defines BIM as:

A Building Information Model (BIM) is a digital representation of physical and functional characteristics of a facility. As such it serves as a shared knowledge resource for information about a facility forming a reliable basis for decisions during its life-cycle from inception onward. A basic premise of BIM is collaboration by different stakeholders at different phases of the life cycle of a facility to insert, extract, update or modify information in the BIM process to support and reflect the roles of that stakeholder. The BIM is a shared digital representation founded on open standards for interoperability.

In the UK the Construction Project Information Committee (CPI) has produced the following definition based on the above as:

Building Information Modelling is digital representation of physical and functional characteristics of a facility creating a shared knowledge resource for information about it forming a reliable basis for decisions during its life cycle, from earliest conception to demolition.

However, Crotty (2012) takes the view that BIM is an approach to building design and construction in which:

- A reference model of the building is created using one or more parametric component-based, 3D modelling system.
- These systems exchange information about the building in one or more agreed standard file formats, with each other and with other systems which conform to the agreed formats.
- These exchanges are regulated by a set of protocols which establish the particular types of information to be exchanged between different members of a team, at different points in the project life-cycle.

From these definitions clearly sharing "common" data, is a fundamental part of BIM but this alone is not enough to achieve a successful construction project. In order to handle the amount of data and understand the implication of data on other aspects of the project, data should be integrated and unified in a way that allows efficient analysis of the decisions and changes made by any one discipline. Data should stream through and simulate the results on the project as a whole and allow project stakeholders to respond early in the process and inform the project team with their decisions. Using 3D modelling, dedicated 4D simulation tools, or glue-ware solutions in which a set of products is stacked together without real integration, provides only partial solutions and promotes a "lonely BIM" approach. To reap the full benefits of BIM – as an individual, a company, or an industry – a bigger change is required; a truly integrated BIM technology and a new set of processes. A researcher of the 1970s (Morris, 1974) examining the then relatively new concept of project management, stated:

The building process is heavily differentiated and is likely to become even more so as technology becomes more sophisticated, yet at the same time there is an increasing need for it to become more integrated.

This was written at a time when computer technology in construction was in its relative infancy and Morris' belief that Project Management could provide a method for integration was perhaps visionary. BIM together with the services of a BIM Manager may indeed provide a method where all members of the construction team can work together with the same sets of non-conflicting data. In

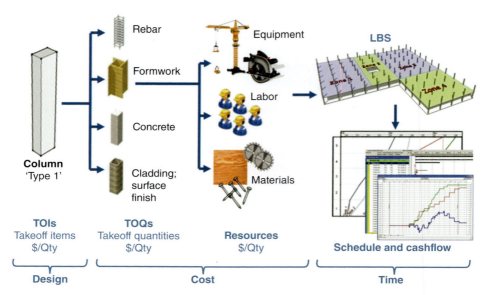

Figure 21.1   Design, cost and time integration. Source: Vico Software.

this sense BIM provides a synergy that the previous attempts of procuring construction via the many and questionable procurement methods, has singularly failed to achieve.

21.3   Integration and dataflow

A truly integrated solution should be created in order to break the traditional silos that prevent the flow of information. A unified BIM database integrates input from design, estimating, scheduling, and on-site production data. This live dataset provides immediate feedback and analysis whenever new information becomes available and allows stakeholders to better understand the impact of new data on the project as a whole.

The following section outlines the process using VICO software. Case study 1 and Case study 2 illustrate how "virtual construction" can provide solutions to the traditional problem of variations due to design errors and how "virtual mock-ups" may be used to reveal errors and clashes within a design, and also enable a value engineering approach to be used.

The quantities in Figure 21.1 show how elementary geometry is used to derive construction-caliber take-off quantities, linked to required methods and resources with cost data. The linked methods are aggregated per location and a cost-and-resources-loaded schedule is generated. Due to the tight integration any change to the design, cost or schedule can be immediately evaluated by the project team.

Data streams from design to cost planning are enriched with schedule planning information to generate bid materials. Once a baseline is established, actual cost and schedule data are analysed and results are streamed to accounting. The data loop ends with actual data for facility management and with data analysis that

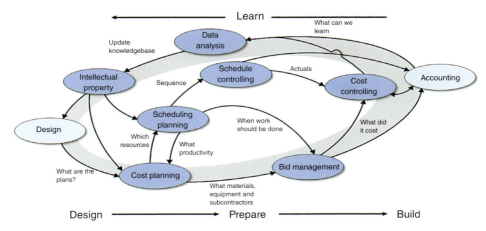

Figure 21.2   Information flow. Source: Vico Software.

enrich the project team's intellectual property. Figure 21.2 illustrates information flow as used by Vico software. This is considered one of the most critical aspects for a BIM project to be successful.

21.4   Model Progression Specification: Developing a common language

An integrated solution requires coordinated input from multiple disciplines. Over the course of project planning, the project's information richness increases over time. It is highly important to set expectations and define what data are needed at each stage of the project and what level of detail is required. This coordination process is essential for supporting effective collaboration and smooth workflow. The Model Progression Specification (MPS) was developed with these needs in minds. The MPS was pioneered by Vico Software and Webcor Builders, refined by the technology subcommittee of the AIA California Council's IPD Task Force, and was adopted by the AIA in late 2008. It is also known as the AIA E202. Since then, Vico Software has refined the solution to further hone virtual construction planning and simplify the MPS structure. The MPS is a language that owners, designers, and builders can use to define every element and task in the building construction process and how it should evolve throughout the project stages. It serves as a coordination point for information about the building, what is being modelled, and to what level of detail it is being modelled, estimated, and scheduled. The MPS provides the needed framework for the project stakeholders – a written checklist that matures from a very schematic level of detail to a high level of detail in terms of 3D geometry, cost, time, and any other required aspect.

Using the MPS the project team can also specify what portions of the project require high levels of detail. Typically some systems will be developed earlier and some later based on the project schedule, the criticality, and the risk involved. This is not a problem as long as all parties understand the expected level of detail at each stage.

The basic Aspects are Model, Cost and Schedule but additional Aspects such as LEED can be added. Each Aspect includes several Classes which reflects level of information. Classes from every Aspect are mapped to LOD levels to create the

Aspects and classes			MPS matrix			
Aspects	Classes	Description	Target LOD	Aspects		
				M	E	S
MODEL 'M'	M0	No model				
	M1	Building/spatial/room massing	000	M0	E1	S1
	M2	Building elements with approximate dimensions (SD)	050	M0	E2	S2
	M3	Building elements with design dimensions (DD)	100	M1	E2	S2
	M4	Construction model (CD)	200	M2	E3	S3
	M5	Fabrication / virtual mockup	300	M3	E3	S3
	M6	As built	400	M4	E4	S3
COST 'E'	E1	Division level ranges	450	M4	E5	S4
	E2	Massing driven ratios & ranges	500	M5	E5	S4
	E3	Element driven – assumption based – resource level & ranges	550	M5	E5	S5
	E4	Element driven – specification	600	M6	E6	S6

Figure 21.3   Aspects and Classes. Source: Vico Software.

project MPS plan. Figure 21.3 shows how aspects include all information categories that the team would like to include in the BIM database.

With the MPS in place, team members move from one project phase to the next, increasing the level of detail in one or more sections of the model specification. The MPS serves as an organisation system for all the information contained in the BIM model. Without a clear and coherent template, the model cannot be exploited for coordination, scheduling, estimating, and on-site production control.

"The Three Buckets": Quality, Cost and Schedule

The fundamental goal of a construction project is to provide a quality product which complies with the target budget and the defined schedule. Achieving this goal is a challenging task. A known axiom is that the further a project is down the road, the lower the likelihood of recovery from errors in quality, cost, and schedule. Hence, the key to success is to identify issues as early as possible, forecast their impact, and fully understand the outcome of possible solutions. In traditional processes, collecting and analysing data is a tedious and lengthy process. The amount of data and the fact that it is delivered in autonomous, discipline-specific chunks makes it hard to track and understand the impact on the other discipline and on the project as a whole. In many cases, when data analysis is available it is already too late to recover without a significant impact. This reactive approach can be eliminated with BIM technology where data-flow between design, cost, schedule and production teams enables a proactive approach and a constant value engineering process. Figure 21.4 illustrates the virtual construction data streams between design, cost, schedule and production teams which allows immediate value engineering and analysis of changes and project status.

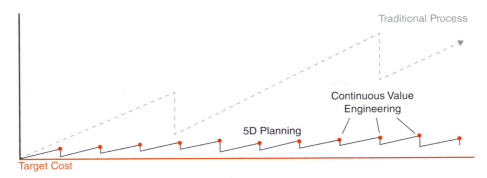

Figure 21.4   The traditional process compared with the BIM process of value engineering.

21.5 Quality

From a high level view, a quality project is one which fulfils the owner's vision and business goals, satisfies the end users, enriches the built environment, complies with the local regulations, and is constructed in an efficient way within the defined cost and schedule constraints. Since major parts of the downstream data are based on the design input, it is highly important to increase the quality of this raw data and help unify the way different stakeholders are interpreting the data. By providing a tightly integrated environment that links together design, cost and schedule data, and powerful visualisation tools, BIM technology helps to achieve the defined quality target. Quality can be ensured thorough effective communication, comprehensive constructability analysis, and better understanding of the design details and the impact of design changes.

Communication

Good and effective communication is a basic requirement for quality. Already in the early stages of the project, stakeholders should understand the various aspects of the design and their effects on the environment, the construction process, the schedule, and the usage of the building. 3D modeling and data visualisation are embedded in the BIM technology. Navigation in 3D is one of the basic ways to explore and experience the design in a virtual environment. In a "learning by playing" process, stakeholders can review and provide feedback about the planned physical space. Loaded with information about the physical characteristics of the building parts, BIM models can also be used for effective simulation. Energy usage, lighting levels, air flow, earthquake resistance, crowd flow, regulations, and essentially almost any real world performance can be analysed and optimised long before putting a shovel in the ground.

Constructability and data integrity

Additional significant aspects of the quality check are constructability and data integrity. The number of drawings in large scale projects can easily reach over 100,000 and the number of versions created during the design phase can exceed three digits. Manually tracking changes, and more importantly, understanding their full impact on the project

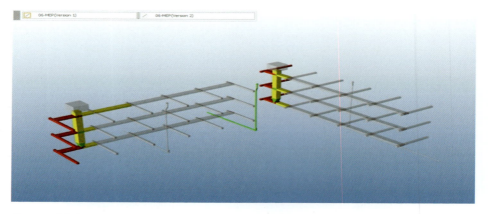

Figure 21.5   Automatic visual analysis of changes between MEP model versions. Source: Vico Software.

as a whole, is a time consuming and error prone task. Using BIM tools including automatic change detection technology and visual analysis algorithms, the project team can identify all changes in 2D and 3D datasets and communicate the outcome with all stakeholders. An important part of the constructability analysis is coordination between trades. With 3D models and automatic clash analysis technology discrepancies can be identified and resolved early in the design stage. This process reduces the amount of changes and delays during the actual construction and also allows greater use of prefabrication to accelerate on-site work and improve build quality.

Colour coding is used to identify deleted, new or changed objects. This is illustrated in Figure 21.5. Typically this quantity assessment and refinement process includes several iterations in which the design is refined and all trades are coordinated.

Using such technology ensures that any change in the design documents is identified as shown in Figure 21.6.

Quality in production

Using BIM technology for quality assurance is not limited to the pre-construction stage. The information captured in the 3D models is sent directly to a robotic total station to ensure the accuracy of complex structural, architectural, and MEP layouts, as well as a streamlined Design-to-Build process. The second step of the process is the Build-to-Design phase which consists of measuring actual construction on site and sending the data back to the BIM environment. This allows a quick feedback loop and immediate response to any discrepancy between design and construction. The B2D and D2B processes improve quality, reduce field rework, help reduce cost, and minimise schedule.

The impact of change orders (variations) on quality

A crucial factor that has a significant effect on project quality is the number and magnitude of change orders in a project. Miscommunication, wrong interpretation, inadequate details, and misrepresentation are only a few reasons for change orders.

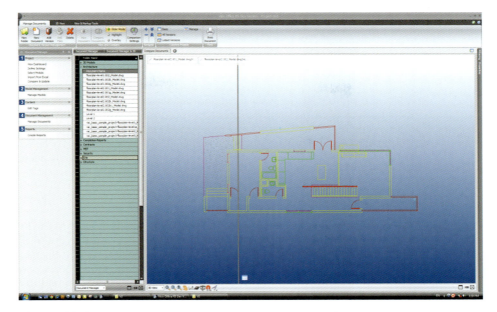

Figure 21.6   Automatic binary and visual analysis of 2D documents. Source: Vico Software.

The segregated structure of data which characterises the traditional process, and the absence of a common integrated data source increase the probability of change orders. In addition to the costly impact, change orders can also cause delays due to loss of productivity and cascading effect on dependent tasks. Using traditional tools it is difficult to assess the cumulative impact of a change on the project. In some cases, radical steps need to be taken in order to keep budget and schedule constraints, steps which in many cases are at the expense of quality and optimised design. In an integrated BIM environment, mainly due to the ability to identify missing data and discrepancies in the design early in the process, the number of change orders is expected to be significantly lower. Additionally, proposed changes ripple through the project database and the impact on all project aspects can be evaluated.

Case Studies

Case study 1 and Case study 2 illustrate the quality issues discussed above.

Case Study 1: In-Wall Coordination and Zero Change Orders

Health care and especially hospitals are among the most complex building types. Coordinating interconnected and critical systems, managing a large number of sub-contractors, communicating with owners and fulfilling requirements of the professional staff, all within a tight schedule and budget constraints are typical challenges that face a project team in this type of project. The owner of the St Jude Medical Plaza in Fullerton, CA, wanted to expand the hospital's wings. To ensure quality and reduce risk for cost and schedule overruns the owner came up with a unique and challenging demand of *zero change orders*. To achieve this goal a detailed design and constructability review was conducted. Using Virtual Construction technology and real-time modeling, the project team analysed and coordinated the design. To include the requirements of the professional staff and avoid last-minute changes, a highly detailed "in-wall coordination" process was conducted with exceptional attention to placement of elements in the walls, on the walls, and in front of the walls including details such as outlets, pipes, caseworks, switches and even paper towel holders. The organisation of these elements in a room is crucial for the comfort and functionality of both the staff and the patients. This technology and process was used to meet the needs of the caregivers, therefore providing the best services possible to the hospital's occupants, while achieving the goal of zero change orders.

Challenges

The walls of medical buildings are congested with equipment from the inside and from the outside of the walls; therefore the placement of equipment in the hospital's rooms must be well-planned and accurately installed. Specifically for this project a tight schedule required that the "in-wall coordination" process would be conducted partially while construction was in progress. The team had to catch up and plan the future steps of the construction in a fast pace process. The ultimate goal of this project was to prevent any change orders that could cause schedule delays or cost overruns.

Solution

Using Virtual Construction technology and an "in-wall" coordination process requires coordination between the general contractor, the owner, the architect, different providers, and the hospital's staff itself. A team of BIM experts used detailed 3D modelling to communicate the design, analyse staff input, coordinated all trades, and ensured a non-disruptive environment for each sub-contractor. During this process every piece of equipment was modelled to ensure a perfect match and find the optimal room layout and the correlated in-wall systems placement. Furthermore, to ensure efficient work environments and avoid last-minute changes, individual models for each doctor's examination room were made according to personal preferences and needs. For example, left-handed and right-handed doctors prefer to stand on different sides of their patients.

Analysing possible solutions using BIM models enabled the project team to provide a comprehensive solution which supported the specific needs and accommodated the different layout of the equipment in each room, including outlets, pipes, and cabinets. Walking inside the virtual environment mimicking the day-to-day tasks helped the nursing team to expose design issues such as obstruction to eye contact with patients in X-ray rooms while standing behind the shielding wall. The "in-wall coordination" process provides visual analysis and allows the project team to match the requirement by changing the size and location of the windows in the shielding walls without interfering with equipment location or in wall systems. The adjusted design was presented for final approval to the hospital team before its execution. The real-time modelling and "in-wall coordination" services proved to be successful and saved both money and time for the hospital. The project was finished as expected with a high quality output and with zero change orders.

Figure 21.7 illustrates how it is possible using 3D models to communicate and coordinate the design. This is highly effective compared with a mere 2D drawing set as presented on the left-hand side of the figure.

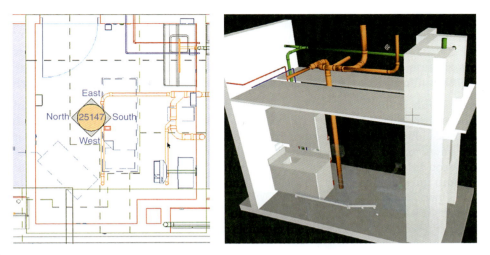

Figure 21.7   Comparison of 2D and 3D models. Source: Vico Software.

Case Study 2: Comparison between Physical and Virtual Mock-ups

The connecting point between building elements or systems is where the complexity lies. A pre-construction physical mock-up of the proposed design is a common way to analyse and identify possible flaws and check the interaction between different systems. Though physical mock-ups are an effective tool, they have some serious limitations. The limited scope of the physical mockup implies that there is still a high risk for constructability issues in neighbouring areas. Additionally, the considerable amount of time required to build the mock-up, the significant cost, and the limited opportunity to analyse design alternative make it difficult to use this solution frequently. BIM technology, on the other hand, provides an ideal platform for this type of analysis. Using BIM, 3D and 4D visual mock-ups can be created and analysed in a fraction of the time and cost required for a physical mock-up. The outcome can be modified for value engineering purposes and the scope can be extended in case there is a need to analyse neighbouring systems. In the case of Mission Hospital in Mission Viejo, CA, the project was already well into construction when the owner decided to build a physical mock-up in order to check the designed façade and detect any problems or risks that the construction may cause.

Challenges

Integration with existing structure, complex design for the building skin, and the need to provide highly accurate data for prefabrication of the metal panels were among the major challenges that the project team was facing. The life-size cross-section physical mock-up with a tag price of US$150,000 was found insufficient to solve the problems. The project team was particularly concerned about the interaction of the skin with the waterproofing and the fireproofing systems which seemed to interfere with the façade components. The project team decided that a more detailed mockup was required to make sure that potential constructability issues were resolved.

Solution

To get a comprehensive analysis within the project time constraints a virtual mock-up was created. This highly detailed mock-up included the precise steel and glass structure with all designed connections and required systems. The virtual mock-up helped reveal major discrepancies. One of the most serious findings was that the exterior skin did not fit over the steel structure. Based on the identified issues, the design team provided an updated set of documents to be used by the fabricator. Working together, the Mission Hospital Team saved the rework of 1,000 panels, 10 project work weeks, and approximately US$1.8 million.

Figure 21.8   3D model showing constructability issues in the steel and glass structure. Source: Vico Software.

Figure 21.8 shows how the virtual mock-up revealed critical constructability issues related to the building skin.

21.6   Cost planning

As one of the most visible and bottom line results, cost is a prominent driving force behind the motivation for a change. There are numerous reasons for cost overruns; a major one is under-estimation of the actual cost during cost planning. Insufficient or impaired data and subjective optimistic bias are common traps when working on an estimate. To reduce the risk and improve the estimating quality, team members need to get the right data on time, fully understand its meaning, effectively compare it with the project targets and with historical projects data, and communicate the output for auditing in a transparent way. There are several crucial elements to facilitate effective cost planning such as the take-off process and the measurement rules, having an evolving cost forecasting system to track changes in design and cost visualisation and value engineering to identify the sources of cost changes and unnecessary cost.

Takeoff (measurement) and cost

Quantities are a basic ingredient for cost calculations. Accurate estimates of materials, labour, and equipment usage are essential for a precise cost planning and require high-quality takeoff data. Getting construction calibre quantities is not a trivial task and handling the changes along the design and construction process is even more challenging. In traditional takeoff processes the quantity

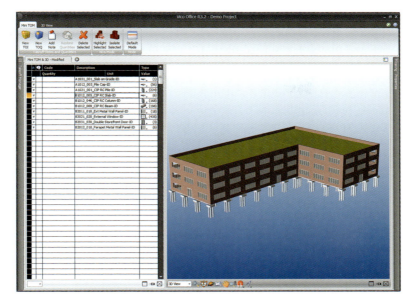

Figure 21.9 Takeoff items in the left pane are automatically generated during the takeoff process. Source: Vico Software.

surveyor is required to interpret large sets of 2D drawings from various sources and to translate them into a spatial environment. When new versions arrive it is sometimes too difficult to identify and assess the changes and the whole process needs to be restarted. Using BIM technology, quantity data from 3D elements geometry can be processed automatically, thus the most time consuming and error prone part of the takeoff can be dramatically improved. As a result of this process, a highly accurate set of pure quantities become available for the estimating work. When changes or more developed stages of the design become available, the system identifies and differentiates between existing new and changed elements. As a result, an updated quantity takeoff is immediately accessible and the variance between the old and new data can be analysed. Another important aspect of this approach is its visual characteristic. As part of the integrated database, 3D elements are directly linked to takeoff data and the project team benefits from a visual 3D takeoff. Auditing, fine tuning, and communicating the takeoff become more efficient as the estimate is founded on solid, trustworthy and fully transparent data and the risk of underestimating decreases.

A visual feedback regarding the takeoff results allows effective quality check and refinement of the takeoff (Figure 21.9).

Evolving cost forecasting

Data are accumulated during the design and preconstruction stages, moving from rough sketches to schematic design all the way through construction documentation and detailed specifications. Maintaining cost planning continuity in this dynamic environment is highly beneficial but with traditional estimating tools it is difficult to keep track and evolve the estimate when new data become available. As a result, large chunks of the estimate are redone when more detailed sets of drawings or major changes are presented. Restarting the estimating is time consuming, error prone, and prevents the team from maintaining a clear history of the cost plan for auditing and clarification of cost changes.

An evolving cost planning process can be maintained with an integrated BIM environment. In this process, the cost plan follows the design process and becomes more and more granular with the availability of more detailed takeoff data. This approach allows team members to track their cost estimate and analyse deviations from initial targets immediately when new and more detailed data become available. In a way this method also protects design intent and removes the risk of late value engineering which might end in reduced scope or non-optimal design changes.

Another benefit of this approach is its explicit and fact-based process which helps avoiding optimistic bias. At the beginning of the project, key figures with assigned value can be used to set the target cost (e.g. number of beds in a hospital or number of hotel rooms). During the early stages of the design, conceptual level 3D massing models are provided and the model-based takeoff output at this stage is merely limited to m^2 or ft^2. In spite of the limited takeoff information, team members can develop a detailed cost plan in which cost items are calculated based on a combination of model quantities such as GIFA (Gross Internal Floor Area), statistical quantity ratio data (e.g. wall/floor ratio and similar ratios), and current resource pricing. The results can be compared with the initial target cost and alternative specifications with significant impact on cost such as finish levels, can be analyzed. As the design is refined, the ratio and m^2/ft^2 takeoff data is replaced with more accurate information derived from detailed 3D model elements. The continuity of this approach matches the evolving nature of a construction project; it provides explicit cost results at the early stages and allows tracking and a clear audit trail of cost data.

Maintaining continuity allows for an immediate feedback about cost variances between different versions. The blue and red arrows indicate the change direction. Figure 21.10 illustrates through the three panes how the hierarchical structure of the cost supports an evolving cost process.

Cost visualisation and constant value engineering

Bottom line cost changes are clear and noticeable. Identifying the specific source for the cost change and when it has occurred is a more challenging task. This is especially true when using traditional cost management tools. The amount of drawings and drawing versions, the lack of effective change management tools, and the fact that the impact of a change is not always clear, make it almost impossible to indicate where and when the deviation from cost or schedule started. To allow effective monitoring of cost changes a more frequent analysis should be performed and better communication tools should be used. When using an integrated database and maintaining continuity as described in the previous paragraph, frequent releases of information are available and an almost constant value engineering process can be obtained. Design and quantity changes, unit cost updates, or reported production rates from the site are captured and their impact is rippled through the database to show clear end results on any part of the cost plan down to the resource level. Maintaining project history in the database enables analysis and comparison between target cost, current cost or any cost version or point in time along the project duration. To communicate the change, the com-

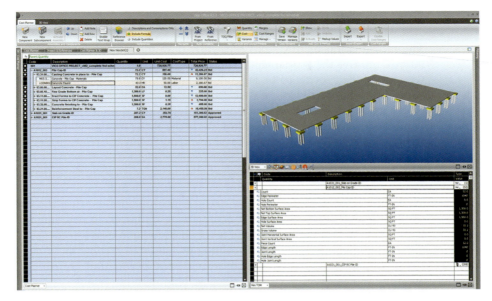

Figure 21.10 Screen shot showing the hierarchical cost structure. Source: Vico Software.

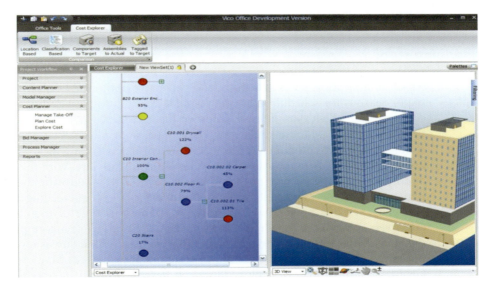

Figure 21.11 Screen shot of the visual monitoring system. Source: Vico Software.

parison can be presented using a combination of 3D and related historical cost plans. This audit trail process supports a proactive approach to cost management and helps to quickly identify and communicate any deviation from the target.

The link between the 3D model and the cost plan helps clarify the variance.

Figure 21.11 illustrates how the VICO software provides a visual monitoring system, from which the project team can easily drill down into the cost hierarchy and identify the cause of cost variances.

21.7   Construction schedule

The introduction of BIM, and especially data integration, provides an opportunity to improve the fundamental scheduling processes and planning quality. BIM-wise scheduling seems to be lagging behind compared with design visualisation, constructability analysis and cost estimating.

One of the major drawbacks of traditional scheduling tools, especially CPM (Critical Path Method)-based solutions, is the focus on the "what" at the expense of the "how." This approach makes the schedule almost irrelevant at the execution stage, and in many projects it becomes a mere picture on the office wall. Another drawback is related to updating the schedule and forecasting the outcome of actual production rates. Deviations from the planned assumptions regarding resource availability, production rates, and other critical factors increase the gap between the plan and the actual progress on-site. Using traditional scheduling tools, updating the plan to match the new data is very difficult and heavily relies on the subjective bias of the scheduler. BIM technology can help resolve these drawbacks by integrating data and providing more efficient production control tools. Using quantity and resource data from the cost plan with physical location information from the 3D models and actual productivity rates, a more reliable plan and a baseline for effective monitoring process can be obtained. Such integration allow team members to achieve significantly compressed schedules without increasing risk. There are two critical elements that are important - the planning and controlling stages - which are discussed below.

Planning stage

Miscommunication starts and stops, and subjective optimistic bias are among the major reasons for schedule delay. In a similar way to cost, the ability to base the schedule on explicit data helps to improve the accuracy of the schedule and prevent subjective bias. To make the most of the integrated data, a proactive approach should replace the common after-the-fact reactive approach in which stakeholders are aware of a delay in the schedule when it is already too late. With this proactive approach, a significant part of the process is devoted to monitoring the schedule, understanding the impact of deviation from the plan on succeeding tasks, and analysing possible preventive measures. To enable this type of monitoring and effectively control the schedule, locations should be added as a core property of the plan. With locations as an additional dimension, resource flow is reviewed and optimised and conflicts between sub-contractors are resolved. Many progressive General Contractors have deployed location-based management system as part of their scheduling expertise. The workflow described in the following is typical.

To avoid crew stops and starts, it is necessary to ensure the flow of work associated with specific task throughout the building location. Task duration and flow is calculated based on quantities per location, loaded resources, and known productivity rates. Quantities per location can be calculated automatically via the integrated BIM database. Productivity rates can be gathered based on historical data or from standard commercial databases. Once the Task durations and flow are set and the basic logic is defined, the team can focus on optimising the schedule by fine tuning the amount of resources, the number/duration of shifts, and the size of buffers between the tasks. The optimised schedule provides continuous

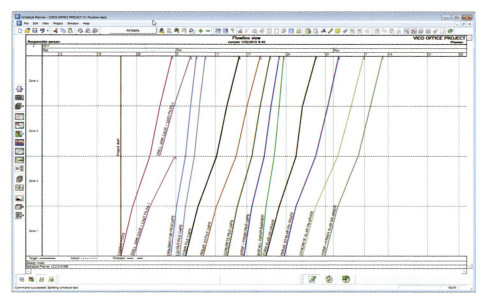

Figure 21.12 Flow line view of a project. Source: Vico Software.

resource flow, prevents conflicts between sub-contractors, and includes the required buffers to allow deviation from planned duration without cascading delays and overall project delays.

Figure 21.12 shows how the flow line view provides a clear picture of the schedule and allows team members to ensure task flow from one location to another, to manage buffers, and to visually identify conflicts, and communicate complete schedules of complex projects on a single page.

To improve communication and present the feasibility of the schedule, a 4D simulation can be produced. The simulation uses the 3D models that are integrated with the location-based and resource loaded tasks. This process allows sub-contractors to thoroughly understand the scope of work and how their work is synchronised with other sub-contractors.

Since the 4D simulation is generated based on actual information from the database its value is beyond a feel-good presentation. Schedule variances can be analysed by changing properties such as the location structure, the task order or the crew size. When actual production starts, the 4D simulation can also be used to clearly present actual progress and schedule delays.

Controlling stage

Creating an optimised location-based schedule is only the first part of the process. Continuously monitoring the actual production on-site is the more critical second part. Weekly reports with information about the status of tasks per location are submitted to the integrated database.

Figure 21.13 illustrates how sub-contractors and project engineers might review the project status using the control chart on weekly site visits and generate forecasts based on actual progress.

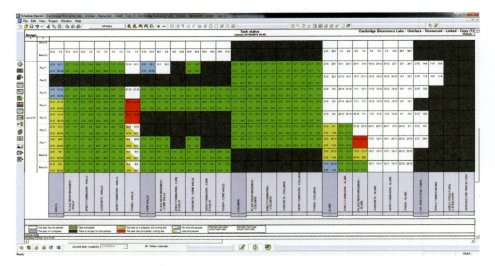

Figure 21.13   Control chart. Source: Vico Software.

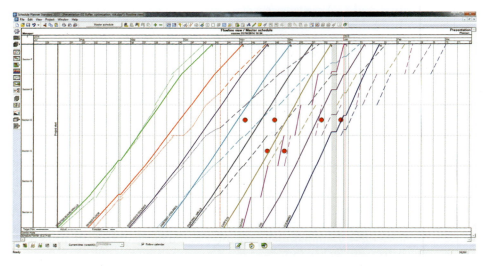

Figure 21.14   Flow chart: the baseline schedule (the solid lines) and the actual productivity rates on-site (the dotted lines). Source: Vico Software.

The data are analysed and the actual progress is compared with the baseline plan. Based on the reported progress and the actual productivity rates from the previous location, a forecast is generated and the completion date of the task is recalculated accordingly. The forecasted line points out possible interferences with other successor tasks which might result in cascading delays.

Figure 21.14 uses the baseline schedule to compare forecast (i.e. the solid lines) to the actual productivity rates on-site (i.e. the dashed lines).

The projected forecast (dashed lines) is based on actual production rate in the previous location and the red warning lights indicate that interference might delay work in that particular location. With a clear understanding about how the project is progressing compared with the planned schedule the team is able to "look ahead" to see and take preventive measures in order to solve problems long before they ever manifest themselves on the jobsite.

21.8   Conclusion and future directions

There is no doubt that BIM technology is likely to prove a major factor in improving the efficiency of the construction industry and has UK Government support. The lack of integration highlighted in the introduction to this chapter and the sheer inefficiency created by "traditional structures" must change. BIM dramatically enhances the understanding of the design and actual construction and provides the ability to perform value engineering to optimise the design. The improvement in quality of design documentation and the effective coordination processes enable greater use of prefabrication which results in a higher quality output and a shorter schedule. The integrated database together with effective visualisation, eliminate waste, grant efficient planning and controlling tools, and provide a highly transparent environment.

As a relatively new technology, BIM has not yet fully matured. Nevertheless, the greater challenge that the industry faces is related to the required paradigm shift and the resistance to change. Internally, for an individual organisation, the adoption of BIM technology and related processes, change established patterns and threatens jobs and organisation structure. To overcome resistance to change, a thorough implementation plan should be prepared and communicated and compatible cross-department processes should be developed. An additional challenge is related to the required adjustment of industry-level processes and legal aspects. Questions such as who owns the BIM model and who is responsible for developing and updating the data should be addressed in the contractual documentation to avoid legal pitfalls about data ownership. Naturally, in a shared environment, stakeholders are also concerned about their Intellectual Property and a satisfactory solution should be generated. Building a more trustful, sharing environment based on new, more integrated business models will help adapt to this challenge.

BIM technology is relatively young and the potential for expansion is beyond the scope of this chapter. Collaboration, communication, and visualisation are major cornerstones of BIM and emerging technologies such as cloud computing and augmented reality can empower BIM capabilities in these fields. Another natural expansion of BIM technology is related to fabrication. The move from 2D drawing to 3D models can be utilised for manufacturing. In a similar way it is done in the mechanical industry, data can be sent directly to manufacturing machines for production. During the controlling phase of the construction, RFID technology can be used to constantly monitor resource availability and actual production. Linking the data to procurement systems supports just-in-time delivery of materials and lean construction. Warranty and maintenance information for building components can be stored in the database and integrated with facilities management software for the owner. The construction industry as a whole is in a process of refabrication and transformation from the traditional linear mode to a parallel and collaborative mode; this change will also be presented in the business models and the definition of roles of the key stakeholders. We can only echo Crotty's (2012) visionary conclusion regarding the future:

> But the most important developments in the global market place will come when all construction, everywhere is carried out using the sort of "perfect information" that the BIM approach makes possible. It is almost impossible to imagine how much of a contribution to the future of mankind and the planet such a form of construction might make.

Just imagine the impact on the world economy and ever growing population of being able, via BIM, to provide 20–30% more construction for the same resource cost.

References

Bowley, M. (1966). *The British Building Industry - Four Studies in Response to Resistance to Change*, Cambridge University Press, Cambridge.

Cabinet Office (2011) A report for the Government Construction Client Group. Building Information Modelling (BIM) Working Party. Strategy paper. March 2011.

Cooney, E.W. (1955) The origins of the Victorian master builders. *Economic History Review*, 8, 167–176.

Crotty, R. (2012) *The Impact of Building Information Modelling – Transforming Construction*, Spon Press, Oxford.

Egan, J. (1998) Rethinking Construction. Report of the Construction Task Force on the Scope for Improving the Quality and Efficiency of the UK Construction Industry. Department of the Environment, Transport and the Regions, London.

Latham, M. (1994) *Constructing the Team*, HMSO, London.

Morris, P.W.G. (1974) Systems Study of Project Management, Parts 1 and 2. Building Magazine, January, pp. 74–80; February, pp. 85–88.

Morton, R. and Jaggar, D. (1995) *Design and the Economics of Building*, E & FN Spon, Oxford.

NIBS (National Institute of Building Sciences (NIBS) (2012) National Building Standard-United States. Version 2.

Chapter 22
Case Study: Value Engineering and Management Focusing on Groundworks and Piling Packages

Richard Powell

22.1 Introduction

Value management (VM) is a creative organised approach to remove unnecessary cost in design and to increase value for the client. In practice a well organised VM process can make a design much better and generate significant savings for clients. This chapter explores how VM theory is implemented in practice using a case study identifying the key cost drivers and value opportunity points. A brief description of the project situation is provided followed by a discussion of the rationale for VM, when, where and how VM was applied in the design and construction process. The key tools used for VM such as issue analysis, open book auditing and workshops to explore ideas and to identify unnecessary costs or cost savings identified are also discussed with examples. The chapter concludes with some reflection on the benefits of VM.

22.2 Why VM?

A single, design and build tender was received for a proposed new build supermarket situated within Kent, which exceeded the client's recently approved budget by +£640, 000 (ca. +22%). Seeking an increase to the budget was not an option. This created a challenging issue, as there were tight time constraints to maintain procurement and construction programme milestones and therefore ensure the client's proposed store opening date and forecast income were not compromised. The client had a clearly defined brief from a quality perspective based on experience of similar retail projects and did not want to deviate from this.

Design Economics for the Built Environment: Impact of Sustainability on Project Evaluation, First Edition.
Edited by Herbert Robinson, Barry Symonds, Barry Gilbertson and Benedict Ilozor.
© 2015 John Wiley & Sons, Ltd. Published 2015 by John Wiley & Sons, Ltd.

Value engineering with the client, contractor and design team was considered the most viable option to address the cost overrun. This was on the basis that the contractor had confirmed that they are able to meet the requested store opening date and there was insufficient time to tender the works on a competitive basis. Whilst competitive tendering may have generated a programme opportunity to offset the increased time associated with a competitive tender, this was not guaranteed; nor was the fact that the competitive tender would mitigate the cost issue on the scheme.

Whilst clients may have different definitions of value and what is important to them, for the retail client on this project the objective was to maximise the earning potential of the supermarket and thus a simple view of value was to maintain an exceptional customer experience. Therefore value engineering was defined such that any aspect that was detrimental to the customer experience was considered cost cutting and not value engineering. The reality of the situation was that the client had many different departments and stakeholders, with their own requirements and different definitions of value. These different requirements and perspective need to be understood, communicated and balanced carefully and therefore creating a situation that was much more complex.

22.3   When and where is VM applied?

VM can be applied at any stage of the design and construction process from pre-brief, briefing, sketch plans, to detailed design and construction to address specific issues relating to concept design, spaces, elements, engineering design, components in a design solution or construction techniques or processes.

In this particular project, value engineering was applied because of significant cost issues and a risk that the project may be abandoned. Value engineering was comprehensively undertaken on a continuous basis throughout the detailed design process. However, with a projected cost overspend of such scale it was clear that the process needed to be resurrected in its entirety. The alternative was for the project to be abandoned by the client, after investing a significant amount of time and effort in getting the project to the tender stage. The client would have incurred a substantial amount of abortive costs. This placed great pressure and expectations from the value engineering process to resolve the issue of cost overrun within this project.

22.4   Value management implemention and tools used

VM was implemented in steps using various tools and approaches. Step 1 focused on identifying value opportunity areas in the design; step 2 focused on investigating in detailed value opportunity areas or design elements with significant cost variances between the pre-tender estimate and the contractor's tender; step 3 involved engaging all relevant stakeholders to identify pre-workshop issues; and step 4 was the value engineering workshop to explore ideas and propose alternative design solutions. Various tools were used such as cost variance analysis, open book auditing, value engineering workshops, face-to-face dialogue with all relevant parties and brainstorming.

Table 22.1   Variances on selected tender packages.

Elements	Pre-tender estimate (£)	Contractor's tender (£)	Variances (£)
Groundworks	2,093,000	2,569,000	+476,000 (+23%)
Piling	764,000	928,000	+164,000 (+21%)
Brickwork	132,261	132,176	−85 (negligible)
Partitioning	134,525	142,856	+8,331 (+6%)
Roof safety system	15,000	14,595	−405 (−3%)
Shop fronts	140,750	156,912	+16,162 (+11%)
Protection	72,500	81,916	+9,416 (+13%)
Lift installations	189,367	162,804	−26,563 (−14%)

Identifying value opportunity areas

Step 1 of the value engineering process was to carefully identify areas to focus the effort of the team on, particularly with the tight timescales involved. Investigating all elements of the project with an equal amount of time and effort was unlikely to yield the savings required to mitigate the cost issue within the timescales. A decision was made to focus on the trades representing the largest cost variances between the pre-tender estimate and the contractor's tender return, which were the groundworks and piling packages. As can be seen from Table 22.1, the extent of these variances was largest for the groundworks tender package (+23%) and the piling tender package (+21%).

Investigating value opportunity areas with significant cost variances

Step 2 in the value engineering process was to make the project team aware that the groundworks and piling trades were the areas of focus given the significant cost variances identified in the two packages. All relevant tender return documentation relating to these packages were sent to the architect and structural engineer for review and comment. This consisted of sub-contractor quotations, assumptions, exclusions, qualifications and the analysis the main contractor produced comparing the detailed tender returns received. This was done as a result of the requirements for negotiation stipulated in the employer's requirements in the tender documents. It was requested that the sub-contract packages of the project should be tendered on a competitive basis (with a minimum of four tenderers per trade), using open book and the main contractor was to provide all tender return information to the quantity surveyor.

Within a fortnight of forwarding the information, responses were received from the structural engineer and main contractor on a few proposed value engineering options, many of which required sub-contractor input. It was therefore decided to schedule a value engineering workshop for that same week (2 weeks after the cost issue was identified) to bring the team together with an agenda consisting of: considering the ideas proposed to date in more detail in order to establish if they were viable; the repercussions (if any) if they were implemented; the risks in implementing them; and further opportunities. It was considered that a workshop and face-to-face dialogue with all relevant parties would expedite the whole process rather than communicating through email correspondence and telephone dialogue. This was particularly important when considering the tight timescales at hand.

Participants requested to attend the workshop included the quantity surveyor, main contractor; the groundwork and piling sub-contractors; the structural engineer and architect; and the employer's agent. Within the 2-week period of steps 1 and 2 elapsing, the main contractor (with assistance from the quantity surveyor) had commenced a tender adjudication on the piling and groundworks trades. Whilst this was not fully complete it had identified who at that time was most likely to be the most competitive, compliant tenderer for each trade. It was these parties that were invited to the value engineering workshop.

A value engineering example raised prior to the workshop was by the piling contractor. This was in relation to the design of the load bearing piles beneath the store. The contractor advised that they had priced what was shown on the tender drawings, for piles designed to a load of 850 kN compression and up to 200 kN shear. They were however advised that other design information produced by the engineer indicated that the maximum compression and shear loads could be 550 kN and 20 kN, respectively, and if they could be designed to this, then this would offer a substantial cost saving. They agreed that they would be willing to resubmit a price based upon confirmation of individual pile loads.

Inviting relevant stakeholders and identifying pre-workshop issues

Step 3 involved all of the parties preparing for the value engineering workshop. Participants were asked to prepare for the agenda items listed. For the quantity surveyor, this entailed reviewing the next level of cost detail within the piling and groundwork trades to establish significant areas of difference to the pre-tender estimate. Whilst undertaking this process, it was identified that for the groundworks the pre-tender estimate assumed that there was sufficient site storage space to facilitate a large cut and fill process to be undertaken, without hindering the contractor's progress in completing the piling and other groundwork activities. However, it was discovered that there was a substantial variance in the contractor's price, with much larger quantities for disposal of excavated material off site and a new item for importing of new material for building up levels.

The contractor's proposals were reviewed to check the position. The contractor also advised that there was insufficient storage space to store the excavated material temporarily on the site and therefore had allowed for disposal off site, followed by importing of new material. This issue was identified as an item for value engineering, as it was not considered by the other team members and would be discussed at the meeting. The quantity surveyor extracted the sums from the main contractor's tender price for these items and added a sum back, based on the contractor's rates within the tender for disposal on site, together with an allowance for additional handling. This created a saving which the quantity surveyor communicated to the main contractor for consideration ahead of the value engineering workshop.

One example of the team members' preparation for the value engineering workshop was the main contractor opening up dialogue with the structural engineer on the load bearing pile issue. The contractor forwarded the advice from the piling contractor on the load bearing piles to the structural engineer, for review and comment. The engineer responded by stating that in order to meet the tender programme, they had produced general loading requirements for the load bearing piles and at that stage they had not sufficiently designed the store steelwork to provide the actual column loads. It was requested by the main contractor that the substructure design information should be produced prior to the structural frame design information. The engineer had subsequently finalised the steelwork requirements and updated the foundation design information to reflect the actual column loads. They identified that from the pricing information they had been given, the piling contractor had not incorporated the final load checks within the tender price. This therefore increased the likelihood for this saving to be realised, but would be subject to discussion at the value engineering workshop.

Value engineering workshop

Step 4 was the value engineering workshop, which was scheduled for three hours focusing on the groundwork and piling trades. The key participants present, with three representatives each from the groundwork and piling sub-contractors. The agenda was to focus on the items previously mentioned, but it was recognised that there was a need to concentrate first on a review of the soils report and ground conditions; followed by methods and sequencing. The contractor also requested a review of adjacent site availability for storage purposes as an agenda item, in response to comments prior to the workshop on the cut and fill works and methodology. Availability of materials on the adjacent site was also requested as an agenda item by the main contractor. The quantity surveyor was not aware of this item but the main contractor had indicated that it was brainstormed as part of their preparation to reduce the cost of importing material to form piling mats.

The quantity surveyor chaired the workshop and set the tone of the meeting by emphasising the importance of the day and the seriousness of the cost issue on the scheme. It was felt that this was important to ensure that all participants were aware of the required output and the magnitude of savings to achieve. The quantity surveyor also informed the team that whilst the cost issue on the groundworks and piling trades amounted to +£640,000, the potential cost savings associated with the value engineering opportunity identified should exceed this amount as there was a risk that some proposals may not materialise in full, due to the client rejecting certain proposals. The team concurred with this approach.

Whilst the quantity surveyor chaired the meeting particular items were designated to those parties best placed to present them. For instance, a request was made for the main contractor to reconfirm the proposed methodology and sequencing by the sub-contractors, together with the basis of the latest design information. This was simply because the contractor had managed the sub-contract tender process, owned the methodology and had committed to a work programme. The main contractor was taking on all design risk under the contract and wanted to ensure that the latest available information was priced by the sub-contractors, to reduce the main contractor's post contract risk. This was essential to ensure that the main contractor and sub-contractors were fully coordinated and had an agreed baseline to work from for value engineering discussions.

22.5   Practical benefits and savings

The output the quantity surveyor wanted to achieve from the workshop was a schedule categorising items as follows: (1) those ready for client approval; (2) items that required further action prior to confirming if they were viable, to which designated owners and timescales would be set; or (3) items considered by the team but had to be rejected. The critical aspect was to ensure items were considered in as much detail as was required and to set owners and timescales for aspects of the review that could not be completed during the meeting. This was particularly important considering the tight timescales to mitigate the cost issue and reduced the risk of not being aware of what actions they or other team members were responsible for reviewing after the workshop.

The workshop format worked very well, with all participants working proactively together, particularly the designers and sub-contractors. For instance, the design information to be reviewed was laid out within all areas of the room. Telephone calls were made to offices where information was not readily available or queries could not be answered within the room. The quantity surveyor tasked the main contractor to produce a schedule recording the value engineering items, together with anticipated level of savings. Items were also recorded, to keep track of key steps, owners assigned with specific tasks and timescales to get agreement from the relevant parties. Examples of value opportunities identified, potential savings, status and key participants are shown in Table 22.2.

An example of a new opportunity that arose at the workshop was an alternative waterproof solution to the upper deck of the two-storey car park, adjacent to the store. The tender information specified an asphalt surface but the groundwork sub-contractor proposed an alternative such as a waterproof sealant product. The groundwork sub-contractor suggested that there were some competitive solutions available and the structural engineer commented that this would reduce the dead loading of the car park structure, thus offering potential opportunities on the car park structural steel frame and substructure. The groundwork sub-contractor's proposal was not accepted on the basis it would be moving away from the client's standard specification for car park finishes. However, the employer's agent advised that the client may accept to deviate from the standard specification for this particular item, due to the potential cost savings which may not impact negatively on the customer journey and experience. Due to the nature of this item, it was clear that a number of criteria needed to be considered further prior to submitting it for client approval. It was therefore agreed by the team that this was not an item that could be presented for immediate approval. Actions were initiated by the contractor to explore suitable alternatives based on the contractor's experience of previous schemes, including supplying the client with information on appearance and current condition in relation to when the schemes were built. In order to offer protection to the client, the quantity surveyor also tasked the contractor to investigate the availability of product warranties from suppliers and the level of cover they would provide. Due to so many varying factors, the team were unable to quantify the extent of savings offered by adopting such a solution. This item was noted as a post meeting action.

With regard to the methodology and sequencing of the cut and fill process, the contractor held a detailed discussion with both the groundworks and piling sub-contractors on their proposed programmes against the planned programme.

Table 22.2   Value opportunities identified, potential savings and status.

Items	Anticipated level of savings and potential problems	Status (e.g. rejected, accepted, further investigation required)	Participants
Alternative waterproof solution to the upper deck of the two-storey car park	−£95,000 but not client's standard specification for car park finishes	Further investigation required initially and subsequently accepted	Groundwork sub-contractor, main contractor, structural engineer, architect, employer's agent, quantity surveyor, client
Sequencing of the cut and fill process		Accepted	
Disposal off site	Substantial saving (ca. −£150,000) if adjacent land becomes available		Groundwork sub-contractor, main contractor, employer's agent, quantity surveyor, client, third party
Load bearing pile alternative solution	Proposed saving, in conjunction with the main contractor (−£50,000)	Accepted	Piling sub-contractor, main contractor, structural engineer, employer's agent, quantity surveyor and client
Rationalising the piles from the structural engineer's specified design	Further saving (−£40,000) but enhanced site investigation was required at a cost of +£10,000	Accepted	Piling sub-contractor, main contractor, structural engineer, employer's agent, quantity surveyor and client
Use of one single, larger diameter load bearing pile in lieu of three smaller diameter load bearing piles	Proposed saving (−£80,000)	Accepted	Piling sub-contractor, main contractor, structural engineer, employer's agent, quantity surveyor and client

Improvement was made on the proposals after the sub-contractors offered advice on activities that could be done in a different sequence or faster on the constrained site. Whilst this reduced the volume of material to be taken off site and import of new material allowed within the contractor's tender, there was still a substantial volume of material that required disposal off site, together with the associated cost premium. The employer's agent was made aware of this issue by the quantity surveyor prior to the meeting and had contacted the owner regarding the availability of temporary storage on the adjacent land. No response was received at the time of the workshop or when the employer's agent attempted to contact the party on the day of the workshop. Notwithstanding this, the quantity surveyor asked the main contractor and groundwork sub-contractor to review the item in as much detail as necessary during the workshop; on the basis that the adjacent land would become available. Based on the review and calculations undertaken and discussed, a substantial saving (ca. −£150,000) would be generated should the land become available.

In connection with the above item, the groundwork sub-contractor had also advised that there was sufficient room on the adjacent land to treat contaminated excavated material rather than dispose the material off site and import new material in lieu of treating it. The groundwork sub-contractor made a telephone call during the workshop to a specialist contractor experienced in treating contaminated material and advised them of the volume required together with the results of the soils report. The remediation contractor advised that subject to further review, limestone stabilisation was the most viable solution and provided budget rates for treating the material. The groundwork sub-contractor therefore forwarded the relevant information to the remediation contractor and requested for a review urgently, so that the saving could be confirmed at the meeting. The remediation contractor responded and further to more discussion with the main contractor a proposed saving of approximately £120,000 was offered, should the adjacent land become available. The quantity surveyor challenged those present and asked if there was an option on whether the excavated material could be reused without treatment. The quantity surveyor also advised that from the calculations prior to the workshop, the level of saving could have been in the region of £200,000 in lieu of the £120,000 proposed. However, the structural engineer and groundwork sub-contractor both agreed that this was not a viable option. The quantity surveyor advised that it should be recorded nonetheless and categorised as an item that was considered but rejected. It was important to demonstrate to the client that this option was explored (which the client may well have asked) but it was not viable.

Once the dialogue had concluded, the quantity surveyor steered the team on reviewing the items that could be taken immediately with emphasis placed on where risk allowances could be removed and the pricing of the load bearing pile as an alternative solution. For the piling, the piling sub-contractor requested the structural engineer to present the latest loadings available, in hard copies. The piling contractor confirmed that the information was received but was not aware that they had to price it, due to a misunderstanding with the main contractor. The piling sub-contractor requested their estimator to review the position and following a number of calls arrived at a proposed saving of £50,000, in consultation with the main contractor. The quantity surveyor was pleased at this result and agreed with the team that this item could be categorised as ready for immediate client approval, subject to checks later in the office that the level of saving was fair and reasonable.

Following a discussion with the structural engineer, the piling sub-contractor also suggested that there was a medium probability to save a further £40,000 on rationalising the piles from the structural engineer's specified design. To facilitate this, however, an enhanced site investigation was required at a cost of £10,000, to provide further information on the soil condition before the savings could be realised. The quantity surveyor emailed the client and followed it up with a telephone call to request for confirmation to proceed. The quantity surveyor recommended to proceed on the basis that the benefit of the potential cost saving materialising would outweigh the consequence. Pursuing this option would mean incurring about £10,000 of additional cost compared with £30,000 of additional gain. The client agreed with this approach and confirmed it in writing given the potential savings that could be realised.

Whilst reviewing the position on the piling, the piling contractor had also proposed the option to the main contractor and structural engineer that for every point beneath the building there was potential to use one single, larger diameter load bearing pile in lieu of three smaller diameter load bearing piles. Following further investigation, calculations and telephone calls by the piling contractor during the workshop, it was found that this would not offer a reduction from the piling contractor's price. The structural engineer however advised that this proposal would not require pile caps therefore offering a reduction from the groundwork sub-contractor's price. This sounded positive and therefore the main contractor worked with the groundwork sub-contractor to identify the level of saving by calculating, via an 'add and omit' exercise of items from the ground-work sub-contractor's bill, a proposed saving of £80,000. The team decided that this could be categorised as an item ready for client approval.

At this point, the quantity surveyor reflected on the overall situation and the level of opportunity. The current position of £130,000 of saving was ready for client approval with a further £300,000 of opportunities requiring further action which had varying levels of risk attached to them. Even with the best case scenario of (–£430,000) of opportunities materialising, this would not mitigate the +£640,000 cost issue on these trades and fell (–£210,000) short of the required saving. Further review was therefore required.

The quantity surveyor focused the team on reviewing the final area of potential cost savings which were the retaining wall proposals. The groundwork sub-con-tractor had advised that the specification offered by the structural engineer was high considering the function required and proposed an alternative solution offer-ing a £50,000 saving. This could not be presented to the client for immediate approval, as it was a retaining wall adjacent to a boundary and therefore required highways approval. As a follow-up, it was agreed that the structural engineer should liaise with the local authority highways department to review the proposal and likely acceptability.

22.6   Reflection and concluding remarks

To conclude the workshop, the team were made aware that the level of savings discussed had not reached the expectation at the outset of the meeting. An intense period of work was required thereafter with the potential for a further workshop if required. Notwithstanding this, there was a consensus that good progress was made in responding to the workshop agenda and expected outputs. A level of sav-ing (–£130,000) was presented for immediate client approval with more opportu-nities for saving (at least – £350,000) requiring further action. Those that required further work were assigned owners to close out the actions over the course of the next fortnight.

Prior to presenting the proposals to the client, the quantity surveyor carried out some checks on the level of savings offered by the contractor the day after the workshop, by reviewing their calculations against the proposed contract sum analysis, to ensure the full value of a particular item had been omitted. As there was no cost data available on the project for new items added back in lieu of those items being omitted (for instance the single, larger diameter pile), the quantity surveyor consulted in-house cost data and contractors for budget rates to ensure that the level of savings was reasonable. The quantity surveyor also identified a number of queries and forwarded them to the main contractor for review and comment. The queries generally resulted in increased level of savings; for instance the benchmark rate identified for the single diameter piles was more competitive than that proposed by the piling sub-contractor. However, the output of this process did not generate any further cost savings and the contractor confirmed this the same day after forwarding the quantity surveyor's queries. Notwithstanding this, the quantity surveyor decided to include an assessment of the savings proposed within the report to be forwarded to the client.

The most important aspect when drafting the report was the format. The quantity surveyor had a conference call scheduled the day after the workshop with a number of client representatives and therefore needed a format that was easy to understand and to facilitate discussion over the telephone. The quantity surveyor decided to use an elemental summary of the latest cost position versus the budget, showing the level of projected cost overspend for each trade. This clearly demonstrated variances for the piling and groundworks trade packages. A section was inserted scheduling each value engineering item against trade, with a description of the proposed change, the saving proposed by the main contractor, an assessment as a benchmark and relevant commentary on next step and timeframe. To simplify the summary, each opportunity was colour coded with a green, amber or red status, to reflect whether it was ready for client approval, required further action or was rejected. If an opportunity was approved by the client, it would be given a blue colour code. This simplified the report and made it easily understandable. Once the client had made a decision on an opportunity (whether through an immediate response or following stakeholder consultation) the colour coding and sums within the financial summary could then be changed accordingly. The aim was for all green and amber opportunities to be converted to blue, to reduce the project forecast and variance to the client's approved budget.

Prior to drafting the report, the quantity surveyor also requested the contractor to provide an assessment of a saving for the waterproof car park finish option, on the basis that even with a best case scenario of all green and amber opportunities being approved, the project would still be over budget. The contractor advised that this was not provided, as noted at the workshop the day before. The quantity surveyor requested the contractor to propose a saving making assumptions on the specification proposed and impacts of implementing the change. The contractor proposed a saving (−£95,000) in respect of the groundworks tender, giving a best case savings total (−£575,000) in relation to the +£640,000 projected cost overspend.

The client was pleased to see the level of progress made from the workshop and was not expecting the results presented within such short timescales. The client immediately approved the opportunities coded 'green' with minimal questions asked other than if the structural engineer had reviewed and approved the proposals and why the items were not identified at an earlier stage. For the latter question, the quantity surveyor advised that for the pile loads, an error was made in the piling contractor's pricing and for the pile caps this was not considered by the team as it required an environment with the relevant parties in one room to brainstorm such an item, which only happened the day prior to reporting to the client. These two items approved reduced the variance from the budget from +£640,000 to +£510,000. On the pending opportunities, the client agreed with the next steps and timescales and requested a further conference call once the actions are completed. The client particularly requested for more information on the proposed specification changes on the car park surface alternative from a maintenance perspective with regard to failure, leaks and the forming of stalactites into the floor below. The quantity surveyor noted these requests and forwarded them to the main contractor for comment. Following the call, the schedule was updated and forwarded to the client for review and agreement. Confirmation was then received in writing that the changes made were acceptable and the value engineering items were taken on board. The project team were made aware of the latest position. The contractor informed the design team, who subsequently commenced the updating of the tender drawings to reflect the revised solutions proposed.

Good progress was made on all of the pending value engineering opportunities by the team and they had all changed from amber to green status. The adjoining owner had agreed that the site could be used temporarily for storage purposes at a cost, for which the value of the savings was reduced by. The enhanced site investigation proved to be worthwhile and facilitated a further reduction in the size of the load bearing piles. The actions on the specification change for the car park finish were resolved earlier than expected and the client was consulted ahead of the next conference call. The highways department of the local authority agreed in principle to the specification change on the retaining wall and the main contractor offered to take the risk on subject to full approval. Thus all of the pending items were approved formally before the next conference call, reducing the project cost variance further by 65,000. Again, a similar process was followed following the last conference call. The remaining +£65,000 was secured as saving through the negotiation process between the quantity surveyor, the main contractor and their proposed sub-contractors, via the tender adjudication and analysis process.

Upon reflection the value engineering process was successful on this project mainly due to the collaboration and ownership of actions between the team. The team were fully aware of the repercussions should the cost issue not be mitigated and therefore actions were completed ahead of planned dates. Long lead procurement dates were met and the client gave confirmation for the project to proceed to the construction phase, after the cost overrun was addressed. Whilst the client was pleased with the outcome, the issue should not have arisen in the first instance and should be prevented on other projects by closer communication during the earlier stages. This requires the involvement of the main contractor in negotiation with the client's advisers, together with greater and early engagement of sub-contractors on the efficiency of the proposed design. The cost of value engineering was limited to the cost for the enhanced site investigation and the cost of the workshop, as there were no additional fees from the design team, main contractor or planning applications.

Chapter 23
Case Study: Value Engineering of a New Office Development with Retail Provision

Paul Ullmer

23.1 Introduction

A brief description of the project situation is provided followed by a discussion of the rationale for value engineering (VE), when, where and how VE was applied in the design and construction process. The key tools used for VE such as issue analysis, function analysis, brainstorming and workshops to explore ideas and to identify unnecessary costs or cost savings identified are also discussed with examples of specific VE opportunities. The chapter concludes with some reflection on the benefits of VE.

23.2 Why value management?

The central London office scheme consists of the demolition of an existing office tower and podium structure, to be replaced by the construction of a 15-storey new build office including retail provision (gross internal area, GIA: 42,601 m²; net internal area, NIA: 30,673 m²) using a Design and Build procurement route.

Due to the current economic climate at the time, it was important that the client agreed a let with a tenant as soon as possible. In order to do this, the client wanted to create a significant commercial office development to maximise the lettable area and its commercial viability whilst delivering a development that significantly adds to the quality of the environment.

A cost plan figure of £114 million (£2,676/m²) was submitted at stage D to the client which was benchmarked against similar schemes and considered acceptable. However, due to the client's reduced budget as a result of not acquiring the expected level of funding, the design team were requested to re-evaluate the scheme and provide a revised stage D cost plan to achieve £10 million worth of savings. The

Design Economics for the Built Environment: Impact of Sustainability on Project Evaluation, First Edition.
Edited by Herbert Robinson, Barry Symonds, Barry Gilbertson and Benedict Ilozor.
© 2015 John Wiley & Sons, Ltd. Published 2015 by John Wiley & Sons, Ltd.

client insisted that the project would not proceed without clear indications that this new budget figure could be met. Some of the savings could be offset by gaining value in other areas – such as net floor area increase – if these could be proven and then utilised. A new figure was submitted to the client within a month in order to maintain the programme and not to compromise existing and ongoing tenant negotiations. The client was not too concerned about how the savings were achieved as long as the revised scheme adhered to the existing brief and requirements.

As a team it was important to consider the options available to find the necessary savings. It was agreed that there are a number of ways of achieving the savings. First, the team discussed carrying out a general VE exercise to get best value for money from the design by eliminating unnecessary costs. By carrying out a general VE exercise, the quantity surveyor (QS) or cost consultant would take advantage of VE processes such as brainstorming, team-working and clarification of the brief. Meeting with other members of the design team encouraged creativity and ideas that may not have been considered previously. There were some disadvantages by taking a general approach to VE, the design team could not utilise suppliers within the market due to the time constraints. However, this did not facilitate the identification of alternate innovative solutions and construction methods in great detail. The signature architect may also be protective of their design vision at all costs.

23.3   When and where is value management applied?

Value management (VM)/VE was carried out at stage D design stage. The design team assessed the viability of Target VE on specific packages such as cladding and landscaping to get best value for money from the package design and eliminate any unnecessary costs. A more detailed evaluation was undertaken looking at elements in terms of quantity, quality and cost. Also, the workshops were shorter in length, and more focused on addressing the client's requirements. Looking at specific packages created a more definitive process in which design could be evaluated. The QS consultant thought that it was important to focus on the high value packages as they can take up a large percentage of the overall cost and therefore can offer the greatest opportunity. This option gave the team specific challenges to tackle and provide more specific actions as opposed to a general approach. By concentrating on high value packages, the QS consultant explored a range of construction methods while maintaining an acceptable quality level, achieving the required cost reductions and increased value. There was an opportunity to meet with specialist sub-contractors to obtain their input into design, materials and costs. Ultimately this created a better understanding of the function of the individual parts that make up the building. A disadvantage of undertaking such a process however was the time it took to complete and, with such a tight deadline to adhere to, the QS consultant was unsure whether the level of detail required would be achievable.

It was agreed that to get the best results, the design team would need to undertake extensive VE on specific packages. This option was considered to be the most viable for resolving the key issue of reducing the cost without sacrificing the function of the building. The senior partner agreed that due to the large size of the scheme, it would be broken down into individual packages with each package examined separately by members of the team.

Table 23.1   Summary of the key packages selected for VE.

Package	Cost (£)	% of Stage D cost figure	Brief description
External walls	20,068,000	18	• Standard unitised curtain walling comprising fixed glass/solid panels • Associated glass reinforced concrete and terracotta cladding features
External works	5,643,000	5	• Planters excluding topsoil, including geotextile membrane, waterproofing, plinth, brick with flint cladding • Water features, including waterproofing, reinforcement, plinth, concrete construction with flint cladding • Landscaping generally, York stone paving to public realm. • Associated public realm features such as bollards, cycle racks, and so on

The VE workshops created the opportunity for a 'second' look at the packages by the QS and the design team which ultimately fulfilled the client's requirements. By adopting this method the team was able approach suppliers and sub-contractors to obtain a greater understanding of costs and what the market could provide. This resulted in a complete assessment of each package whilst fulfilling the consultant's commitment to the client brief. As a result, the QS consultant was satisfied that the client achieved the best value for money on each package for the scheme.

Such detailed assessment requires time and there were concerns that the exercise could not be completed within a month. The QS consultant suggested looking at the high value packages (external walls and landscaping) as the team could achieve significant savings in these areas. With the cladding and the external works package (Table 23.1) representing a significant cost to the overall scheme (23% of the overall cost) it was important to maximise value from these packages and generate cost savings.

It was suggested to the client that this was the safest option to proceed with as it would allow the current programme to be maintained. Further VE workshops could then take place at a later date on the smaller value packages and any further design developments could be integrated into the revised stage D cost plan. The consequence of this action meant that the design team undertook a series of initial VE workshops focusing on large value packages. The QS consultant was assigned the cladding and external works having worked on them during the preparation for the stage D cost plan.

23.4   Value management implementation and tools used

The process began by organising a workshop with the relevant parties. Focusing particularly on the external works package, the QS consultant invited the Architect and the Structural Engineer to make sure they would be represented in the workshop. A landscaping architect was also invited by the Architect as they were consulted during the initial designs. VE workshops gave the design team the opportunity to meet with suppliers and designers as it was important to take advantage of this if the best results were to be achieved. In this particular instance, the landscaping architect had far greater appreciation of current trends in design of outside space, which was invaluable in achieving the best results. Where possible other experts were available to consult and to provide various inputs during the VE process.

During the meeting, the QS consultant highlighted the areas where it was felt that savings could be made. These were then set out in an agenda issued to the participants ahead of the workshop. The agenda gave the workshop focus and members had a clear understanding of topics in advance of the meeting so that they could make the necessary preparations. It also gave the workshop a structured approach rather than an open ended approach with people simply discussing options randomly which can be inefficient and time consuming. Other participants in the workshop also had the opportunity to add to the agenda relevant topics which required discussion and development before the meeting.

The duration of the workshops – depending on the package to be discussed – ranged from a couple of hours to 2 days. For example, for the external works package, a day was set aside so that the agenda items could be completed and the issues extensively examined.

Working through the agenda, participants brainstormed and shared their collective experiences to find solutions to specific problems. In this example, it was imperative to find savings on the overall package as it was too expensive. During the workshop the design team focussed in particular on the tree planters and water features. As a team, they bounced ideas off each other to find solutions which best suited the needs of the design as well as fulfilling the requirements of the client. The design team needed to find savings so the quantity of tree planters and water features was discussed with respect to the possibility of reducing them. It was felt that the project would not suffer by loosing a couple of tree planters and water features in the design of the building as the unique surroundings required them in specific areas around the site.

The total cost/amount for the external works package submitted at stage D was about £6 million. Having worked on the package for the cost plan, the QS consultant was aware that the design was realistically still at stage C and not developed as far as the other packages. It was therefore agreed with the senior partner on the project that it was not unrealistic to try and find a 10–15% saving on this package.

The design team looked at reducing the heights of the features of the planters. This proved to be a good way of finding savings whilst still adhering to the design without sacrificing its function. Lowering the wall heights would use less material, require less sub-structural work and reduce external finishes providing significant savings. This also adhered to the architect's wishes for a complementary design so all the parties were able to achieve their respective goals through the VE process.

Following the workshop, minutes of the meeting were issued to highlight actions to be taken after the workshops. Also, a report was published to highlight the areas and points which required further development. A timeframe had to be agreed to complete any revisions so the new design could be priced efficiently and quickly. Once this was completed, the QS consultant was able to price the new designs and report to the rest of the team about the potential savings which could be achieved within the package. These findings were put together in a report with the other assessments done on the external walls package, sub-structure and superstructure, and submitted to the client to confirm the intentions of the team on how they intended to find the savings.

23.5　Practical benefits and savings

The QS consultant was able to contribute over £900,000 of savings on the external works package which meant hitting the target comfortably of between 10% and 15% for the package (Table 23.2).

On the other packages the savings varied. For example, there was very little scope to find savings in the external walls package as it was well developed by the design team as they looked at ways of maximising the NIA (Net Internal Area). Overall,

Table 23.2　Examples of value opportunities identified, savings and participants.

Items/Packages	Anticipated level of savings and potential problems	Status (e.g. rejected, accepted, further investigation required)	Participants
External works	Savings of 10% and 15% for the package (over £900,000)		
Omit York stone paving up to and including extended pavement along London Wall, include for Marshalls Perfecta Square Paving	£263,910	Accepted	QS Architect Landscape Architect
Reduction in quantity of cycle racks	£5,000	Further investigation required	QS Architect Landscape Architect
Omit green wall to existing garage	£20,000	Accepted	QS Architect Landscape Architect
Omit 2nr water features to site	£120,000	Accepted	QS Architect Landscape Architect
Omit flint cladding to water features and planters and replace with alternate cladding synthetic material – likely metal	£122,500	Further investigation required	QS Architect Landscape Architect
Reduction in planter wall heights, including reduction to wall thickness, omit minor foundations into existing GF slab, reduction in waterproof membrane and other associated works	£295,001	Accepted	QS Architect Landscape Architect Structural Engineer
Reduction in water feature wall heights, including reduction to wall thickness, omit minor foundations into existing GF slab, reduction in waterproof membrane and other associated works	£114,723	Accepted	QS Architect Landscape Architect Structural Engineer

the team was able to find the £10 million of savings (within the deadline) over the majority of packages whilst maintaining the client's brief and requirements.

The practical benefits of the process meant that designers developed a better understanding of the design and so removed waste from that design without sacrificing the function of the design elements of the building. In this particular case study, cost savings were achieved and there was confidence that the new proposals would fulfil the client's requirements.

The VE process was found to be an extremely useful and successful part of the pre-contract stage. Through the VE process the team acquired a better understanding of materials and their design and through meeting with suppliers, the team developed a better understanding of the market in terms of cost and availability.

23.6   Concluding remarks

For the lump sum fee agreed as part of the pre-contract, VM/VE services were carried out and the benefits were significant. In conclusion, the key issue was resolved with a successful outcome. Not only did the design team provide the savings required by the client within the project timeframe but more significantly they were able to adhere to the client's initial requirements to provide a high quality design. Furthermore, by undertaking VE workshops, the design team were able to show stringent budget control, explore design options and examine alternative construction methods and materials, agree solutions and therefore demonstrate value for money to the client. Value engineering is key economic tool focusing on exploring the cost of alternative design and materials to reduce waste in construction and enhance value which is central to the sustainability debate.

Chapter 24
Case Studies: Sustainable Design, Innovation and Competitiveness in Construction Firms

Arthlene Amos and Herbert Robinson

24.1 Introduction

Construction firms are experiencing greater demands for reducing their carbon footprint due to sustainability. As a result, there is gradual shift away from traditional methods to environmentally friendly design and construction processes. This chapter examines the challenges faced by construction firms as a result of sustainability and its impact on innovation and competitiveness. Using a case study strategy, the sustainability challenges of five organisations in various positions of the construction supply chain are discussed. The chapter also explores how sustainability in construction firms is driving changes in their strategy, which has triggered process, design and product innovation influencing competitiveness and profitability both in the short and long term. Following this introduction, the concept of sustainable development and key drivers influencing sustainability strategies of construction firms are reviewed. The case study findings, analysis and discussion of the key issues based on the experience of five construction firms are presented and the lessons learned with the implications for construction firms are also discussed.

24.2 Background and context

The construction industry uses a significant amount of energy and generates an astounding amount of waste both from excess materials, which cannot be reused, and from new materials not stored properly on site resulting in wastage. Construction firms have been put under increasing pressure in recent years as a result of sustainable development driven by international agreements such as the Kyoto Protocol, EU Emission Treaty and various government legislation and initiatives aimed at reducing environmental impact and carbon emission. These agreements have led to profound changes in the behaviour of construction firms such as changes in design,

Design Economics for the Built Environment: Impact of Sustainability on Project Evaluation, First Edition.
Edited by Herbert Robinson, Barry Symonds, Barry Gilbertson and Benedict Ilozor.
© 2015 John Wiley & Sons, Ltd. Published 2015 by John Wiley & Sons, Ltd.

project processes creating challenges as well as opportunities for the construction supply chain. However, for sustainable design to be effective a balance between addressing the environmental concerns about design and construction (environmental objectives), needs of society (social objectives) and profitability and competitiveness (economic objectives) is required. If firms are not competitive and profitable then their sustainability efforts would fail as firms exist to maximise profits to reflect market risks and to reward their shareholders. Jiménez and Lorente (2001) noted that firms can contribute individually towards sustainable development by innovation in products and processes during design and construction. Innovation in design and construction is therefore necessary to respond to the increasing pressure to adopt sustainable development and to maintain or enhance their market position, level of competitiveness and increase profitability. Fergusson and Langford (2006) noted that companies who willingly show environmental concern, command a strong market position because they are driven by more than just legislation and client requirements. Previous studies have not adequately addressed the impact of sustainability on design, process and product innovation, competitiveness and profitability.

Sustainable development based on the seminal Bruntland Report is defined as "development that meets the needs of the present generation without compromising the ability of future generations to meet their own needs" (cited in Atkinson, 2008). In any sustainable development strategy, three important dimensions have to be addressed. These are the environmental (the planet), social (the people) and economic (profit) aspects. Cooper (1999) and many other researchers recognised this tension between protecting the environment, social obligations and economic development, and the difficulties in operationalising the concept of sustainable development (Kaatz et al. 2005). Edum-Fotwe and Price (2008) used the "Triple Bottom Line Sustainability Model" to demonstrate that sustainability lies at the core of the three dimensions, when they are combined. The Forum for the Future (2009) argued for sustainable development to be incorporated through an organisation's strategy. Responding to the challenge of sustainable development is important to ensure the survival of construction firms in the market. There is now a greater demand for adopting a sustainable development strategy increasingly driven by government legislation, clients and other stakeholders affected by construction activities.

24.3   Key drivers of sustainability in design and construction

There are a number of drivers influencing the rate of adoption of sustainable development strategy. First there are the legislative requirements. To support the government's sustainable development policy there are a number of legal instruments, directives and Acts passed to address environmental issues such as energy usage, water, waste, wildlife and land use and carbon emissions (Ministry of Defence, 2009). The Climate Change Bill, which became law on 26 November 2008, serves as a legally binding target to reduce the UK's carbon emission by at least 26% in 2020 and at least 60% in 2050 (Defra, 2008). The Government is required to publish 5-yearly carbon budgets from 2008 onwards and a Committee on Climate Change has been formed. The Committee is expected to advise the Government on the levels of carbon budgets to be set, the balance between domestic emission reductions and the use of carbon credits, and whether the 2050 target should be increased. In a recent article, the Secretary of State for Climate Change for the UK Government sets out the low-carbon plan for 2050 (Harvey and Stratton, 2011).

Secondly, there are specific planning and building regulations such as the 'Merton Rule', Energy Performance Certificates (EPCs), BREEAM assessment and Code for Sustainable Homes, which affect design and construction activities. The planning rules, building regulations and assessment tools are aimed at protecting the environment by providing guidance on design approaches and construction procedures which obviously affect the activities of construction firms. For example, the 'Merton Rule' originally introduced by the London Borough of Merton has been adopted by other boroughs throughout Britain (Merton Council). This rule is a prescriptive planning policy that requires new developments to generate at least 10% of their energy needs from on-site renewable energy equipment. An EPC is also required when a building is constructed, sold or rented. The EPC gives home owners, tenants and buyers information on energy efficiency of properties using a standard energy and carbon emission efficiency grade from 'A' to 'G', where 'A' is the most efficient (Directgov, 2008). The BREEAM assessment process was introduced in 1990 and there are various versions updated regularly to comply with UK Building Regulations. BREEAM looks at a broad range of environmental impacts from management, health and wellbeing, energy, transport, water, material and waste, land use and ecology and pollution (BREAAM, 2007). Credits are awarded in each of the above areas according to performance and a set of environmental weightings then enables the credits to be added together to produce a single overall score. A Code for Sustainable Homes was also introduced and from May 2008 it became mandatory for all new homes to have a rating against the Code.

Thirdly, there are the economic arguments as one of the most powerful triggers for changing the behaviour of construction firms and their clients. The Stern Review focused on the economic costs of the impact of climate change, and the cost benefits of action to reduce the emissions of greenhouse gases (HM Treasury, 2009). Carbon Trust developed carbon management to help companies including construction firms to recognise the business opportunities associated with climate change (Carbon Trust, 2006). A major economic driver of sustainable construction is improving building performance and reducing maintenance and operational costs (Khalfan, 2001). An argument sometimes put forward is that the 'least sustainable is the more profitable' as it avoids the environmental cost. Sir Jonathon Porritt, Chair of the Sustainable Development Commission, was quoted as saying:

> You have occupiers saying we want to live in green buildings, but there aren't any. So the contractors say we can build them but developers don't want them. Developers say we want them but investors won't pay for them. Then the investors say we would pay for them but there is no consumer demand (Pickard, 2007).

The misconception of increased costs of sustainability and lack of a market value discourages both developers and contractors (Cole, 2000). Johnson (2000) found that sustainable buildings could produce more economical benefits for their owners/operators than more traditional designs. Yates (2001) explored the business benefits of sustainable construction and concluded that the benefits are diverse and potentially significant. He identified the business benefits such as capital cost savings, reduced running costs, increased investment returns and image/ marketing spin-offs. There are various other economic incentives or disincentives to adopt sustainable solutions such as the capital allowances scheme, landfill tax, climate change levy and aggregate levy (ICE, 2009).

These legislative and planning instruments affect construction firms in a number of ways. The construction industry draws materials from natural resources, uses highly energy intensive processes, removes land from other uses and is responsible for designing and making products that have lasting effects on the environment and users [Chartered Institute of Building (CIOB), 2008]. Materials extraction, production and recycling interfere with complex ecological and socio-economic systems (Steen, 2005). Construction firms also contribute to pollution of water, dust, noise and toxicity. The CIOB argued that construction firms account for a third of the nation's waste through demolition and excavation processes (CIOB, 2008). Clients, both public and private, are demanding changes and expressing preference for construction firms that are up-to-date with their design and construction procedures. Changing design and construction processes to adhere to legislation and planning requirements and to avoid potential economic and environmental costs as a result of non-compliance or to benefit from economic incentives such as enhanced capital allowances or tax allowances can drive an organisation to become innovative and steer away from traditional practices. There are significant opportunities for innovation to respond to the challenges of sustainable development. However, design and construction firms often invest relatively little in research and development to drive innovation compared with other sectors. Ivory (2005) noted that due to a client-focused construction industry the innovation process may be in jeopardy as clients are in general not keen on taking any risks with new designs or ideas as their main concerns are budget and time completion. Reichstein *et al.* (2005) argued that "construction firms have become inherently risk averse and many construction firms do not need to innovate to remain successful". Innovation can be delivered through change of processes or creation of new products, often referred to as process or product innovation (Jiménez and Lorente, 2001). However, if successful, innovation can reduce cost and increase revenue for an organisation thus increasing competitiveness and profitability.

24.4   Case studies

Case study A: managing consultant

Case A is a managing consultant firm, which serves the public and private sectors worldwide from over 152 offices. They currently operate in all sectors and specialise in building, environmental, health, communication, water, energy, transport, oil and gas, tunnels, bridges and power stations. They have an annual turnover of £800 million with about 13,500 employees worldwide. The corporate responsibility for sustainability is driven through their Quality, Environmental and Safety (QES) teams, which are spilt, into various management units operating in different sectors of the business.

Case study B: main contractor

Case study B is an international company that operates in all sectors besides property and residential. They have an annual turnover of about £800 million and employ approximately1,500–2,000 people. Sustainability has always been a key part of what the organisation does and is driven through a Corporate Responsibility (CR) team, which consists of 15–20 members, who constantly measure the organisation's sustainability performance.

Case study C: main contractor

Case study C is an international company that was formed from an acquisition, with an annual turnover of £3.24 billion and about 30,850 employees. The sectors in which they operate are diverse and include lifestyle, social infrastructure, business and transport. They have a specific department (CR team) which deals with sustainability and it is a key part of the responsibilities of the Procurement Director, Human Resources Director, the Design Management team, the Environmental Manager and Community Managers.

Case study D: specialist contractor

Case study D operates in both the rail and highways sectors locally in the UK and internationally with an annual turnover of £240 million. The company's sustainability commitment is directly linked to the requirements set out in their section 61 documents, issued by their environmental team. Section 61 is a legal document which comes from the Control of Pollutions Act (COPA) 1974 issued by the local council to construction companies to govern and control interface works with the public.

Case study E: subcontractor

Case study E forms part of a group company, which operates internationally. They have an annual turnover of £1.2 billion worldwide and employ just over 500 engineers in the UK and 8,000 worldwide. Sustainability is relatively new on the agenda and is currently being progressed through the quality and environmental side of the business. The appointed department so far consists of 10 key personnel with an additional 5 in a peripheral setting.

24.5 Findings and discussions

Key findings from all the five case studies are summarised in Table 24.1, Table 24.2, Table 24.3, Table 24.4 and Table 24.5 and discussed in relation to their sustainability strategy and the key drivers, effects of sustainability on processes, types and nature of innovation and competitiveness.

Profile of Companies and Sustainability Strategy

Case studies A, C and E operate in all sectors, with specialities in a number of areas and Case B in all sectors except property and residential while Case D, being a specialist contractor, only operates in two sectors – rail and highways (see Table 24.1 and Table 24.2 for their sustainability strategy).

Table 24.1   Company background.

	Case Study A	Case Study B	Case Study C	Case Study D	Case Study E
Type of firm	Managing Consultant	Main Contractor	Main Contractor	Specialist Contractor	Sub-contractor
Geographical focus	National – UK-based and International with over 152 offices	National – UK-based and International	National – UK-based and International	National – UK-based and International	National – UK-based and International
Operational sectors	All sectors	All sectors besides property and residential	All sectors	Rail and highways	All sectors
Annual turnover	£800 million	£800 million	£3.24 billion (March 2008)	£240 million (for 2008)	£110 million in the UK and £1.2 billion world wide
No. of employees	13,500 worldwide	1,500–2,000	30,847–11,482 in Europe, 17,326 in Middle East/Asia and 2,039 in Australia	1,350	8,000 world wide (including 500 engineers in UK) operating out of 60 different countries

Table 24.2 Sustainability strategy.

	Case Study A	Case Study B	Case Study C	Case Study D	Case Study E
Organisational ethos towards sustainable development	Committed to promoting a strong culture of corporate sustainability through their values, practices and projects	Sustainability has always been a key part of what the organisation does and is driven by different departments and committees	Sustainability within the organisations is dependent on who is responsible for the long-term cost of operating asset. If they are building and operating, they are happy to invest upfront to gain long-term savings	Highways division is driven by being "environmentally responsible in all activities" Rail division's commitment stemmed from, requirements set out in Section 61 documentation but if there is no profit involved then there is no point in doing it	Important – seeking to gain a better understanding
Corporate strategy and department responsible	Integrated into company's culture through QES teams	Managed by a specific department (CR department/ team) for reporting and e-collation of data on sustainability, that is CR KPIs (KPIs set by the highways agency)	Managed by specific department with belief that there is no point in pricing something in if Client is not willing to pay Central CR cross functional team of Champions ensures that sustainability is intertwined in the business	Highways – integrated into company's culture, focus on management of their carbon footprint through energy, fuel and waste projects to deliver their strategic aims Rail – no specific department that dealt with sustainability, as there was no budget. Highways – not specified Rail – environmental team	Implementation stage – relatively new on the agenda, and as a result they are now in the process of developing their strategy further. Also seeking to gain a better understanding of what each client within the sector requires in terms of sustainability Developed a sustainability quality statement as part of their management strategy

Members in team	QES split into various management units with representative for each division	15–20 (from across different sectors of the business)	Thousands directly involved in sustainability strategy Champions within each business unit meet regularly with central sustainability team to discuss action plans	Ten people in highways division whose main focus is sustainable development with resources and budget	Consists of 10 key people with an additional 5 in a peripheral setting
Leadership	Leaders of various management units from different sectors of the business consisting of 200–300 people	Two CR managers specially appointed and one is a member of the Executive Board	Project leader reports to the Director responsible for sustainability, supported by the team of CR Champions from functions and business units – HSE Director – Procurement Director, HR Director – Design Management teams Environmental and Community Managers, and so on	Not specified	Currently being managed through the environmental department

CR, corporate responsibility; KPI, key performance indicator; QES, Quality, Environmental and Safety.

Key Drivers of Sustainable Development Strategy

Case studies A and B adopted a personal stance on sustainable development and the need for it (Table 24.3). Case study B is currently in the top 20 based on the Construction News top 100 contractors, which shows that they are among the industry leaders in terms of their market position. Case study A was ranked in the top 5 by the Sunday Times in the 20 'Best Big' Companies to work for. Both Case studies C and D (specialising in rail and highway sectors) view profit as their main driver. However, Case study C recognises that while sustainable development may cost more at the beginning they can benefit from long-term profits if they invest in it. The Project leader for Case C (main contractor) argued that on projects where they are building and operating, they are willing to invest in sustainability costs up-front. Case study D (rail division) does not see the need for sustainable development if there are no profits whilst Case study D (highways section) views it as an environmental responsibility. Case study E's (sub-contractor) main drive for sustainability stems from the client's requirements. Case study C's approach is to implement sustainable development as a long-term objective (i.e. paying up front costs for long-term benefits).

Case studies A, B, C and D (highways) are aware of BREEAM and have implemented it on a number of projects. They however did not express knowledge of the key legislative requirements such as the Climate Change Bill, Clean Air Act and others including major reports such as the Stern Review. Case study D (rail) and Case study E had little or no knowledge of BREEAM. Unlike the others, Case study B was the only organisation voluntarily signed up to a sustainable development programme that does not carry any legislative requirements.

Effects on Processes

Case studies B and C experienced changes at the tendering stage and believed that clients are seeking contractors with commitment to sustainability. As a result, they changed their procedures for tendering (Table 24.4).

Case study B also reported that sustainability related questions have increased at tender stage, especially with public sector clients. Case studies A, D (rail) and E changed their procedures based solely on clients' requirements. For Case study D (highways), no change was made to their procedures as sustainability is always as part of their package. The tendering process for a main contractor is usually more stringent than for the others along the supply chain, upstream and downstream. For this reason, Case studies B and C as main contractors have experienced increased pressure at tender stage to adjust their procedures.

Case studies B, C, D (highways) and E changed their processes to incorporate sustainable development. Case studies B and C both changed some of their key account and management plans and adjusted the way they manage their supply chain. Case study D (highways) experienced change as they have to manage their carbon footprint while case study E incorporated changes in processes such as their procurement and waste disposal methods. In Case studies A and D (rail) changes were directly driven by client requirements (Table 24.4).

Table 24.3 Key drivers of sustainable development.

	Case Study A	Case Study B	Case Study C	Case Study D	Case Study E
Key drivers	Driven by the 'if it can be done, then it would be' attitude not only on local but international projects as well Always seeking new ways to do things but believes that innovation is difficult in traditional sectors Profit and increasing competitive advantage are key factors	Recognises the need for climate change and reduction in carbon footprint Increased innovation from sustainable development, for example innovation on health and safety has increased significantly. The SHE newsletter highlights innovations at project level Profit and increasing competitive advantage are key factors as they see a direct link between innovation and competitiveness	Increased innovation as a result of pressure from sustainable development Profit – investments in upfront to gain long-term savings. Profit is therefore a key factor	Different drivers. For Rail its profit and for Highways its environmental responsibilities In Highways, there is increased innovation from sustainable development but none in Rail. "Innovation Scheme" introduced to stimulate and encourage innovation or improvement of ideas by providing an approach that ensures proper examination, approval, recognition and award Profit is a key factor	Client requirements Major drive to increase innovation As piling contractors their biggest spend is on concrete and steel so constantly looking for ways to reduce cost and improve efficiency. Profit is therefore a key factor Monetary prizes are awarded to employees who come up with good ideas, which consist of its office as well as site based, so if someone comes up with an idea as simple as printing on both sides of paper, it would be awarded.

(continued)

Table 24.3 *(continued)*

	Case Study A	Case Study B	Case Study C	Case Study D	Case Study E
Awareness of tools, reports, legislative instruments and implementation	Aware of Bruntland Report and BREEAM adopted on all projects to improve design and management process	Adopted BREEAM on some building projects and signed up to voluntary requirements of Civil Engineering Environmental Quality Assessment – specific to rail and road projects Client's approach differs greatly. For example, highways – highways agency would expect greater resources, reports on waste KPIs every month and energy consumption KPIs, and they also look at carbon accounting	Adopted BREEAM and often build to its requirements. However, the Project leader believes the system does not always encourage most sustainable solution. For example, a contractor does not receive BREEAM credits for cement replacement in concrete	Highways – aware of BREEAM but has not specified Sustainable development implemented in all highway projects Rail – none	Limited knowledge of BREEAM and its requirements. Felt BREEAM is not very clear as to which requirements are statutory and voluntary Main contribution to the environmental aspect of sustainability at the moment is by adhering to all the requirements in their Section 61 documentation

KPI, key performance indicator.

Table 24.4 Effects on processes.

	Case Study A	Case Study B	Case Study C	Case Study D	Case Study E
Procedures and methods	Change is influenced by client requirements when tendering for projects but difficult to achieve in sectors such as oil and gas Generally seek solutions that will not have an adverse impact on natural environment, for example rely on local manufacturing and locally sourced materials rather than importing	Experienced change when tendering – more questions asked on sustainability and how it is dealt with in project Changes in their design/construction methods. Required to check credentials at the design stage using BREEAM Use of local labour is now a mandatory requirement on all international projects. Projects employ a percentage of local labour	Change their methods and procedures for tendering, particularly for public sector clients to meet government targets Design/construction methods have also been affected. Product selection to drive down long-term energy costs, locally sourced materials to reduce transport miles and costs Procurement team working with suppliers to reduce packaging delivered to sites, buy-back unused materials, support local business, and so on	Promote sustainable solutions where specification and client approval has allowed it Highways – no changes required as it has always been part of culture for the highway sector Rail – changes introduced based on client requirements at tender stage	Currently have not changed tendering methods but will in future depending on clients' requirements. Noted that requirements for sustainability are increasing All offices are video linked and meetings are carried out via videoconference, which reduces transportation costs, and makes people more efficient because they do not have to travel long distances Changed their design/construction methods mainly in the procurement section and on site in terms of their waste disposal methods

347

Nature and Types of Innovation

Each organisation reported increased innovation due to sustainable development, except for Case study D (specialist contractor, rail division) who argued that sustainability has had no effect on innovation within that sector of their organisation (Table 24.5). Case study E (sub-contractor) argued that innovation directly related to sustainable development is now on the increase within their organisation. However, they have always viewed innovation as an important part of the organisation's development and a way to increase their profitability.

The results in Table 24.4 suggest that sustainability has played a positive role in increasing innovation. Case studies B and C cited examples of innovation in processes, product and design. Case studies D and E only reported innovation in processes whilst Case study A cited examples in both processes and product. They all argued that the key driver for innovation was profit. However, Case studies A and B argued that innovation is a way to increase their competitive advantage and competitiveness.

There were a variety of examples of process, design and product innovation to respond to the challenges of sustainability. This includes the use of polystyrene instead of piling in hard stone for constructing railways embankment (Case study A), use of timber from more sustainable sources for design solutions (Case study B), use of new materials for cement replacement, recycling by-products and development of "Waste Tracker" to quantify waste generated, recycled, reused or disposed of (Case study C). Case study D (highways) provided an example of the use of a surfacing product, which was used on a South African project. Case study E also developed "Screwso" which reduces the amount of spoil that comes up when a pile is dug and use a lot of replacement mixes including ground granulated blast furnace slag and pulverised fuel ash to reduce their cement content.

Effects on Competitiveness and Profitability

Case study A (consultant) as one of the major suppliers of oil and gas has experienced no effect in competing for work. They argued that they have filled a niche in this market and do foresee some competitive benefits. However, in other sectors where there are strong key players tendering, they have experienced some effects (Table 24.6).

Case study B reported that the change in their level of competitiveness was project specific. They argued that in sectors such as highways where the clients (Highways Agency) are more sophisticated compared with the retail sector the requirements are different and thus will have to show more commitment at tender stage. Case study C reported an effect on their level of competitiveness in all sectors, because of the increasing demand to construct more sustainably from all sectors. Case study D experienced no effects in both their highways and rail sector business whilst Case study E only noted a minimal effect.

The ICE (2009) stated that the active management of sustainability performance can deliver significant improvements in business, efficiency and profitability. However, some organisations (Case studies A, B and C) found it difficult to quantify the impact of such changes in their profits at the organisational level but argued that any change positive or negative in profitability was more project specific. Case study D experienced no change in their profitability in both their rail and highways

Table 24.5 Nature and examples of process, product and design innovation.

	Case Study A	Case Study B	Case Study C	Case Study D	Case Study E
Process innovation	Seek to implement BREEAM in business processes Key processes introduced focusing on: – Minimisation of power and maximization of their products – Minimising pollution to the local environment and emissions (oil and gas engineering)	Key account plans introduced and appointed specific key managers to manage this process Changes in downstream supply chain management project performance assessment on projects. Audit carried out by business systems Sustainability key performance indicators Introduced new site waste management process and plans to cut landfill waste, encourage recycling aggregate and recycling on site	Changed management processes to incorporate sustainability. For example, there is a system called "Waste Tracker" to quantify waste generated, recycled, reused or disposed of Introduced new business code of conduct and supplier code Changed business process – each business unit and business function has an action plan. Introduced effective downstream supply chain management Lean construction techniques (DfMa), offsite construction, local sourcing of materials and resources and waste management	Highways introduced new process for management of carbon footprint through energy Rail – none	Use a lot of replacement mixes including ground granulated blast furnace slag and pulverised fuel ash to reduce their cement content Concrete and steel represents 80% of their material spend, and are limited in their use of timber and plastic products. They are constantly looking at ways they can save on concrete. If they can reduce the concrete volume by 20% and still have it carry the same loading then it is more feasible to do this; it would also save 20% of the material taken out of the ground Developed 'Screwso', which reduces the amount of spoil that comes up when a pile is dug. Introduced CSM, which is better known as a remix soil to reduce a wall

(continued)

349

Table 24.5 (continued)

	Case Study A	Case Study B	Case Study C	Case Study D	Case Study E
Product and design innovation	Use of polystyrene instead of piling in hard stone for constructing railways embankment. Polystyrene able to take the same loading as piles	Use renewable energy as part of their energy mix for a school project. Use of timber from more FSC and sustainable sources for design solutions. Use of natural ventilation instead of air condition units. Taking a more in depth look at the design stage to show whole life costing	Use new materials for cement replacement, recycling by-products. Pre-manufactured units, use of recycled materials, carbon management DfMa, specification of sustainable materials, design out waste, reduce, reuse and recycle	Rail – not specified Highways – not specified	None specified
Innovation influence from international projects	Yes. Innovations highlighted in company newsletter. Use multi-phase pumping, adopted from an Abu Dhabi project	None	No influences known	Highways – yes, use of surfacing product, adopted from project in South Africa Rail – no	Yes but no examples given

DfMA, Design for Manufacture.

Table 24.6 Effects on competitiveness and profitability.

	Case Study A	Case Study B	Case Study C	Case Study D	Case Study E
Competitiveness	No effect in niche markets/sectors (e.g. oil and gas) Experienced more competition at tender stage in other sectors such as rail and building projects	Change in competitiveness is sector specific	Change affected all sectors and their competitiveness when tendering. They believe that it can provide a competitive advantage where it can be demonstrated in their offering to clients	Experienced no change in level of competitiveness, as they have always had sustainability at the heart of their business Believe that effects on competitiveness would be different for each sector Highways – not affected Rail – not affected	Minimum effect. Noticed a small difference in their level of competitiveness and were not able to report any noticeable effects. They believe that the level of competitiveness in each sector is different and they are currently involved with innovations to try and improve their sustainability; although predominantly the innovations are mainly there to increase their profit margin rather than improve sustainability

(continued)

Table 24.6 *(continued)*

	Case Study A	Case Study B	Case Study C	Case Study D	Case Study E
Impact on profitability	Not affected at organisational level, benefits are more project specific	Difficult to quantify at organisational level Profitability is more project specific. For example, in a demolished building being able to use/ recycle materials on site saving on lorry journeys and material costs	Not quantified at organisational level Sustainable development affects short-term profitability but it is for long-term gain Profit margin is the same for a construct only project as they build what is specified and charge a percentage on top of the costs. It is only when they go onto operating a building or asset that they see the payback on investment	Profitability for each sector is different and it is believed that sustainable development has affected their profitability due to improved management of their carbon emissions Highways – not affected Rail – not affected	Has not been quantified but has had an impact on their profitability although very small. For example, they have not seen paramount benefits such as winning a job because they were showing commitment to sustainability The difference in profit levels for each sector has not been quantified to date, but the Project Manager is convinced that it would exist later, particularly in the housing sector where there are requirements for builders to produce more efficient and carbon neutral properties

divisions. Case study E had taken no steps to quantify the effects on profitability at a project or organisational level. The direct impact of sustainability for construction organisations would be in the form of increased cost or savings due to changes in processes, procedures and methods in complying with sustainability strategy. For example, Case study A has implemented a multi-phase pumping for oil, gas and water from an Abu Dhabi project. The usual process for pumping involves separating the oil, gas and water, pumping the oil and water and compressing the gas, which involves the use of three different pieces of equipment (a separator, two pumps and a compressor). With multi-phase pumping however, one piece of equipment is used which carries out all the processes. Using this method of pumping rather than the traditional method provided savings in their capital and manufacturing costs. These savings can be aggregated for each type of innovation and assessed at the project level, which can directly improve profitability at an organisational level.

24.6   Concluding Remarks

There are several key findings. First, main contractors, higher up in the construction supply chain, need to demonstrate their commitment to sustainability during the tender and bidding stages more than those in the lower part of the supply chain. Sub-contractors in the lower end of the supply chain do not necessarily have to demonstrate the same level of commitment unless they are involved in major projects where sustainability is top of the client's agenda. Secondly, whilst sustainability is deemed important there are still questions about the level of awareness in the construction industry. Requirements for sustainability set out in BREEAM was the most common source for increasing awareness. Action on sustainable development should be taken or required at the tender stages, especially when bidding for work with public sector clients. Thirdly, as a result of the pressure of sustainability, some organisations adopted new processes in the tender phase resulting in changes in design, construction and different types of process, design and product innovation. The innovative ideas were not limited to UK projects as there were examples of innovation adopted for international projects. Specific examples included innovative ideas to reduce their waste, energy consumption and carbon footprint. In many instances such innovative ideas have had a direct positive impact on competitiveness and/or savings due to a reduction in for example, the higher fees associated with waste disposal at landfills.

Whilst there were obvious examples of how innovations can lead to savings in the cost of processes, design and products, the level of profitability from innovation at a project or organisational level has not been assessed or explicitly quantified in monetary terms. Sustainability is more than just complying with legislative or planning requirements but provides the opportunity for increased innovation, competitiveness, and more significantly, profitability which can be quantified to strengthen an organisation's market position. For an organisation to effectively manage sustainability, a specific department, dedicated team and resources are required with clear leadership and authority. A quality driven agenda allows for easier quantification of the benefits and savings resulting directly from innovation. Monitoring of sustainability progress is also crucial to identify areas for innovation to facilitate continuous improvement in design and construction processes. Construction firms should seek to adopt sustainability, increase awareness, and develop a strategy as it can have a positive effect on innovation in processes, design and products and their level of competitiveness and profitability.

References

Atkinson, G. (2008) Sustainability, the capital approach and the built environment. *Building Research and Information*, **36**(3), 241–247.

BREEAM (2007) A Record Year for Carbon-cutting BREEAM. http://www.breeam.org/newsdetails.jsp?id=530 (accessed 11 January 2009).

Carbon Trust (2006) Carbon Footprints in the Supply Chain: the Next Step for Business. https://www.carbontrust.com/media/84932/ctc616-carbon-footprints-in-the-supply-chain.pdf (accessed 2 August 2014).

Cooper, I. (1999) Which focus for building assessment methods – environmental performance or sustainability. *Building Research and Information*, **27**(4/5), 321–331.

CIOB (Chartered Institute of Building) (2008) Sustainability and Construction. www.ciob.org.uk (accessed 21 October 2008).

Cole, R.J. (2000) Cost and value in building green. *Building Research & Information*, **28**(5/6) 304–309.

Defra (Department for Environment Food and Rural Affairs) (2008) Climate Change Act 2008. http://www.legislation.gov.uk/ukpga/2008/27/contents (accessed 2 August 2014).

Directgov (2008) Energy Performance Certificates for the construction, sale and let of non-dwellings. https://www.gov.uk/government/publications/energy-performance-certificates-for-the-construction-sale-and-let-of-non-dwellings--2 (accessed on 2 August 2014).

Edum-Fotwe, F.T. and Price, A.D.F. (2008) A social ontology for appraising sustainability of construction projects and developments. *International Journal of Project Management*, **27**(4), 313–322.

Fergusson, H. and Langford, D.A. (2006) Strategies for managing environmental issues in construction organizations. *Engineering Construction and Architectural Management*, **13**(2), 171–185.

Forum for the Future (2009) Sustainability Concepts: Natural Step. http://www.gdrc.org/sustdev/concepts/19-n-step.html (accessed on 2 May 2009).

Harvey, F. and Stratton, A. (2011) Chris Huhne Pledges to Halve UK Carbon Emissions by 2025. The Guardian, 17 May 2011. http://www.theguardian.com/environment/2011/may/17/uk-halve-carbon-emissions (accessed 2 August 2014).

HM Treasury (2009) Stern Review: The Economics of Climate Change, Executive Summary (Short). http://webarchive.nationalarchives.gov.uk/+/http:/www.hm-treasury.gov.uk/sternreview_summary.htm (accessed 2 August 2014).

ICE (2009) Sustainable Development Strategy. http://www.ice.org.uk/downloads/Sustainable%20development%20strategy%20July%2007(1).pdf (accessed 21 January 2009).

Ivory, C. (2005) The cult of customer responsiveness: is design innovation the price of a client focussed construction industry? *Construction Management and Economics*, **23**, 861–870.

Jiménez, J.B. and Lorente, J.J.C. (2001) Environmental performance as an operations objective. *International Journal of Operations and Production Management*, **21**(12), 1553–1572.

Johnson, S.D. (2000) The economic case for high performance buildings. *Corporate Environmental Strategy*, 7, 350–361.

Kaatz, E., Root, D. and Bowen, P. (2005) Broadening project participation through modified building sustainability assessment. *Building Research and Information*, **33**(5), 441–454.

Khalfan, M. (2001) Sustainable Development and Sustainable Construction – A Literature Review. Loughborough University.

Ministry of Defence (2009) Acquisition Operating Framework for Sustainable Development. https://www.gov.uk/government/uploads/system/uploads/attachment_data/file/27613/1_20091120684SubmissionstoMinistersNewRequirementsAttachmentFinalPUS91U.pdf (accessed 2 August 2014).

Pickard, J. (2007) Green Credentials under Scrutiny. The Financial Times, 22 May 2007.

Reichstein, T., Salter, A.J. and Gann, D.M. (2005) Last among equals: a comparison of innovation in construction, services and manufacturing in the UK. *Construction Management and Economics*, **23**, 631–644.

Steen, B. (2005) Environmental costs and benefits in life cycle costing. *Management of Environmental Quality: An International Journal*, **16**(2), 107–118.

Yates, A. (2001) Quantifying the Business Benefits of Sustainable Buildings – Summary of Existing Research Finds. BRE Centre for Sustainable Construction.

Chapter 25
Case Study: Retrofitting Building Services Design and Sustainability in Star Island

Victoria Hardy

25.1   Introduction

This chapter presents a case study of how one non-governmental organisation (NGO) is confronting the costs of fossil fuel utilisation in an isolated location with an innovative approach to funding the costs of moving to an alternative energy solution. The case study starts with an overview of the background and location of the project, the key challenges and issues faced by the Board of Directors and leadership of the NGO, which is followed by an in-depth analysis of the problems and the options explored for resolving the problems in terms of various design solutions and their cost implications.

Project background

Star Island is a 40-acre (ca. 16 ha) barrier island 10 miles (6 km) off the New England coast of the US. The island has been occupied with a wide variety of historical uses since the mid-1660s. It has been a fishing camp, a fishing village, a nineteenth century resort, and finally, is currently in use as a retreat and conference center. The current owner of the island is the not-for-profit (NGO) Star Island Corporation (SIC), which has been operating the island for "religious, educational, and kindred" conferences, gatherings, and meetings since 1916.

The twenty-first century has not been kind to the island's management, with the intrusion of dozens of regulatory agencies and a rapidly aging infrastructure. For example, there have been a number of changes in building regulations relating to building services such as requiring new exterior fire escapes on historic buildings with little or no regard for the building's historic attributes; requirements to replace existing zone-based fire alarm systems with addressable units (even if the old systems

Design Economics for the Built Environment: Impact of Sustainability on Project Evaluation, First Edition.
Edited by Herbert Robinson, Barry Symonds, Barry Gilbertson and Benedict Ilozor.
© 2015 John Wiley & Sons, Ltd. Published 2015 by John Wiley & Sons, Ltd.

were still functional); and substantially increased reporting requirements for the water generation and wastewater treatment systems.

Key challenges and issues

The Board of Directors and CEO of SIC have recognised that the sustainability of the historic property and its continued use as a retreat venue must address several core issues: the age of the buildings, which range from 50 to 250 yr old; the harsh environment; the current dependency on diesel for power; and the costs of maintenance/repairs and construction in a remote marine environment that typically adds 25–30% to the cost of capital projects.

In addition, SIC was confronted in 2007 with another set of facility issues. In June 2007, one week before the island was to open for its 110th season, the local fire inspector refused to grant the certificate of occupancy citing more than 12 pages of violations in the fire notification and suppressant systems; in addition, the local building inspector also found numerous electrical code violations. With many buildings more than 100 yr old, this level of violations was not difficult to find. The resolution was that the SIC spent more than US$2 million over the following 6 weeks to make all the necessary repairs and upgrades. The subsequent loss of more than US$1 million in lost bookings and reservations almost bankrupted the organisation, and most certainly created substantial operating challenges.

25.2   Initial study or analysis to identify problems

In the following months, SIC embarked on an intensive analysis of why this event happened, and what could be done to prevent its reoccurrence. In the first step of the process a community-wide assessment was completed, which included interviews with more than 100 individuals. Among other findings, the resulting report revealed a lack of communication and candor between the CEO and the Chief Facility Executive (CFE); facility budgets were presented but not accepted during the annual budget process; and there was no respect for the facility function. The Board of Directors and the volunteer leadership were also not in the communication loop; the advisory committee met infrequently and their reports were buried. Overall, there was a lack of understanding of the critical value of the facilities to the core business; facility budgets had been underfunded for years, and preventive maintenance was almost non-existent; and depreciation was minimised on the balance sheet.

The second step in this organisational reengineering was to redesign the staff, and to empower the CEO and the CFE with a level of authority to match their responsibilities. The Board of Directors hired a CEO with professional expertise to support these changes. They authorised the CEO to hire a CFE with professional expertise, and they promoted the volunteer Strategic Facilities Planning Committee to a board-level committee with regular reports on their efforts.

In the final third step, the Board of Directors authorised the CEO to implement an intensive facilities planning process, which included the first ever complete Building Assessment for every building on the island; this document laid the foundation for the 5-yr Capital Improvement Plan, and lays out more than US$3.5 million in expenditures to bring the island buildings up to par.

25.3   Funding for capital improvement plan

SIC also embarked on a capital campaign to raise US$5 million to fund the Capital Facilities Improvement Plan, re-build the endowment fund (which was used for the repairs), and establish a US$500,000 capital reserve for the future (and natural disasters endemic to this coastal region).

It was with this backdrop that the newly configured senior management team was confronted with the issue of power generation and the source of fuel: diesel. Under current operational modes, island energy is produced by diesel-powered generators, with a season total of approximately 154,000 kWh used during the busiest months. The cost of transporting the diesel, and storing a highly volatile material, drives the per kilowatt hour costs to extraordinary heights. The island costs of energy generation in 2011 were US$0.74/75 per kilowatt hour, while on the mainland just the average cost for a similar operation runs at US$0.14/15 per kilowatt hour. The island costs are five times the average on land, with additional hidden costs as well.

The supplier is the sole source of diesel by tanker in the New England seacoast region, with additional transportation charges sometimes arbitrarily assessed if weather conditions are less than ideal. With this power position, the supplier refuses to consider advance purchase contracts and charges the maximum going rate at the time of delivery. In addition, the island is subject to increased scrutiny from the regulatory agencies, both state and federal, that have oversight in this domain. As a result, SIC was ordered in 2009 to rebuild the secondary storage structure under the diesel tanks that are designed to catch any possible spills. This project cost more than US$100,000 to design, construct, and certify. Among other challenges in this project environment was the transportation to the island of two cement trucks on a commercial landing craft to produce the concrete needed for the containers.

25.4   Evaluation of design options and the cost implications

Researching possible solutions to this design dilemma revealed several possible paths. The first option considered was using geothermal wells and the inspiration for the solution came from a university project. In 2009, Columbia University in New York City retrofitted Knox Hall, a "century-old building of landmark quality" that ultimately received the Gold certification under the U.S. Green Building Council's Leadership in Energy and Environmental Design (LEED) rating system. The virtual gutting and outfitting of the building with new heating, ventilation and air conditioning systems relied upon four 2,000 ft deep geothermal wells that "descend through Manhattan schist and draw water that is being used for heating and cooling the building's mechanical systems". Management anticipated that the utilisation of these wells will save the institution 22% in energy costs and reduce city water consumption by 47% (Columbia, 2009). However, this solution was not viable for Star Island as the costs of drilling wells into New Hampshire granite was prohibitive, and more likely than not would produce brine, not fresh water.

The second option that was discussed was the development of the John H. McCormack Post Office and Courthouse in Boston, MA, USA where reference was made to the energy systems incorporated into the plans. This project was another renovation of an existing building that resulted in the awarding of

LEED Gold certification. The key element in the McCormack project included development of roof gardens to reduce energy use and thus costs. A number of "energy-saving measures such as insulation behind the building skin and occupancy sensors" were installed to reduce overall energy costs (US EPA, 2010). However, in reviewing this project, it was clear that the engineers and designers of the building infrastructure were focused on reduction of energy, not the development of alternative sources for the building.

Other options considered included wind power but earlier studies on Star Island completed by the University of Massachusetts in 2009 indicated that at best, wind turbines could be used to offset the energy needs of the island's Reverse Osmosis Desalination unit, but consideration for energy for the entire island operation would require substantially larger turbines (Henderson *et al.*, 2005). Unlike Europe and Canada, the placement of very large wind turbines in visible off-shore environments in New England and other coastal regions of the US has met with stiff public and regulatory resistance. More importantly, the study revealed that wind velocities were greatest in the winter, which is when the island is closed.

The final option under consideration to solve this conundrum is the execution of a Solar Power Purchase Agreement with a regional vendor experienced in this type of alternative energy operation. It is also understood that prior to the signing of such an agreement, a long-term plan for alternative energy must be developed for all aspects of island operation. It does not benefit SIC if the cost per kilowatt hour of energy generation on the island drops to coastal levels but diesel transportation costs skyrocket and individuals must pay exorbitant surcharges for ferry service to the island.

25.5 Proposed design solution and costs

The proposed solution was a Solar Power Purchase Agreement. A Solar Power Purchase Agreement is a financing mechanism that facilitates the installation of solar energy systems at no upfront cost. The vendor builds the power generating system, and operates it, charging the client (SIC) a fixed rate per kilowatt hour for a specified term of years. At the end of the Power Purchase Agreement (PPA), the client takes ownership of the generation facilities. Under a PPA, the host agrees to purchase the electricity at a predetermined rate for a set period of time and at the completion of the contract has the option to purchase the system. With this arrangement, the vendor is able to access significant US federal tax incentives (30% of the total system cost) that Star Island is ineligible for due to its non-profit status. The benefits of a PPA are summarised in Table 25.1.

The vendor would design, install and maintain a complete solution for Star Island at no upfront cost to SIC. The foundation for this system is a completely revamped renewable energy grid controlled by high-efficiency, redundant inverters that are fully integrated into a 325 kWh battery storage system, a 125–250 kW solar array and new high-efficiency small generators for minimal charge support to supply 100% of the island's energy needs. Additional upgrade recommendations will address strategies for completely replacing the island diesel needs for steam generation for the kitchen and propane for the domestic hot water for the hotel. With this system, diesel demand can be lowered by 80% immediately and by 100% within 1–2 yr with a new system.

Table 25.1   Benefits of a PPA.

Key benefits	
Performance, efficiency and reliability	• Reliable, non-explosive and warranted technology • Increases the quality of electricity available on the island – a constant problem under the existing system • Increases in generator efficiency from conversion from load following to battery charging expected to be 10–15% – resulting in significant fuel savings • System produces the most amount of electricity when needed by Star Island – closely follows occupancy and needs of the island
Capital costs	• No upfront capital costs or down payment required • Receive the benefits of renewable energy today – no time spent trying to consider alternatives based on limited finances
Operating and maintenance costs	• No operations and maintenance concerns (vendor operates the system) – reduces operating expenses for Star Island
Environmental and sustainability	• No noise pollution and reduced risk of spillage during fuel transport, delivery and operation • The inverter will adjust the island's power factor closer to unity, meaning the amount of electricity produced will more closely equal the amount of energy actually required. The island's actual load will decrease by up to 11% – resulting in less energy needed and increased cost savings
Energy usage charge	• 15-yr flat rate PPA of US$0.380 per kilowatt hour with 0% escalation charge to SIC – compared with the current rate of US$0.74/75 per kilowatt hour

If the full system is not installed, the project can be built out in stages and incorporate other renewable energy technologies as desired. For example, one section of the solar array could be built first to generate power for just the hotel or the water generation plant (a very high energy consuming process). New technologies for heating water and compacting trash could also be integrated in phases.

In summary, the 15-year flat rate PPA would include these commitments:

- No down payment.
- US$0.380 per kilowatt hour with 0% escalation charge to SIC – compared with the current rate of US$0.74/75 per kilowatt hour.
- The solar array is projected to produce 275,000 kWh annually – close to 100% of the island's energy needs.
- An estimated annual PPA payment during contract of US$87,000–93,000.
- System can be purchased at Year 15 for the Fair Market Value of 5% of total system cost (US$81,250) or PPA contract can be extended.
- Initial back-up to be provided by small (25 kW) generator running at optimal efficiency.

Another major issue for the Corporation is reliability: in this remote location, it is critical to have a reliable power source that does not vary in output. The proposed system will be internally redundant, as it will include batteries, and have its own set of small generators to maintain battery charge levels during extended inclement weather and during any maintenance-related outages. It is also recommended that the existing generators in the current power plant remain on the island to provide for another 100% level of system redundancy to ensure seamless operation. The existing generator can be integrated into the battery bank, which will allow the generators to run at their peak efficiency while recharging the batteries. With the current arrangement, the generators are required to be "load following" forcing them to run less efficiently. The 10–15% increases in efficiency from the generator will reduce overall fuel costs and subsequent operating expenses for Star Island.

The island Facilities Superintendent and the vendor are also proposing upgrades to current equipment; it is paramount to identify any "low-hanging fruit" efficiency improvements that can be incorporated into the overall project. The kitchen has the highest energy use on the island (other than the wastewater treatment plant and the reverse osmosis fresh water generation operations) and is a logical place to look for efficiency improvements from refrigerators, freezers, coffee machines and other kitchen equipment. Most appliances are relatively inexpensive to replace and can offer significant decreases in electrical load, which certainly adds up quickly with an electrical rate.

25.6 Concluding remarks

Finally, the financing of this type of capital improvement has many advantages to an NGO.

As the vendor is a for-profit company that will have full ownership and maintenance obligations for the system, they will be able to access tax-based incentives that can lower the overall energy cost to a point that is lower than Star Island's current burden (with no maintenance or fuel costs). The time to gain these benefits is today to insure the best protection from failed equipment and increased diesel prices. The SIC Board of Directors and the CEO are considering the next steps in this process. A consultant has proposed the development of a 10-yr plan with the goal of being completely energy independent within the next 6–8 yr. To summarise, the challenges will never go away; it is all in how the continuing absurdity of limited resources, high expectations, aging infrastructures, and the twenty-first push for sustainability in all operations are coped with; and incorporating the acceptance of challenge and the absurd situations that are confronted on a daily basis in the fluctuating business environment.

References

Columbia University (2009) Columbia University Conversion of Century-old Building for Academic Classrooms Meets Today's Energy Standards. http//facilities.columbia.edu/knox-hall-receives-green-preservation-and-renovation-work (accessed 10 May 2011).

Henderson, C.R., Manwell, J.F. and McGowan, J.G. (2005) A Wind/Diesel Hybrid System with Desalination for Star Island, NH: Feasibility Study Results. www.sciencedirect.com (accessed 15 September 2011).

US EPA (2010) Sustainable Facilities at EPA: John W. McCormack Post Office and Courthouse. www.epa.gov/oaintrnt/documents/boston_508.pdf (accessed 15 September 2011).

Chapter 26
Case Studies: Maximising Design and Construction Opportunities through Fiscal Incentives

Paul Farey

26.1 Introduction

In Chapter 11 some of the key sustainability drivers for building owners and occupiers, as well as some of the incentives that exist to complement the required expenditure are examined. The landscape regularly shifts, in response to new legislation, funding challenges and technological developments that appear at an almost exponential rate. The resulting level of specialist advice is increasing accordingly and any proposed solution is further burdened by the cost of associated professional advice and impacts of compliance on systems and overhead. Therefore, in order to reduce abortive expenditure, it is imperative that any decision taken considers as many issues as practical, along with the results of their impact on the asset and cashflow.

This chapter reviews some of the strategic considerations, focusing on how opportunities are explored during the design, construction and operation stages drawing on the principles (regulatory and statutory framework, etc.) discussed in Chapter 11 and demonstrating using case studies how opportunities are maximised for building owners/occupiers or investors.

26.2 Strategic considerations

The design and construction process offers plenty of opportunity to embed significant value through the early consideration of available tax allowances, particularly where low-energy/carbon solutions are a driver. Early recognition of the potential benefit has the ability to reduce capital and operating costs,

Design Economics for the Built Environment: Impact of Sustainability on Project Evaluation, First Edition.
Edited by Herbert Robinson, Barry Symonds, Barry Gilbertson and Benedict Ilozor.
© 2015 John Wiley & Sons, Ltd. Published 2015 by John Wiley & Sons, Ltd.

as well as improving cashflow. Some of the key factors and considerations include:

- **Tax status of project stakeholders** – for example, landlord and tenant. Which parties pay tax and if so, at what rate? Are any of the parties UK companies, such that they could benefit from payable tax credits for unrelieved losses?
- **Key performance indicators (KPIs) and measures of success** – inputs for development appraisals; is success measured by performance against capital budget or running costs over a particular period?
- **Procurement strategy** – how much influence does the owner or occupier have? Is product specification and selection determined by the contractor under a design and build contract, whose interest is relatively short term?
- **Funding** – are budgets set based on capital cost? If so, is initial cost the only consideration? How are alternatives considered, where the capital cost may be higher, but the whole-life cost is lower? Can a top-up budget be created to fund the initial capital cost?
- **Timescales** – what are the proposed holding periods for assets: sale on completion or longer term investment/occupation?
- **Holding structure and eligibility issues** – what impact does the holding structure have on the tax position? Is the property held within a special purpose vehicle for ease of sale and Stamp Duty Land Tax in particular? Can group relief be used on allowances? Does the funding or tax domicile prevent the claiming of Land Remediation Relief or allowances?
- **Appetite of the parties** – what appetite actually exists? Will 'greener buildings' generate higher rental values or returns? Do shareholders expect corporate real estate to be low energy/carbon?

The above list is clearly not exhaustive and no particular weight or preference is given to the particular points highlighted. Stakeholders will have their own priorities for any given project and early consideration will ensure that sufficient data are available as part of the decision-making process. Opportunities to change are enhanced when considered early in the development process (Figure 26.1).

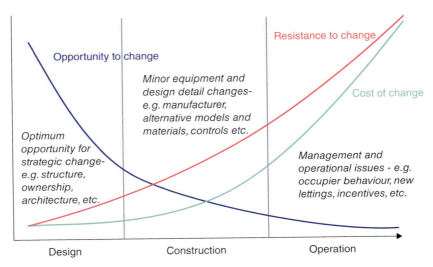

Figure 26.1   Key opportunities for tax planning.

Some of the key design considerations for property owners and occupiers are discussed in the following sections.

26.3 Capital allowances planning

The valuation of plant and machinery is a subjective exercise involving many variables. The process is complicated by the fact that there is no satisfactory legal definition of 'plant', with substantial grey areas. Achieving a maximised claim requires a detailed understanding of the legislation, case law, HMRC guidance, the construction process and procurement methods against an appropriate recognition of the risk factors.

Allowances are available for the capital cost incurred installing qualifying plant and machinery assets, including additions for associated builder's work, profit, overheads, fees and other appropriate on-costs. Problems are often experienced when trying to identify the true cost of an item, particularly where the exercise is carried out on a retrospective basis. The challenge is compounded because building contract documentation is not designed to capture information to substantiate claims for capital allowances. Engagement with the design team is essential, as consideration of various design issues and data collation at an early stage can be beneficial to any claim (Figure 26.2 and Case study 1). Analysis post-completion requires forensic review of documents without a real understanding of the reasons or nuances behind design decisions on installed assets. Some of the key opportunities include:

- Minor design changes that do not impact on building function can sometimes increase claim values with no extra construction cost.
- The correct wording of design and other documentation can help in substantiating claims for specific items.
- Changes in the format of contract documentation and placing an obligation to providing information will improve the identification and evaluation of qualifying items.
- Collating supporting information and monitoring of works while the project is in progress improves the claim level and subsequent agreement process with HMRC.

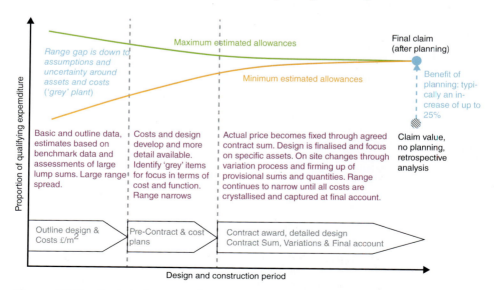

Figure 26.2 Capital allowances benefit through proactive planning.

■ Early identification and risk assessment of items which may be challenged by HMRC gives time to compile evidence and increase the chances of a successful outcome. The value of allowances can be factored into financial models and can assist with optional appraisals.

Case Study 1 – City Centre High-Rise Offices

Camilla Ltd, a UK property company, had plans to develop a city centre site for a large high-quality office building. The project was one of the most ambitious undertaken by the company to date and considerable planning went into the holding structure and funding over the development period. Because some of the equity was provided by investors subject to the higher rate of income tax, the company was keen to ensure that the capital allowances position was maximised. The development also had a strong 'green theme' with Camilla's stated desire for low energy underpinned by planning and building regulation requirements.

The company had benefitted from capital allowances on previous developments and employed a special advisor prior to the tender process. The advisor worked closely with the project team to provide the following support through the design, procurement, construction and letting process:

■ Engagement with the designers – established the basis for claiming allowances with designers and impactive factors for a successful claim with HMRC. Stressed importance of linking impact of business requirements upon design solution to support the concept of 'function' for plant tests and energy-saving assets for Enhanced Capital Allowance (ECA) purposes.
■ Creating information routes with the Quantity Surveyor and Main Contractor – creating systems to allow for the request, supply and breakdown of sensitive cost information to support claim.
■ Identifying and addressing claim risks – highlight assets and treatment that might be contested by HMRC. Identify further information or other necessary evidence to support the resulting claim. Reject potential assets where no defensible position can be established. Review potential risks and approach with Camilla.
■ Regular reporting – at key project and accounting milestones, provide updated reports on levels of estimated or identified allowances for management reporting, forecasting or tax payments on account. Highlight actions taken to date, areas of risk and opportunity, impact of potential legislative change and planned future actions along with delivery dates.
■ Value engineering input – contribute to proposed cost reviews and potential variations by highlighting the impact on available allowances. During this stage, the advisor was able to differentiate between two options for a planned equipment change to demonstrate that one marginally more expensive option would generate a cashflow benefit through first year allowances for ECA purposes.
■ Lease incentives – as the development progressed, the advisor was able to work with Camilla and the letting team to model the impact of potential lease incentives to the post-tax positions of landlord and tenant to determine the optimum position based on a variety of inputs.

The overall result was a maximised, low-risk claim for capital allowances that was agreed in full with HMRC. Since all assets claimed were fully justified in terms of eligibility and quantum, the advisor was able to demonstrate that the claim was some 18% higher than for similar projects where a retrospective forensic analysis of expenditure was made.

26.4   Enhanced capital allowances (ECA)

In addition to the regular capital allowances planning opportunities identified above, sustainability requirements can give rise to additional areas for maximising the cashflow position of proposed capital expenditure. Eligible technologies are determined from the Energy Technology List (ETL) and are typically product specific. Therefore, without planning, only a forensic review of installed assets will determine whether ECA compliance has been achieved.

However, where combined heating and power installations are to be installed, ECAs are only available where the system complies with the Combined Heat and Power Quality Assurance (CHPQA) Scheme which ideally requires review at design stage. Retrospective certification is possible but only where particular performance and efficiencies are delivered, something easier to tackle before the equipment is procured and installed.

Engaging with the services consultant, main contractor and mechanical and electrical (M&E) sub-contractor gives building owners and occupiers greater tax efficiency in the procurement period. Late changes to design or products result in potentially expensive variations to the project cost; the optimum window of change arises earlier in the process, meaning that consideration of the impactive factors needs to occur earlier in the development period.

In practical terms, the short-termism of capital cost is likely to be a deterrent to sustainable design; a sophisticated, longer term approach of whole-life cost will provide a more balanced picture reflecting economics, energy and carbon. Proactive planning should take into account the following key metrics:

- **Design workshops** – early focus with the design team; raising awareness and understanding with key decision makers; involve client representatives; stimulate open discussions where tax, energy and water efficiencies can be achieved through ECA qualifying technologies; address client specific issues and ensure ECA savings are a measure of project success.
- **Implementation of ECA strategy** – implement a strategy that fits comfortably with the project brief and the philosophy and objectives of the design team.
- **Work with consultants and suppliers** –discuss qualifying technologies and establish any areas where the design can be more tax efficient; incorporate recommendations into M&E specifications; draft appropriate clauses for tender or procurement documentation.
- **Manufacturers** – work with the manufacturers to ensure that sufficient detail is available to support a claim for ECAs.
- **Monitoring** – as the project progresses, monitor ordered and installed equipment to ensure qualifying technologies are installed.

The ECA agenda should be driven as an integral part of the project. To ensure claims are maximised, an open dialogue needs to be maintained between all the key stakeholders in the project. This is especially important during value engineering exercises that could remove ECA qualifying equipment without considering the post tax position. An example of the benefit is shown in Case study 2.

Case Study 2 – Chas Ltd

Chas Ltd is a company with owned and leased properties throughout the country. In recent years, the company has experienced difficult trading conditions, resulting in tax losses. The company has a strong corporate social responsibility (CSR) agenda, with a stated aim to shareholders to reduce carbon consumption. A change in legislation recently required the company to replace R22 refrigerant within its air-conditioning systems across the estate.

The company was struggling with cashflow and had concerns as to how it would balance its CSR obligations to shareholders as well as funding the enforced intensive capital programme.

Chas Ltd appointed an advisor to determine how best to manage the process and mitigate the costs associated with this statutory compliance. As part of this process, the advisor convened a workshop with all the key stakeholders in the project: Sustainability Manager, Estates Team, Procurement, Engineering and representatives from the Treasury team to cover finance and tax.

Some of the equipment solutions under consideration were on the approved ETL, such that ECAs would be available. The group quickly realised that, as they were unable to benefit from the 100% first-year allowance, the 19% payable credit could be quite valuable. Equipment costing £1,000 would generate a credit of £190, resulting in a net cost of £810. This is in addition to the running cost savings as a result of the equipment being energy efficient in the first instance. The group agreed that as a policy they would pursue low-energy machinery and equipment that was eligible for ECAs as part of the project.

The workshop then focused on ensuring that the supply chain systems delivered the appropriate support for the tax team to claim ECAs on the capital expenditure incurred. During this discussion, it became clear that a trip hazard existed at the procurement stage: the buyer advised that their KPIs were measured around purchasing at the lowest initial capital cost and that given the choice between two units that were seemingly compliant, they would always opt for the cheapest one, even if this was not on the approved ETL.

Further, as only capital cost was measured, no consideration could be given to any efficiencies through running costs or available tax relief.

The solution was for the board of Chas Ltd to create a 'green fund' as part of its commitment to low energy and sustainability. This fund is now used to support the differential of more expensive capital items subject to a number of qualifying criteria. Approval for funding is given only where the additional capital cost is more than offset through improved operational costs and the cashflow of ECAs/payable credit over ordinary Special Rate pool relief. Any funding also specifically mitigates the impact of purchasing dearer 'green' equipment for the purposes of the buyer's KPIs.

26.5 Land remediation relief (LRR)

The main challenges taxpayers face are around the definition and extent of eligible expenditure and, in particular, the interaction with sub-structure solutions and normal site preparation costs. Further difficulties arise around the nature of remediation and whether replacement materials qualify: each solution is dependent on the particular nature of the development and that no two projects are ever the same.

The nature of the pollution is also an issue and since 1 April 2009, HMRC have acknowledged that LRR is also available for the removal of contamination arising from:

- naturally occurring arsenic and arsenical compounds;
- radon;
- Japanese knotweed.

These are the only naturally occurring contaminants permitted and are defined within the corporation tax legislation [CTA09/S1145 (2)(b) & (3)]. The application of this is strict and does not extend to other contaminants even 'by analogy'.

There is no definitive list that prescribes a list of items that qualify for LRR. The legislation provides a very broad definition against which expenditure can be claimed. In development terms, it might typically include:

- Excavation and disposal costs for contaminated soils
- Backfill to excavated contaminated areas
- Onsite remediation techniques such as bioremediation
- Capping layers
- Gas protection measures
- Preparatory costs
- Related professional fees.

Due to the specialist nature of the works, it is not always clear from contractors' documents which items relate to remediation costs. The fundamental test applied should always determine whether the cost would have been incurred, had the land not been in a contaminated state. It is highly interpretative and the determination and extent as to whether a cost is eligible depends upon site-specific circumstances.

The compliance demands of corporation tax self-assessment put particular burdens of proof upon the taxpayer. Typically a claim for LRR will need to include:

- Site investigations and/or remediation strategies
- Company expenditure records
- Actual cost data for relevant items of work.

A claim for LRR will comprise the necessary evidence extracted from this documentation; any omission of these crucial data may adversely impact on the claim success. Table 26.1 and Table 26.2 highlight how the benefit is claimed by developers and investors.

26.6   Value added tax

Value Added Tax (VAT) is effectively a tax determined by the EU, albeit administered locally. As such, the Government's ability to utilise the legislation as part of its arsenal towards low carbon is difficult without breaching state aid rules. The opportunities to treat expenditure as reduced or a zero rated supplies of VAT generally apply irrespective of the attention paid to sustainable design.

Table 26.1   Example of LRR for developer undertaking land remediation prior to sale.

A developer buys contaminated land in 2012. He remediates and develops the land and sells the development in 2014.	EXAMPLE 1	£000s	£000s
	Sale Proceeds		20,000
	Purchase price of land	500	
	Qualifying remediation	5,500	
	Other infrastructure costs	1,000	
	Building cost	10,000	
	Total cost in 2012		(17,000)
	Taxable profit before LRR		3,000
	Additional 50% relief on LRR costs		(2,750)
	Taxable profit after LRR		£250

Cash saving as a result of LRR = £2,750K × 21% tax rate
= £578K (10.5% of LRR cost of £5,500K)

Table 26.2   Example of LRR for investor undertaking land remediation works.

A landlord buys land to build an office to rent out. He remediates the land in 2013 and sells the investment in 2018.	EXAMPLE 2	£000s	£000s
	Purchase price of land	500	
	Qualifying remediation	5,500	
	Other infrastructure costs	1,000	
	Building cost		(10,000)
	Total capital cost in 2013		17,000
	Investment sale in 2018		30,000

2013   Tax relief claimed: £5,500k × 150% × 23% tax rate= £1,898k (35% of LRR cost)
    Capital account reduced by £5,500k to £11,500k.

2018   Additional Capital Gains Tax paid: £5,500k × 21% tax rate= £1,155k

Benefit is additional 50% at higher rate of tax, plus cashflow

For residential or certain charitable developments, consideration should be given to the following energy-saving materials subject to a reduced rate of VAT (VAT Act, 1994):

- insulation for walls, floors, ceilings, roofs or lofts or for water tanks, pipes or other plumbing fittings;
- draught stripping for windows and doors;
- central heating system controls (including thermostatic radiator valves);
- hot water system controls;
- solar panels;
- wind turbines;
- water turbines;
- ground source heat pumps;
- air source heat pumps;
- micro combined heat and power units;
- boilers designed to be fuelled solely by wood, straw or similar vegetal matter.

Expenditure incurred on any of the assets identified above qualifies for the 5% reduced rate of VAT. This will be a factor where the overall supply of the installer is exempt with irrecoverable input tax, resulting in a 15% cash saving.

26.7   Taxation anti-avoidance

As a general rule, the tax position will be determined by the nature, extent and timing of expenditure incurred in construction works. Whilst pre-planning is an acceptable part of a taxpayer's right to organise their finances, where tax treatment drives the design, there is a risk that this may be construed as avoidance. The remnants of loopholes and potential lucrative planning opportunities in the UK tax system are red-flagged with extensive anti-avoidance provisions added through various statutory instruments and legislation rewrites. Some of the areas to look out for include:

- Proposed general anti-avoidance rule (GAAR) – where the activity is structured to avoid tax (Gauke, 2011).
- Prevention of claiming a 100% first-year allowance through ECAs, where a Feed-in Tariff (FIT) or Renewable Heat Incentive (RHI) payment is taken.
- Restrictions on tax losses generated for income tax payers. This may impact on equity investment which is geared around high and accelerated levels of tax relief, say through capital allowances.
- Restrictions on qualifying expenditure for connected parties' transactions or sale and leaseback arrangements.
- Inability to claim plant and machinery allowances on certain leased fixtures under the long funding lease rules. This does not affect most commercial property.
- The 'polluter pays' principle to restrict claims for LRR where the tax payer polluted or has contributed to the contamination of the site.

In many cases, the client and design team will not come across these provisions, however for particularly complex projects, especially where there are large levels of relief, professional advice should be obtained.

26.8   Conclusion

Sustainable design is clearly a permanent addition to the construction agenda. With real momentum behind it in terms of Government policy, legislation, planning requirements and the focus of economic performance; building owners and occupiers cannot ignore the resulting impact upon the design, construction and operation of property. When rising energy costs and other impactive factors such as corporate social responsibility and shareholder expectations are taken into account, then clearly more thought and rigour will be required when incorporating low-energy solutions into buildings and real estate.

Buildings are intricate and active assets; the days of fitting mechanical and electrical services into an architectural box have passed; although the process has not turned full circle, the services requirement now significantly affects the architectural form, for example the impact of solar gain on a building's thermal mass and the requirement to keep cooling demands as low as possible.

It is important not to forget that installing renewable and low-energy solutions actually produces operational benefits on running costs. Aside from the reduction in cost of energy used, there are further economies in terms of:

- Reduction in indirect taxes paid – for example through Climate Change Levy for commercial users and the CRC Efficiency Scheme.
- Positive income generated through FIT and RHI.
- Reduced income or corporation tax liability through allowances or reliefs: sustainable solutions often (but not always) give rise to enhanced or accelerated levels of relief.
- Payable credits from HMRC where enhanced allowances are unable to be utilised in the short term.

Whilst design and construction is challenging enough, securing funding is critical to the development process. The pressure on designers and advisors to comply with regulatory and economic demands for low-energy solutions has to be balanced with budgetary constraints, usually on the basis of short-term capital expenditure. In addition occupier demand puts further load on the drivers. Against the backdrop of balancing commercial 'wants' with green 'needs' the client has to ensure that the business case satisfies the requirements of lenders and investors alike.

The challenge is compounded as a result of the shifting landscape of legislation, taxes and energy sources. In order to fully balance the equation, taxpayers need to take into account fully burdened cost, reflecting timing and magnitude of cashflow as well as the potential to earn income or mitigate outgoings as a means of preserving real profit.

Clients and advisors who do not recognise the risks and opportunities associated with the fiscal demands of sustainable design are in real danger of losing out, in terms of economics and reputation.

References

Gauke, D. (2011) Written statement 21 November 2011. http://www.publications.parliament.uk/pa/cm201011/cmhansrd/cm111121/wmstext/111121m0001.htm#1111212000181 (accessed April 2012).

VAT Act (1994) Schedule 7A, Group 2 – Installation of Energy-saving Materials. http://www.legislation.gov.uk/ukpga/1994/23/contents (accessed April 2012).

Chapter 27
Mapping Sustainability in the Quantity Surveying Curriculum: Educating Tomorrow's Design Economists

Chika Udeaja, Damilola Ekundayo, Lei Zhou, John Pearson and Srinath Perera

27.1 Introduction

The climate change debate has generated considerable interest in the sustainable development agenda throughout the world. The UK, like most other nations, is becoming increasingly aware of the significance and value of having a sustainable environment policy (Khalfan, 2006). In the built environment, the challenges are massive, given the size of the construction industry which accounts for 8% of the UK's gross domestic product; consumes an enormous amount of resources with a major impact on the manufacturing industry that creates products for construction projects and society at large. The built environment's contribution to the economic well-being of a country, the social well-being of people, and the impact on the environment is hugely significant (Cowling *et al.*, 2007; BERR, 2008). Theron (2010) estimated that the built environment in its widest sense is responsible for 40% of CO_2 emission, as well as 40% of all energy used. The Kyoto Protocol, EU Emission Scheme, recent changes in building regulations, and Climate Change legislation are a growing recognition of the need to minimise the consequences of human activities on the environment. These initiatives have created the need for a major reform in the UK construction industry with significant implications for the educational systems.

The green agenda and construction education are intricately linked (see e.g. Walton and Galea, 2005; Cotgrave and Alkhaddar, 2006; Hayles and Holdsworth, 2008; Theron, 2010; Ekundayo *et al.*, 2011). The rationale, therefore, for embedding green issues within the construction curriculum is a powerful imperative for change. This is mainly as a result of policy drivers and in some cases existing research but the response from the academic community is less clear. However, it is increasingly recognised that the curriculum should incorporate sustainability, in order to produce graduates that will confidently take care of the environment and not damage it for future users. Hayles and Holdsworth (2008) argued that the twenty-first

Design Economics for the Built Environment: Impact of Sustainability on Project Evaluation, First Edition.
Edited by Herbert Robinson, Barry Symonds, Barry Gilbertson and Benedict Ilozor.
© 2015 John Wiley & Sons, Ltd. Published 2015 by John Wiley & Sons, Ltd.

century is seen as the time for UK universities to embrace new ways of working. This is especially important if the educational system is to continue to be competitive and also meet the needs of its ever demanding stakeholders. A major challenge for the universities is the ability to provide products, and to an extent services, that meet stakeholders needs and aspirations, especially in relation to the sustainability agenda.

This chapter addresses how sustainability is incorporated into the education curriculum, to address professional competencies in construction related programmes with specific reference to quantity surveying (QS). It proposes a strategy to support the development of desired competencies in sustainability.

There is a growing interest in the sustainability agenda and to identify the quality and quantity of sustainability-related materials within the QS curriculum. This chapter focuses on QS degree programmes to identify broad and specific changes needed to develop competencies relevant to QS practices. First, it attempts to map the sustainability activities within the QS programme. To achieve this, a review is undertaken to determine the main areas of interest in sustainable construction particularly in QS, design or construction economics. Primary data collected using case studies are used to qualitatively map the extent of sustainability-related features within the curriculum in the QS degree programmes. The findings are analysed to determine the extent of sustainability-related topics or areas within the curriculum.

27.2   Literature review on sustainability issues

The terms "sustainability", "sustainable development" and 'sustainable construction' are words that have become common currency in recent years. They are phrases that are interpreted in different ways but the underlying principle is one of doing things differently to safeguard the environment. Numerous definitions have been proposed but there is no universal agreement of what exactly sustainability is meant to be within the curriculum. This section will review the fundamental change of Sustainable Development in Higher Education and identify the challenges the QS programme will face if the sustainability agenda is not addressed in the curriculum.

Greening the curricula

This section explores the views of some academics on sustainability-related education within the built environment curricula. There are a number of studies carried out to explore the opportunity to embed sustainability agenda into the built environment curricula (Perdan et al., 2000; Fenner et al., 2005; Cotgrave and Alkhadder, 2006; Murray et al., 2006; Cowling et al., 2007; Hayles and Holdsworth, 2008; Sayce et al., 2009; Iyer-Raniga et al., 2010). These studies have been carried out to encourage staff to make commitment to sustainability by making changes to their modules or provide new modules for student learning. As early as 2000, Perdan et al. attempted to adopt a multidisciplinary approach to teaching sustainability for engineering students at the University of Surrey and they developed IT-based learning materials and case studies to facilitate understanding of concepts of sustainability and how solutions could be developed. Fenner et al. (2005)

reviewed the education for sustainable development in the Department of Engineering at the University of Cambridge and encouraged students' self-reflective learning processes to obtain solutions for the challenges of sustainable development.

Cotgrave and Alkhaddar (2006) reviewed the undergraduates' construction management curricula at Liverpool John Moores University and established that sustainable design and technology was superficial during the final year. Murray et al. (2006) implemented a full curriculum review to identify the gap in provision of sustainable construction education at Plymouth University. The study found that although discipline-specific environmental aspects were included in the curriculum, few generic aspects of sustainability such as citizenship or poverty were also covered.

Cowling et al. (2007) argued that education for sustainable development has become increasingly significant within the built environment curriculum at Kingston University. They explored students' familiarity, understanding and interest in sustainable development and how these developed over their time at the university. The university's emphasis on sustainable development provided an opportunity to contribute greatly to the students' awareness of the subject given that they are enrolled on courses with interest but often with a low knowledge base. Hayles and Holdsworth (2008) conducted an action research project at RMIT University, Australia to embed sustainability agenda into the core curriculum of the undergraduate programme at the School of Property, Construction and Project Management. The results showed how sustainability issues were embedded into three new modules. Iyer-Raniga et al. (2010) conducted research on construction management students at RMIT to compare sustainability activities between Melbourne and Singapore. The findings showed that there are no significant differences in the perceptions, knowledge and understanding of sustainability issues between the two sets of students.

While the list of previous research in the area of sustainability education is not exhaustive, it does indicate the wide range of challenges faced in incorporating sustainability-related education within the QS programme.

Challenges facing the QS professional

Previous research provided some understanding of the meaning and significance of QS (Lee and Hogg, 2009; Perera et al., 2010; Simpson, 2010). The role of the quantity surveyor as suggested by the Royal Institution of Chartered Surveyors (RICS, 1971) cited in Nkado and Meyer (2001) is associated with measurement and valuation. They argued that quantity surveyors provide cost management services for construction projects in the context of forecasting, analysing, planning, controlling and accounting. Others suggested that competent quantity surveyors must have a range of skills and knowledge which can be applied in a range of projects and organisations. What is clear is that the roles of quantity surveyors have become extremely diversified, to match the changing needs of employers (Ashworth and Hogg, 2007). In the UK, a number of construction companies have rebranded themselves to respond to the needs of the sustainability agenda.

Achieving progress towards sustainability is critical to the future well-being of society which has long being recognised by the Higher Education Funding Council for England (HEFCE, 2010). They have placed sustainability as a major objective both organisationally and within their sphere of influence and activity. It is suggested that universities have a major role to play in tackling the sustainability agenda (Jones *et al.*, 2008). The universities and colleges are in a unique position to lead the way and change the awareness of sustainability agenda (HEFCE, 2010). It is therefore expected that universities will be at the forefront of embedding sustainability both within their own institutional values and the curricula that they deliver.

Dixon (2009) argued that there has been progress made in recent years in linking sustainability into professional practice globally but suggested that the key barriers are lack of knowledge and expertise from graduates and experienced professionals. At the EcoBuild conference in 2010, Paul Morrell, the then government adviser on construction, stated that the government's greatest concern is how to satisfy the carbon and green agenda. He went on to state that the construction industry does not have the capacity to meet the sustainability agenda because the universities are not producing graduates with adequate knowledge.

It is therefore crucial for the construction sector to make significant contributions to sustainable development. Architects and engineers are providing leadership in sustainable construction in the world. However, there is lack of evidence showing that QS professionals are demonstrating sustainability leadership in the business environment. It is therefore of paramount importance to identify what types of new skills are required by quantity surveyors to tackle the sustainability agenda. The RICS (2007) review identified competencies and new skills required for QS to provide sustainability services through the life span of a building project. The areas identified are: value for money, whole life costing, cost of alternative materials, renewable energy schemes, recycled content schemes, the ethical sourcing of materials and labour. Other key elements also discussed in the literature include: sustainable procurement and sustainability performance measurement. Furthermore, the RICS also identified specific responsibilities for QS in sustainable development:

- Protecting and enhancing the natural environment
- Encouraging the sustainable use of resources
- Reducing waste generation and responsible disposal of waste
- Reducing energy consumption
- Promoting community development and social inclusion
- Minimising any negative social or environmental impacts of development
- Promoting sustainable land use and transportation planning and management
- Promoting sustainable design, development and construction practices, including whole-life costing.

However there is a huge knowledge gap for those studying QS in higher education. Embedding sustainability within the QS curriculum will require an exploration of its three spheres: economic, environmental, and social dimensions. In addition, knowledge of regulatory and technological issues is important as cross-cutting themes. Dale and Newman (2005) argued that the key to achieving these skills is adaptability, and the ability to change, particularly in an evolving economic climate with threats of climate change. Clearly universities operating in the built environment have a vital role in shaping the future pattern of practice and policy in relation to the sustainability agenda. So, it is essential to map the curriculum to capture the sustainability content. This will enable staff to educate and, inspire and influence the new generation of quantity surveyors or design economists to be tomorrow's leaders in sustainable development.

27.3   Development of the Sustainability Framework

In developing an appropriate strategy that will embed sustainability education within the built environment curriculum, a case study approach was adopted using three methods of data collection: use of published sources to identify the key components of sustainability education; structured interviews with academic staff involved in the decision making to establish the key categories of sustainability relevant to built environment professionals; and document analysis to determine the extent of sustainability topics in the curriculum based on the module descriptors. The coverage of sustainability in the current QS curriculum was identified and ideas on how to improve sustainability education in the QS degree programme were suggested.

The case studies include four RICS accredited QS degree programmes. The curricular of the programmes (module specifications) were mapped at detailed level using amount of time spent, that is module credits (as a depth measurement). The ensuing mapping was then verified for accuracy and consistency with programme directors responsible for delivery of these programmes. The four case studies selected were leading QS honours degree programmes in the UK all accredited by the RICS. There are a total of 360 Credits equivalent to 3,600 hours of learning in a degree programme. The depth measure reflects the amount of time spent on achieving a competency. In degree programmes, time spent on achieving module outcome is stipulated as Credits. Where 10 hours spent is considered as 1 Credit, a typical 20 Credits point module reflects 200 hours of learning by the student. This constitutes direct contact with formal teaching; lectures, seminars, tutorials as well as students' expected study time on the module content (time spent on their own). The depth measure is only indicated at competency level and not at topic level as it is impractical to stipulate expected number of study hours spent at a detailed topic level. A percentage score is used to indicate the proportion of time spent on each competency to provide a valuable measure to understand the relative time spent for each competency. The depth vector scale mappings of the four case studies were initially carried out using the respective module specifications of programmes. The results were sent out to the programme leaders of the degree programmes for necessary adjustments and validation. Descriptive statistics such as mean and percentage scores were used to analyse the results of the case studies as a conceptual benchmark.

The literature findings, document analysis and the interviews led to the development of the final sustainability framework which identifies the knowledge areas relevant to the QS degree programme and the profession. The framework has been developed in the light of the current and future roles of the professional quantity surveyor as informed by the sustainability agenda. According to the findings from the research, QS graduates will need to have awareness and knowledge of the issues identified in the framework (though to differing levels of detail) to be capable of delivering their professional responsibilities in the built and natural environments now and in the future. The refined framework (Table 27.1) categorises the sustainability-related knowledge areas relevant to QS education into six main categories (high level categories) with several sub-categories (low level categories).

Category A – background knowledge and concept

The low level categories identified in the framework under high level category A are relevant to QS education and should be taught but it is questionable whether they are being taught under the current education system. There is a need to understand the principles of sustainable development, to have a background understanding of climate change and global warming issues which most students are already aware of. It is suggested that the latter should not take more than half a lecture and should not be taught in too much detail. Students need to be taught the history that led to sustainable development, and how it linked to climate change and global warming issues. The sustainable construction concept should have greater emphasis in QS education. Students need to know the link that exists between all the identified sub-categories and the roles of QS, why they are learning these and their application in the industry. It is agreed that the identified sub-categories are exhaustive and the advisory role of QS in sustainable development should be taught in more detail and greater depth under background knowledge and concept. All the sub-categories should always be linked to QS roles.

Category B – policies and regulations

The only module that will address the sub-categories identified is the Sustainable Development module taught at Level 4 or possibly any of the Technology modules which will not be addressed to a satisfactory depth. Building regulations and Code for Sustainable Homes are not covered although Energy Performance Certificate (EPC) might be covered in the Building Services optional module taught at Level 5. Category B is vital and should be covered in the QS education especially the Building regulations and Code for Sustainable Homes which are central to the degree programme. QS students have to be able to advise the client accordingly. Students should also be familiar with the sustainable construction strategy and sustainable procurement action plan. However, the students are never examined on these topics and there have been possibly no exam questions in the past. Testing their knowledge on how they can advise clients in these areas is therefore important. Even though there is a piece of course work on the Sustainable Development module taught at Level 4, it is not QS specific.

Category C – environmental issues

Most of the sub-categories identified under this main category are not covered and at best it tends to receive a superficial treatment. Part-time students are sometimes aware of these issues based on their experience at work. Moreover,

Table 27.1 Sustainability framework relevant to QS degree programme.

SUSTAINABILITY FRAMEWORK

	CATEGORY A – BACKGROUND KNOWLEDGE AND CONCEPT	CATEGORY B – POLICIES AND REGULATIONS	CATEGORY C – ENVIRONMENTAL ISSUES	CATEGORY D – SOCIAL ISSUES	CATEGORY E – ECONOMIC ISSUES	CATEGORY F – TECHNOLOGY AND INNOVATION
HIGH LEVEL CATEGORIES	Sustainable development overview and principles	Changes to Building regulation, e.g. Part L (energy efficiency) and Part F (means of ventilation)	Protecting and enhancing the built and natural environments	Corporate Social Responsibility (CSR)	Cost planning and management	Renewable energy technologies (Photovoltaic, Wind Turbine, Geothermal, Biomass, etc.)
LOW LEVEL CATEGORIES	Climate change and global warming issues	Code for Sustainable Homes	Environmental Impact Assessments (EIA)	Ethical issues such as ethical sourcing of materials and labour, for instance	Value management or engineering (cost of alternative materials and designs)	Green Building Materials
	Impact of the construction industry on the environment	Energy Performance Certificate (EPC)	Environmental Management Systems: ISO 14001	Equity and social justice	Sustainable procurement strategies	Rain water harvesting and Grey water collection systems
	Sustainable construction concept	The Kyoto protocol	Environmental Assessment Methods: BREEAM, LEED, Green Star	Community development and social inclusion	Feasibility studies	Professional and management software packages such as BIM, etc.
	Role of QS in sustainable development	Relevant EU Directives such as the EU climate policy, EU ETS, etc.	Reducing energy consumption, that is, emitted and embodied	Health and safety	Whole-life appraisal/Life cycle costing	Modern methods of construction: offsite production, use of precast material, lean construction, etc.
		Climate Change Act	Reducing greenhouse emission such as methane, carbon, nitrous oxide and refrigerant gases	Employment, training and education	Financial incentives (such as subsidies, climate change level, aggregate tax, carbon credit, Brownfield land tax, etc.)	Passive design methods such as day lighting, intelligent facades, carbon storage and offsetting, etc.
		Sustainable Construction Strategy	Carbon Agenda (Carbon footprinting, Zero Carbon, Retrofit)	Social assessment methods (e.g. Design Quality Indicators, KPIs and benchmarking, etc.)		Supply chain management
		Sustainable Procurement Action Plan	Waste reduction principles (recycling, reduction, reuse, effective design)	Cost Benefit Analysis (i.e. impact of human factors on the community)		Effective information control and management (using e-business)
			Brownfield development			
			Natural resources, renewable and non-renewable materials			
			Water usage and Sustainable Transportation Plan			

how much a quantity surveyor has to know about this category is questionable. Having general knowledge and awareness of the issues in this category may be enough. For example, Environmental Management Systems and Environmental Impact Assessment are relevant to a contractor, architects, other designers and clients who tend to be more aware of these issues. QS students need awareness of some of these issues in the Technology lectures but the traditional role of QS which involves the economic aspects still needs to be brought to focus. A minimum amount of understanding is required for this category so that a quantity surveyor could be effectively involved from project inception to completion.

Category D – social issues

Public sector clients are more likely to be interested in Category D, as private sector clients are generally more interested in cost. A participant in the study argued that "Everyone knows the importance of local sourcing of labour and materials to support the local economy for instance, this is very obvious, but I am not interested in the level at which we teach this to QS students". Cost Benefit Analysis and other Social Assessment Methods are important and would be of interest to QS. It was further argued that the environmental issues, polices and regulations are more important aspects than social issues. Even though there is a sympathetic view on the impact of social aspects sustainability, the other areas are considered more relevant to QS education due to the time factor. Private sectors will need to be educated when it comes to social issues as they are more interested in cost, not the social impact of their development.

Category E – economic issues

These are far more relevant and very important to QS education; cost planning and management, value engineering, feasibility studies, life cycle costing and financial incentives. Quantity surveyors should be fully aware of the financial incentives available so they can encourage their clients to use different sustainable technologies. Good examples include households that received some financial incentives as a result of using ground source heating. New graduates having this sort of up to date knowledge to take into the industry will be very beneficial for the companies they work for and their clients. The sub-category list for the Economic issues includes all relevant topics that should be taught under economic aspects of sustainability.

Category F – technology and innovation

All the sub-categories identified here are relevant and should be taught in line with Category E using the analogy of Measurement and Technology module where the students are taught the relevant construction techniques and later taught how to measure them. The different technologies that could achieve sustainable development should be taught to QS students in parallel with how to measure and cost them. This way the quantity surveyor will be able to advise clients on the life-cycle cost implications of sustainable design technologies, which is the ultimate goal.

27.4   Mapping of Sustainability Education in QS Degree Programmes

A method of mapping sustainability education to curricular was developed as there is no standard method to compare the level of attainment of sustainability. A scoring approach was devised to systematically analyse the extent of mapping of sustainability education to individual modules of four RICS accredited QS degree programmes (Case studies A, B, C and D). These results are presented in the following section.

Level of Sustainability Education in QS degree programmes

The analysis undertaken was to establish the extent of sustainability education in the different QS degree programmes. The analysis of the four case studies revealed that there are considerable variations in the degree programmes on how sustainability education is taught and delivered (Figure 27.1). Figure 27.1 shows a massive difference between the highest (Case study C) and the lowest (Case study B) coverage. The reason for this significant variation is attributed to the fact that Case studies C and D are considered as research intensive universities. Incorporating sustainability in their curricula was much easier and less demanding because the majority of the staff are active in research. Individual module contents also varied in greater detail than the generic comparison presented here.

Framework mapping in QS degree programmes

This research also considered how the formulated framework was mapped within the degree programmes (i.e. Case studies A, B, C and D). These also revealed massive variations in the various categories (Figure 27.2). Figure 27.2 shows that Category F (i.e. Technology and Innovation) received the most

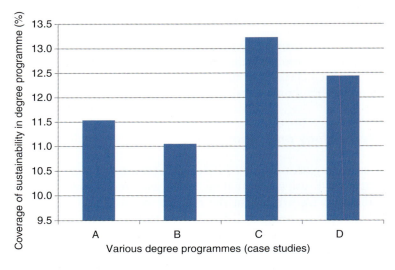

Figure 27.1   Coverage of sustainability in QS degree programmes.

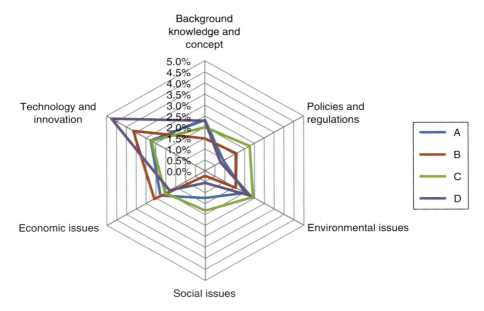

Figure 27.2   Sustainability framework mapping in the four case studies.

attention and understanding by the staff and the programmes. This can be attributed to the argument that there is some synergy between sustainability and technology/innovation in the construction field. Figure 27.2 also reveals that Category A (i.e. Background Knowledge and Concept) was also high on the agenda of most universities. However, it seems limited in Case study B which is a less research intensive university, and staff do not feel confident to teach contemporary issues because of lack of up-to-date knowledge and skill. This is one of the factors that hamper and prevent adequate coverage of sustainability education in the curriculum.

Mapping of sustainability education at various Stages (i.e. Levels 4, 5, 6)

It was considered imperative to consider at various stages of students' development how the QS degree programmes mapped the sustainability agenda.

The next section presents a summary of the sustainability education, QS degree programme and sustainability framework, extent of coverage and ideas on promoting sustainability education in QS degree programme based on the literature review findings and content analysis of the interviews conducted.

Summary of key findings

- There is no prescribed threshold benchmark standard for achieving a decent level of sustainability education at undergraduate level.
- There is a clear lack of understanding or common agreement of what sustainability construction is or should be and this is hampering a structured agenda with the QS programmes.

- Although this research work has formulated a framework, in practice (i.e. universities) there are no detailed specifications to indicate what content should be covered to achieve a reasonable knowledge of sustainability for graduates.
- Different universities aim to achieve sustainability education at different levels, based on their own interpretation.
- There is no standard way or formal mapping process of sustainability education available for universities in curricula development or revision.
- There is no formal training for staff who consider themselves out-of-date to acquire current and contemporary knowledge and skills on sustainability.

27.5   Discussion and conclusions

Awareness is growing amongst the general public, encouraged by politicians at local, national and international levels, of the need to reflect upon the relentless consumption of resources by the growth of the built environment.

The industry will not survive by mere reflection and associated promises but, rather, by making best use of those scarce and sometimes non-renewable resources and by seeking sustainable solutions for the long-term future. Those involved in the construction industry will be key players, particularly the quantity surveyors with their perspective on the economics of design or construction which will be crucial. However, this "economic perspective" must be expanded beyond capital cost of construction (the traditional boundary of their skill) to embrace life time cost of the buildings, districts, cities, together with the infrastructures which serve and link them in the built environment.

This chapter has examined the core body of knowledge currently taught to students on four QS degree programmes. It is held that these graduates, together with other construction professionals, will be responsible for shaping and managing the built environment in the coming decade. Through a series of interviews with key staff, that is those responsible for developing and directing the programme at universities, a set of criteria was created, within broad categories referred, which could be tested on programme leaders from a sample of four universities from across the UK, each offering QS degrees. The study revealed the role which sustainability plays in the undergraduate studies at all four, and thus the importance attached to it. Six key categories were identified (A–F) which extended beyond background knowledge and the purely technical to cover political, social, economic and general environmental issues – in line with the "global" and cross-discipline nature of the problem of sustainability.

The study indicates that there is quite a large sustainability-related void in the education of student quantity surveyors, and quite possibly those in other construction disciplines. The total percentage of the curriculum devoted to Categories A–F range between 11% and 13%. As may be expected within a QS tradition, technology and economic issues (in this case chiefly a reference to cost-related capabilities) tend to be the areas where there is most concentration of teaching. Only two of the four universities focus on the broader environmental issues. Policies, regulations and social issues appear to receive the least attention.

Discussion with participants has indicated two possible causes of the void. First, it appears that realisation of the very real threat of sustainability is only just becoming apparent to those in academic institutions and to the professional bodies, who to a significant extent direct the pattern of the curriculum. Secondly, a limited number of academics have enough detailed knowledge of sustainability-related issues to incorporate into the subject confidently within the materials they deliver. They themselves were educated when the sustainability debate was not a top priority in the academic agenda. To some extent, education on matters of sustainability needs to be extended up the chain, to those academic programme leaders planning the curriculum, staff doing the teaching as well as down to the students who will be the future leaders.

One practical reason cited for the apparent failure to recognise and address the significance of the sustainability issue, often given in apology and sometimes as an excuse, is the lack of spare time or space within the existing curriculum. "So much to teach, so little time within which to teach it" is the cry. Indeed, this is already apparent from the varying emphasis placed upon the categories referred to above, the most time being afforded to the technical. However, awareness of the sustainability agenda and its importance is vital for the survival of the QS profession. Social and environmental issues drive the broader agenda. Therefore, whilst it is not suggested that academics should talk of nothing else, the research implications suggest that they might plant an awareness of its relevance to most things, emphasising to a greater or lesser extent its importance across the whole of the existing curriculum. To certain subjects such as Law and Management it may indeed seem and be somewhat peripheral but to Construction Technology and Construction Economics, for example, it must surely be of fundamental importance.

The examination of the existing curriculum and of curriculum leaders' perceptions of its content and delivery at one institution suggests some uncertainty as to exactly where, and how, sustainability-related issues should be delivered. It is hoped that eventually it will be possible to produce a template, illustrating the relevance of sustainability to each key subject area, and ways even by which it may be effectively incorporated. A number of specific suggestions were made both as to the general direction teaching might take, and on specific areas worthy of increased emphasis within the syllabus. There is a general consensus and agreement of the appreciation of the sustainability agenda which should be a thread visible through all teaching at all levels. It was suggested that where a multidisciplinary School set-up existed, every opportunity should be taken for students of differing disciplines to work through the sustainability issues together, as they will have to in their professional careers. There was also an agreement that, where possible, classroom work should take as its model, data from local schemes which exemplified good practice in sustainability. Also, current research has much to offer and it is agreed that the technological and cost implications were crucial, together with the ability to transmit these concepts effectively to clients.

Participants agreed that: '[whilst] quantity surveyors are not there to advise on designs for sustainable development, which is the designers' job really [they] should be trained as design economists to understand the technologies involved and their implications more in terms of costs'.

The current research supports the findings of RICS research by Perera and Pearson (2011), where sustainability was ranked low in terms of the content of the curriculum at present, although the same research shows that a growing body of professionals in practice do recognise the part it must play in their future workload. Surely academic institutions must do better to equip the quantity surveyors or tomorrow's design economists for what will undoubtedly be a pivotal role, in terms of the management of time, cost and quality in deciding the future costs to society of sustaining the built environment.

As one interviewee remarked: "Sustainable development is not going to go away... students are going to go out there in the next couple of years upon graduation to confront these issues which [are] out there and [are] not going to go away". To echo an earlier statement, we shall not survive, as educators, by mere reflection and associated promises. Our delivery and content must change to address this challenge of our age.

This chapter has presented the results of a study carried out in QS degree programmes to establish the content of sustainability education within the curriculum. The research is part of larger research which aims at diffusing sustainability into the curricula of all built and natural environment programmes. This research and other research have established that a holistic understanding across the disciplines is needed to accommodate the evolving concept of sustainability. Consequently, future research is needed to extend or map the sustainability education within other construction related programmes to enable decision makers to have a better understanding of the situation. Also, it is of paramount importance for this research to consider and explore the link with other stakeholders. Thus, a key strategy for incorporating sustainability education within the construction related programme would be to include professional bodies, industry and students in the research (Figure 27.3).

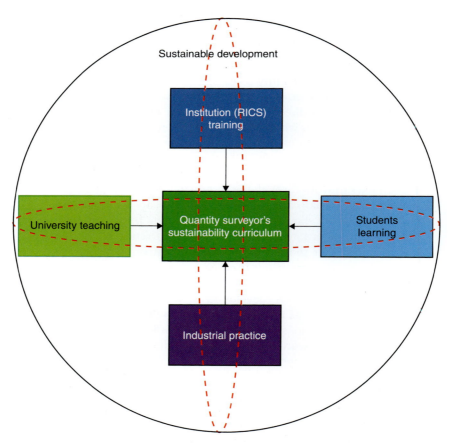

Figure 27.3   A holistic view of the QS sustainability research strategy.

Figure 27.3 shows that for any meaningful strategy, the input from the various stakeholders is necessary to establish what is required and how the strategy will be implemented. Finally, it is anticipated that this strategy will lead to the development of a methodology that schools or universities generally can use to incorporate sustainability education within their curricula it is expected that there will be profound changes in the QS curriculum. This will ensure that the education and professional training of design economists is fit-for-purpose to continue to play a leading role in achieve sustainable design and cost effective solutions for the built environment.

References

Ashworth, A. and Hogg, K. (2007) *Willis's Practice and Procedure for the Quantity Surveyor*, 12th edn, Blackwell Publishing Ltd, Oxford.

BERR (2008) *Strategy for Sustainable Construction*, Department for Business, Enterprise & Regulatory Reform, HM Government in association with Strategic Forum for Construction, pp. 1–64.

Cotgrave, A. and Alkhaddar, R. (2006) Greening the curricula within construction programmes. *Journal for Education in the Built Environment*, 1(1), 3–29.

Cowling, E., Lewis, A. and Sayce, S. (2007) Exploring the changing nature of students' attitudes and awareness of the principles of sustainability. Built Environment Education Conference, CEBE, 12–13 September 2007, London.

Dale, A. and Newman, L. (2005) Sustainable development, education and literacy. *International Journal of Sustainability in Higher Education*, 6(4), 351–362.

Dixon, T. (2009) RICS Green Gauge 2008/09: RICS Members and the Sustainability Agenda. RICS research, summary report.

Ekundayo, D., Zhou, L., Udeaja, C., Pearson, J. and Perera, S. (2011), Mapping of sustainability education to construction related curricula: a case study of quantity surveying degree programme. Proceedings of RICS Construction & Property, COBRA, 12–13 September 2011, School of the Built Environment, University of Salford, UK.

Fenner, R.A., Ainger, C.M., Cruickshank, H.J. and Guthrie, P.M. (2005) Embedding Sustainable Development at Cambridge University Engineering Department. *International Journal of Sustainability in Higher Education*, 6(3), 229–241.

Hayles, C.S. and Holdsworth, S.E. (2008) Curriculum change for sustainability. *Journal for Education in the Built Environment*, 3(1), 25–48.

HEFCE (Higher Education Funding Council for England) (2010) Carbon Reduction Target and Strategies for Higher Education in England. Policy development statement, January 2010.

Iyer-Raniga, U., Arcari, P. and Wong, J. (2010) Education for sustainability in the built environment: what are students telling us?. Proceedings of 26th Annual ARCOM Conference, 6–8 September 2010, Leeds, UK, pp. 1–10.

Jones, P., Trier, C.J. and Richards, J.P. (2008) Embedding Education for Sustainability Development in higher education: A case study examining common challenges and opportunities for undergraduate programmes. *International Journal of Education Research*, 47, 341–350.

Khalfan, M.M. (2006) Managing sustainability within construction projects. *Journal of Environmental Assessment Policy and Management*, 8(1), 41–60.

Lee, C. and Hogg, K. (2009) Early career training of quantity surveying professionals. RICS COBRA, 10–11 September 2009, Cape Town, South Africa.

Murray, P., Goodhew, S. and Turpin-Brooks, S. (2006) Environmental Sustainability: Sustainable Construction Education: a UK case study. International Conference on Environmental, Cultural, Economic and Social Sustainability, 8–12 January 2006, Hanoi, Vietnam.

Nkado, R. and Meyer, T. (2001) Competencies of professional quantity surveyors: a South African perspective. *Construction Management and Economics*, **19**(5), 481–491.

Perdan, S., Azapagic, A. and Clift, R. (2000) Teaching sustainable development to engineering students. *International Journal of Sustainability in Higher Education*, **1**(3), 267–279.

Perera, S. and Pearson, J. (2011) Alignment of Professional, Academic and Industrial Development Needs for Quantity Surveyors: Post Recession Dynamics. RICS Education Trust funded research report. http://www.northumbria-qs.org/RICS_Alignment/Reports/Alignment_of_views_final_report.pdf (accessed 29 June 2011).

Perera, S., Pearson, J. and Dodds, L. (2010) Alignment of professional, academic and industrial development needs for quantity surveyors. RICS COBRA, 2–3 September 2010, Dauphine University, Paris.

RICS (Royal Institution of Chartered Surveyors) (2007) Surveying Sustainability: a Short Guide for the Property Professional.' Report, pp. 1–24.

Sayce, S., Clements, B. and Cowling, E. (2009) Are Employers seeking Sustainability Literate Graduates? C-SCAIPE, The Higher Education Academy. Final report, pp. 1–41.

Simpson, Y. (2010) Twenty first century challenges for the professional quantity surveyor. RICS COBRA, 2–3 September 2010, Dauphine University, Paris.

Theron, C. (2010) Surveying the sustainable and environmental legal and market challenges for real estate. RICS COBRA, 2–3 September 2010, Dauphine University, Paris.

Walton, S.V. and Galea, C.E. (2005) Some considerations for applying business sustainability practices to campus environmental challenges. *International Journal of Sustainability in Higher Education*, **6**(2), 147–160.

Appendix A

UK PROPERTY INVESTMENT YIELDS | DECEMBER 2013
KEY THEMES FOR THIS MONTH

- UK GDP now forecast to be around 1.4% for 2013. Consumer and retail figures strong, earnings growth only 0.8%.
- 50 basis point movement in good secondary sectors due to strong institutional interest and wider availability of debt.
- A Happy Christmas and New Year and thanks to all our clients for your support in 2013.

	Dec-12	Mar-13	Jun-13	Sept-13	Dec-13	Trend
HIGH STREET SHOPS	%	%	%	%	%	
Prime	4.85	4.85	4.75	4.75	4.50 (4.75)	Stronger
Good Secondary	7.00	7.00	7.00	7.00	6.50 (7.00)	Stronger
Secondary	11.25	11.25	11.25	11.25	11.25	Weaker
SUPERMARKETS						
Prime (25 yrs,3.5% pa cap RPI)	4.50	4.50	4.25	4.25	4.25	Stronger
SHOPPING CENTRES						
Prime	5.50	5.25	5.25	5.25	5.25	Stronger
Best Secondary	7.50	7.50	7.50	7.25	7.25	Stronger
Secondary	9.50	9.50	9.50	9.50	9.00	Stronger
RETAIL WAREHOUSES						
Park - Open A1 (inc fashion)	5.25	5.25	5.25	5.25	5.00	Stronger
Park - Prime - Bulky User	6.25	6.25	6.25	6.25	6.00	Stronger
Solus - Prime - Bulky User	6.25	6.25	6.25	6.25	6.00	Stronger
Park – Secondary (rebased)	8.00	8.00	8.00	7.75	7.50	Stronger
LEISURE PARKS						
Prime	6.35	6.35	6.25	6.25	6.25	Stronger
OFFICES						
West End	4.00	4.00	4.00	4.00	4.00	Stronger
City	5.00	5.00	4.75	4.75	4.75	Stronger
M25/South East	6.00	6.00	6.25	5.75	5.75	Stronger
Regional Cities	6.25	6.25	6.50	6.00	5.75 (6.00)	Stronger
Good Secondary	8.75	9.00	9.00	8.25	7.50 (8.00)	Stronger
Secondary	13.50	13.50	13.50	13.50	12.50 (13.00)	Stronger
INDUSTRIAL						
Prime Distribution	6.50	6.50	6.50	6.25	5.75	Stronger
Prime Estate (GL ex HTW)	6.00	6.00	6.00	6.00	5.50 (5.75)	Stronger
Prime Estate (Ex Greater London)	6.75	6.75	6.75	6.50	6.50	Stronger
Good Secondary	8.50	8.50	8.50	8.25	7.25 (7.75)	Stronger
Secondary Estate	12.00	12.25	12.25	12.25	12.25	Stable
FINANCIAL INDICATORS						
Base Rate	0.50	0.50	0.50	0.50	0.50	↔
5 Year Swaps	1.03	1.06	1.17	1.73	1.77	↑
10 Year Gilts	1.79	1.96	2.00	2.78	2.76	↓
RPI	3.20	3.30	2.90	3.10	2.60	↓
CPI	2.70	2.70	2.40	2.80	2.20	↓

NB Prime yields refer to the equivalent yield for a prime (well specified, well located and rack rented) property let to financially strong tenants, but not Govt, on a lease with 15 years unexpired and open market rent reviews. Last month's yield in brackets if changed.

Appendix B

An MSCI Brand

IPD/RICS Sustainability Inspection Checklist
2014

The IPD/RICS Sustainability Inspection Checklist has been designed to help valuers identify environmental risk factors which should be taken into account when undertaking on-site inspections. These questions have been taken from IPD's EcoPAS measurement service, a benchmarking service that enables real estate investors to understand the potential environmental risks in their investment portfolios. The data collection process for EcoPAS is shared between IPD, valuers and investors.

Date of inspection:		Surveyor name:		
Full address of asset:			City:	Post Code:
Portfolio Name:		Sector:	Client Ref:	Your Ref:

Is this a listed building? Yes – Grade I / Yes – Grade II / Yes – In conservation area / No

SECTION A: ENERGY			please circle/answer accordingly
A1	What are the EPC ratings for the asset for each lettable unit?	None / Complete in the table overleaf	
A2	In which year was the EPC rating undertaken?	Date (yyyy):	
A3	Is there any evidence of on-site renewable energy generation?	Yes / No / Unknown	

SECTION B: FLOODING Postcode search here: England & Wales: www.environment-agency.gov.uk; Scotland: www.sepa.org.uk; Northern Ireland: www.ni-environment.gov.uk Should a flood assessment be undertaken by a qualified organisation, this may supplant the EA assessment, provided the same categorisation is used.			please circle/answer accordingly
B1	What is the asset's Environment Agency (EA) flood risk rating?	High Medium Low Very low	
B2	What is the asset's Environment Agency (EA) flood zone?	Zone 3 (dark blue) Zone 2 (light blue) Zone 1 (no blue shading)	
B3	Are flood defences in place at the asset – at the site only? (only answer if question B1 reveals a flood risk and/or B2 is zone 2 or 3)	Yes – on site / No – none	
B4	Are flood defences in place that cover the asset's location? (only answer if B2 is zone 2 or 3)	Yes / No	

SECTION C: WATER			please circle accordingly
C1	Is there any evidence of on-site water efficient practices? (e.g. Water recycling, water metering or water efficient fittings)	Yes / No / Unknown	

SECTION D: WASTE			please circle accordingly
D1	Is there any evidence of on-site or off-site waste recycling? (e.g. On-site segregation of waste or contracts for off-site recycling)	Yes / No / Unknown	

SECTION E: FLEXIBILITY			please circle accordingly
E1	Would the physical structure and configuration of the asset allow for different use types, within or without the current use type, subject to appropriate consents and at a reasonable cost?	Yes / No	

SECTION F: ACCESSIBILITY (optional – recommended but not essential)			please circle accordingly
F1	How long does it take to walk to the nearest well-used public transport node?	0 -5mins / 6 -10mins / more than 10mins	
F2	What are the X Y Coordinates (GPS latitude & longitude) of the asset?	X: Y:	

SECTION G: QUALITY (optional – recommended but not essential)			please circle accordingly
G1	Does the asset have a BREEAM rating? (or equivalent such as LEED)	None / Pass / Good / Very Good / Excellent / Outstanding	
G2	What is the date of the BREEAM (or equivalent) rating?	Date (mm/yyyy):	

An MSCI Brand

IPD/RICS Sustainability Inspection Checklist
2014

EPC Rating Table

The following table needs to be completed as follows:
- If there are no EPCs, please note that overleaf;
- If there is one EPC for the whole building, please provide the information for the whole building and not at tenancy level;
- If there is one EPC on part of the building, please provide tenancy level information and EPC information for that part;
- If there are multiple EPCs, please provide all tenancy level information and EPC information.

Unit Ref	Floorspace (net lettable, m²)	Unit EPC rating	Unit EPC Score	Benchmarks (for commercial buildings only)		Unit EPC date of rating	Tenant name in unit
				If newly built	If typical of the existing stock		
Unit 1	200m²	F	130	37	73	2010	J. Bloggs Ltd.

ipd.com/EcoPAS +44.20.7336.9200 @IPDnews

Index

References to figures are given in *italic* type; references to boxes and tables are in **bold** type. Abbreviations: BIM = building information modelling; VE = value engineering; VM = value management

Design Economics for the Built Environment: Impact of Sustainability on Project Evaluation, First Edition. Edited by Herbert Robinson, Barry Symonds, Barry Gilbertson and Benedict Ilozor. © 2015 John Wiley & Sons, Ltd. Published 2015 by John Wiley & Sons, Ltd.